COMPOST TEAS
for the organic grower

ERIC FISHER

Published by
Permanent Publications
Hyden House Ltd
The Sustainability Centre
East Meon
Hampshire GU32 1HR
United Kingdom
Tel: 01730 823 311
International code: +44 (0)
Email: enquiries@permaculture.co.uk
Web: www.permanentpublications.co.uk

Distributed in North America by
Chelsea Green Publishing Company, PO Box 428, White River Junction, VT 05001
www.chelseagreen.com

Designed by Two Plus George Limited, info@twoplusgeorge.co.uk

Printed in the UK by Bell & Bain, Thornliebank, Glasgow

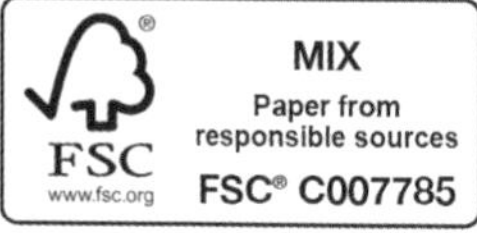

All paper from FSC certified mixed sources.
The Forest Stewardship Council (FSC) is a non-profit international organisation established to promote the responsible management of the world's forests.

Products carrying the FSC label are independently certified to assure consumers that they come from forests that are managed to meet the social, economic and ecological needs of present and future generations.

British Library Cataloguing-in-Publication Data
A catalogue record for this book is available from the British Library

ISBN 978 1 85623 327 9

Prologue

The impetus for this book came about from an interest in composting and what avenues such an undertaking might take me. Whilst many books have been written about this practice, few have been devoted to compost teas after you get past the fleeting and incomplete accounts you will find in popular gardening books.

Doubts are multiplying over conventional agri-chemical style solution based practices in plant nutrition, pest control, disease control and the consequences for human health. This book attempts if not to bridge the gap at least to alert the grower to the possibilities offered by compost teas.

About the Author

Eric Fisher MSc BSc (Hons) was born in North Yorkshire, has a degree in Environmental Science from Plymouth University and a masters in Technology from Cranfield University. He owns a small permaculture orchard where he practices cider making and worm farming. He has an interest in trading cryptocurrencies and a love of being outdoors in nature.

Talk to the Author

If you have a go at anything in this work, I would be very interested to find out how you got on. In the future there may be blogs, webinars, talks and so forth online regarding many of the practices I covered and I can keep you in the loop. Additionally, if you have any other feedback, critiques, questions about this book or any of the topics around it, I would be delighted if you got in touch: efis707@aol.co.uk

Praise for the book

Eric Fisher's landmark and masterful book brings the understanding of the care and cultivation of the plant world back to its proper place by tying it to the cycles of nature. By weaving the biological sciences with the newer sciences of quantum physics and chaos theory, and then finishing up by making a case for ideas inherent in biodynamics, Eric demonstrates in a holistic manner how we can not only repair the plant world, but also repair our relationship to the natural world, and in essence, repair the entire world.

Michael Wayne, Ph.D., L.Ac., author, *The Quantum Revolution: The Power to Transform*, and director/producer of the documentary film, *On the Path to Strawberry Fields*

This book is a dual gem: full of practical applications for organic growers seeking to develop their skills with compost teas, whilst at the same time, scientifically grounded so that it offers a broader perspective looking to shape a better approach to human food production without synthetic chemicals.

Dr Colin Trier, environmental scientist and cider producer

This is a book for the connoisseurs of compost – those that love to delve deep into the detail and dig out nuggets of knowledge. It will tell you everything you didn't know you didn't know about compost and compost teas, from the ancient of days to modern microbiology.

Liz McGregor, nutritional therapist, amateur herbalist, keen composter and organic gardener

A wide-ranging study of the principles and use of compost teas. From soil and plant ecology to bioremediation, pest control and plant defence, topics are covered from scientific, observational and even metaphysical perspectives. It will open your eyes to the possibilities!

Angie Polkey, permaculture and nature conservation teacher and advisor

Compost Teas for the Organic Grower is a practical manual that is both fascinating and, at times, romantic in its exploration of the history and concepts behind this controversial amendment. This book is deceptive. A Tardis of tools and much more than just a book on compost tea!

Simon Parfey, founder of SoilBioLab and Soil Hub International

An informative and refreshing guide to making compost teas, with plenty of practical advice. I really enjoyed it.

Lausanne Tranter, Co-ordinator Transition and Permaculture Hull

Eric's book is both systemic and timely, as evidence of the loss of soil fertility is building and our appreciation of the systemic nature of global ecosystems has never been more apparent as a source of future sustainability. His approach demonstrates that there are many elements affecting soil health and fertility, from insect pests and predators, to the range of nutrients found in the soil and all plants - including the so-called weeds! Other books on plant and vegetable growing skip over the ways in which these latter plants can be harnessed, through a variety of compost teas, to improve the life-giving quality of soil. This book should be an important addition to every gardener's working collection.

Dr Wendy Gregory, chair of a community garden supplying organically grown fruit and vegetables to charities serving Kingston-upon-Hull

Table of Contents

Acknowledgements

I would like to thank the following people for helping me go forward with what was intended only as a small project but developed into something altogether more massive and challenging for me to complete.

Particular thanks to Sarah Morris and Phillip Rhodes who helped with the photos. Also to my long time friend Adrian Gant who rescued my computer with his ninja skills when noone or nothing else could. Thanks to Joanne Wallis for using her networking skills to help me and to David Mache who took the time and effort to clarify my legal concerns.

Special thanks to T. Fleisher (Director at Battery Park City Authority) and Dr Gavin Lishman (Director of Martin Lishman Ltd) whose generosity and goodwill permitted me to conduct two excellent interviews over Skype. Also many thanks to Dr Wendy Gregory for her good humoured help with last minute test runs using Skype. Professor Emeritus George Barron, University of Guelph, Canada, for his good natured support of my endeavours letting me display his stunning SEM of *Arthrobotrys anchonia*. Professor Wilhelm Foissner of Salzburg University, Austria, for being kind enough to let me use his excellent SEMs of *Metopus hasei* and *Polytomella* sp. Dr Manfred Schleyer at the Institute of Flow Sciences, Germany, for his release of the drop pictures shown in Chapter Ten. Freelance artist Amy Edwards, New South Wales, Australia, for letting me use her marvellous Celtic Knot Tree to start off Chapter One. Kalpana Sharma, Permissions Sales Administrator, Cambridge University Press, regarding release in principle of *Notommata collaris* SEM, although it was not finally used. Industry expert Eddie Funderburg, Ed.D. from the Noble Research Institute, Ardmore, UK, for confirming my observations in Chapter Nine regarding E.C.C.E. (see Glossary) and eggshells. David Graham and John Curtis of the Vegan Organic Network (advice@veganorganic.net) for their help in tracking down an important reference associated with their organisation (Growing on Clay Soils). Pat and John from Newland Avenue Allotment who first started me thinking about compost tea in practice with their acrid but marvellous water butts of comfrey. I never got their second names; they moved on and we lost touch. Studying at Plymouth University gave me a great background for this book and Cranfield University helped more indirectly in my practice of describing and depicting different processes. Heidi Tillin, now Dr Heidi Tillin, during one of our many conversations at Plymouth University got me curious about an important and controversial thinker in sustainable agriculture. This figure's impact in this area is introduced in Chapter One, crops up in various places throughout and for interest is extended into the Appendix. Thanks to Paul and Yvonne Fisher for taking the time to read and review my book. I would also like to thank the council official who wished to remain anonymous who permitted me an informal discussion regarding the remediation of contaminated land, and surprised me with his open minded go ahead approach; more of this in Chapter Nine.

Working through my emails and reflecting on all the brilliant people who have helped me develop this book with no gain to themselves, I could not help noticing the significance of the academics and support staff of the American universities who furnished me with so much useful information and encouragement. I cannot mention them all here but some of the most important are: Professor Charles C. Mitchell of Auburn University for taking time to furnish me with extra details of his interesting tillage studies. Professor Margaret Tuttle McGrath of Cornell University for supplying the finer details of her compost tea studies that were not easily available without a lot of digging; pardon the pun. Dr Mark S. Hoddle, University of California, for his generosity and goodwill in allowing a rare photo of *Encarsia formosa* to be displayed. Helen Atthowe (veganicpermaculture.com) for taking the trouble to give me a detailed insight into how she does soil testing in her Biodesign Farm in Stevensville, Montana. Laura Frey, University of Nebraska-Lincoln, for her detective work regarding copyright permission considerations. Alicia Foth, Washington State University, for her help in tracking down an important reference.

Anyone with a computer and internet access can find vast stores of information which would be impossible when we just had physical books. The significance of this may be taken for granted today, but a considerable amount of work has gone into facilitating these remarkable resources and it would be a shame not to acknowledge some of the most significant ones I used here. I find myself in the privileged position of being able to access abstracts, archives, eBooks and so forth from all around the world at zero expense. There were a vast range of repositories that have been applied in this work, but in particular I would like to thank: the Soil Association, the Royal Horticultural Society, Garden Organic (formerly Henry Doubleday Research Association) and the Natural History Museum. Elsevier Scientific Publishing, Springer, ResearchGate, Chelsea Green Publishing, Wiley, CRC Press, Rodale Press, Cambridge University Press, Landlinks Press, National Academy of Sciences, National Academy Press, Global Science Books and Oxford University Press. The University of Manchester, Imperial College London, Stellenbosch University, Princeton University, University of Maine (Orono), Cornell University, Ohio State University, Kansas State University, Iowa State University, University of California, University of Nebraska-Lincoln, University of Wisconsin and Oregon State University. African Journal of Food Science, American Naturalist and Biotechnology, Agronomy Society Environment, American Society for Microbiology, Biocontrol Science and Technology, Soil Biology and Biochemistry, Plant and Soil, Environmental Health Perspectives, International Journal of Advanced Research, American Society for Microbiology. Teagasc Co. Dublin, CABI International, Pesticides Action Network and The Rudolf Steiner Archive.

In collaboration with Wikipedia, the Creative Commons and its contributors although not an organisation as such have also been instrumental in making this book richer and more interesting than it would have been without spending thousands of pounds on lab facilities and photographic equipment.

Finally I'd like to thank Maddy Harland, Tim Harland and Rozie Apps at Permanent Publications for their constructive critiques during the development of this project and thanks to the late Tim Green, their expert reviewer, whose observations guided what I did in the later stages to finish things off.

'Celtic Knot Tree' by Amy Edwards[1] used here as a symbol of life and connectedness.

CHAPTER ONE

Introduction

The use of compost tea was written about as early as 160 BCE by the Roman scholar Cato[2] in his book *De Agricultura*. With today's increased environmental awareness, escalating costs and the banning of certain chemicals that were once considered acceptable for agriculture, greater attention has been given to more natural approaches of supporting crops. Teas come in a variety of names – liquid extracts, tinctures, liquid feeds, compost tea, aqueous extracts, organic liquid fertilisers and black jack to note just a few. They are used by the grower for a variety of purposes including plant nutrition, growth stimulation, pest control, pest prevention and more controversially as a soil tonic and enhancement of beneficial organisms.[3]

Compost tea can be seen as part of the repertoire open to the organic grower looking for an alternative to the ecologically degenerate, toxic and unsustainable practices of contemporary mainstream agriculture.[4] Compost tea has advantages over solid composts in some situations because it can reach parts of the plant in ways not otherwise possible, such as in the use of foliar sprays. There is also the question of transportability where some active liquid composts can be prepared onsite using only a fraction of the compost you would expect to use with a solid compost. If you are an enthusiast of companion planting but have concerns over the amount of space given over to these beneficials, in some circumstances a foliar spray or soil dressing can be prepared which will do the same job but with greater control over the timing and impact.

This book is accessible to anyone with an interest in gardening and no technical knowledge is assumed; any terminology that is beyond day-to-day language is described in the glossary at the back of the book. We take a bird's eye view of the possibilities open to the inquisitive grower who wants to take a closer look at compost teas. Later on in this chapter we will look at radically different approaches to compost teas from Steiner's Biodynamics to Integrated Pest Management.

Before proceeding it might be worth defining what is meant by 'organic' because there is considerable confusion over the term since it is used in different circumstances to mean different but related things. In science it is used to mean the chemistry of carbon compounds. The word 'organic' is derived from the

Latin for 'life force' and it is notable that carbon, because of its remarkable bonding potential, is present in all known life forms. For the grower it has come to mean very generally a way of growing, to improve crops, that attempts to work in greater harmony with nature where industrially produced chemicals are avoided.

A broader more formal definition of 'organic' in a farming context is:

> ... avoids or largely excludes the use of synthetically compounded fertilisers, pesticides, growth regulators, and livestock feed additives. To the maximum extent feasible, organic farming systems rely upon crop rotations, crop residues, animal manures, legumes, green manures, off-farm organic wastes, mechanical cultivation, mineral-bearing rocks, and aspects of biological pest control to maintain soil productivity and tilth, to supply plant nutrients, and to control weeds, insects and other pests.[5]

The distinction between organic and non-organic or conventional growing developed in the 1930s and 40s as a backlash towards the introduction of industrially produced synthetic fertilisers, non-organic pesticides and herbicides. Many commentators have characterised this period up till today as 'the chemical era' but strictly speaking this is a misnomer because all natural interactions can be seen in chemical terms.

The idea of organic agriculture was developed by Rudolf Steiner in the 1920s[6] where he conceived of the farm as an organism. In 1940 Lord Northbourne popularised this conception and brought organic growing further into the mainstream with his book, *Look to the Land*, where he characterised the 'organic farm' and described a more holistic and ecologically balanced way of growing than the dominant regime.[7]

Much of the old knowledge and wisdom passed on from generation to generation was discredited and the value of scientific growing linked to progress was naively accepted and celebrated. The 'green' revolution was underway with vastly improved yields of single crops as the farm was now viewed as an outdoor factory. Progress in this instance built in opportunities for large-scale agro-industry to expand their businesses and lock growers into a dependency on artificial fertilisers, pesticides and fungicides which yielded short term gains at the expense of long-term soil fertility,[8] food quality[9] and ultimately the grower.[10] In this cultural milieu, many of the unsustainable practices of the era filtered down to the small-scale grower. Yet as is always the case, a thread of healthy scepticism remained and continued in defiance. This backlash towards the wholesale disruption of nature by detached decision makers could be seen as part of an ancient tradition dating back to the industrial revolution and romantic movement from the late 1700s onwards.

Over the ensuing years the research community backed by the agri-chemical industry became preoccupied with plant nutrition and monoculture at the expense of soil health and biodiversity. Holes in this approach began to appear with gradual declines in yields and plagues of pests grown resistant to short term chemical fixes. Today greater attention is now paid to organic growing and support for organic food is expanding yearly.[11]

Organic growing, of which compost teas are an integral part, underpins a philosophy of not just feeding the plant but nurturing the soil and the web of life around it on which a healthy plant life depends. In years when oil was cheap, concerns over sustainability were simply ignored and the dominant approach was to bludgeon nature into submission with a policy of high energy and resource intensive practices. Where natural organisms came into the picture was as a potential threat to crops rather than an ally.

In recent years the idea of 'suppressive soils' has begun to be explored where natural microbial communities actively combat weeds and diseases and develop an incredibly fertile growing medium. Such practices involve enhanced organic matter accumulation, increased enzyme activity and a diverse microbial population[12] which facilitate weed suppressive soil communities[13] and could be considered a worthy goal for any dedicated organic grower.

We concentrate on the small-scale horticulturalist: allotment growers, residential garden growers, amateur gardeners and everything in between. The kinds of areas we are aiming the book at primarily are from the smallest urban container garden of a few square metres to a double plot of around 500m^2. In a residential context this would equate to two good sized gardens. Many of these ideas are applicable on a larger scale but after a certain point you might have to consider your level of mechanisation plus innumerable other factors which are beyond the scope of this book.

Teas are described using plants that are readily available locally either popularly in cultivation outside, in the household, greenhouse or are available for you to forage responsibly in nature, rather than as an alternative to goods shipped half way across the world by supermarket chains which erode local business and undermine local self reliance. The reason for choosing this criterion is that it is cheaper, more sustainable in terms of food miles, intellectually stimulating and an opportunity to explore your own potential for self sufficiency.

Today there are serious worries of whether global supply chains can maintain their intensity in the light of fuel shortages. Governments are well aware of the high possibility that a low energy economy may emerge by necessity and have softened their stance on self sufficiency with greater encouragement for forward looking organisations such as the Transition Network.[14] In a sense the scenarios that are being prepared for are not unlike the broken supply chains of World War Two. History is dotted with the remains of once mighty civilisations that have lost contact with the soil and some decision makers have now become acutely aware of this. If predictions of energy shortages and disintegrating food security bear out and we find ourselves in a future of low carbon usage, suddenly from living like emperors with the fruits of every nation at our fingertips, we could move to a civilisation in decline that is unable or unwilling to adapt to the new realities.

With the ascent of social networks across the internet it is easier to find people to share resources and organisations. Some of the preparations included require some basic knowledge of what certain plants look like. If you are uncertain many online user groups and local foraging clubs would be only too pleased to help with plant identification. Alternately you could get hold of one of the many excellent identification keys[15] available. We provide an abundant supply of photographs to help you identify which plants to collect and have made sure that they are not rare or protected species, but even if they are not on the protected list it is still important not to destroy your supplies so the plants can return year in year out and minimise your impact. Better still, propagate some of the more useful plants yourself if this is viable and you can contribute to building resilient, sustainable, local self supporting networks that are not dependent on the vagaries of global trade or the exploitation of less politically strong countries. With this ethos in mind this book aims to exploit the rich and lavish local resources currently available but detach them from their unsustainable source. Such a transition might require a better awareness in the local area of what is available. To take a topical example, aloe vera can be a useful plant as a compound for homemade soap sprays: it is not indigenous to temperate climates but it is well established in a very special microclimate, namely people's homes. If you can get cuttings from someone in the local area, you don't need to get it shipped from wherever via your local supermarket. Whichever networking style you like, internet user groups or word of mouth, your plant will not be far away. From that plant you can propagate as many further plants as you need with little extra expense.

Whilst this book was intended for use in the British Isles, the same information is still relevant to other temperate regions of the world including most parts of Europe, North America and Northern Asia. If you are making compost teas in the tropics you would need to consider using a different range of flora and fauna than many described in this book.

Approaches to making compost tea

Compost teas are either aerated, non-aerated, or a combination of the two. Aeration is simply the process of applying oxygen to a solution. In the absence of atmospheric oxygen, anaerobic processes take over.

Traditionally compost tea was made by steeping various plant and animal products that were thought to help plants grow. Varied amounts of manual stirring were employed and the often foul smelling solution was then applied to crops. Fundamentally this approach aimed to feed the plant with no explicit consideration of the soil. In 1924 Rudolf Steiner introduced a system that blended traditional approaches with his own brand of spiritual holism. Compost teas were developed as part of a system called *biodynamics* that originated from his agriculture lectures.[16] These 'BD preparations' were added to composts and plants in homeopathic proportions with a strong focus on natural cycles. Purely on the basis of nutrient value, a review of some of his ingredients indicated that the plants used had heightened levels of nutrients useful to crops. They were partly fermented and subject to somewhat involved and unique procedures. One of his views was that water had a special vitality that could enhance crop growth. His method of stirring by hand works well for aerating a compost tea even if you cannot accept that some special energy is added by the act of stirring. Whilst the scientific establishment was, and still is very sceptical to his methods, solid evidence supports improved plant growth using his techniques[17] that cannot be accounted for by just the good organic practice advocated. Further strong empirical evidence confirms that in farms applying biodynamics there is an enhanced microbial activity that outstrips both conventional and other organic systems.[18] The emphasis is not on maximising yield but on creating a dynamic balance within the boundaries of the growing area. Although biodynamic systems are generally considered less productive than conventional agriculture they can be more profitable because of reduced inputs and the high premiums attached to the produce.[19] Subjective accounts highlight a less quantifiable impression of biodynamic farms that held a sense of vitality that was lacking elsewhere.[20]

Research has attempted to quantify this sense of vitality in produce grown by holistic methods and asserts differences exist. Particularly interesting is the work in biocrystallisation, illustrated below,[21] where holistically grown vegetables and conventionally grown vegetables were treated with copper chloride and the patterns compared. It must be noted that the science behind these techniques is still poorly developed and may be refined to illustrate further distinctions in the future.

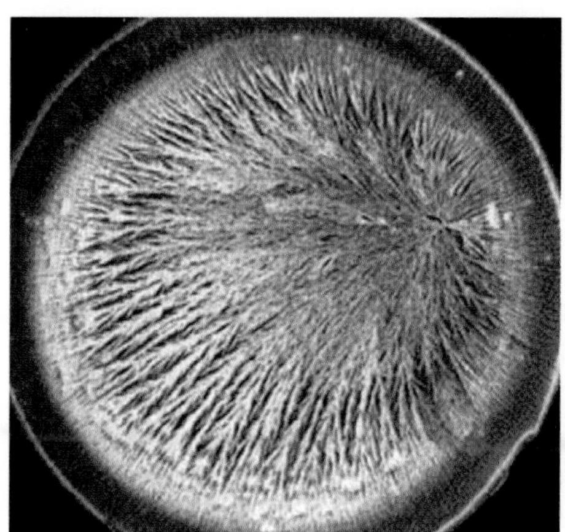

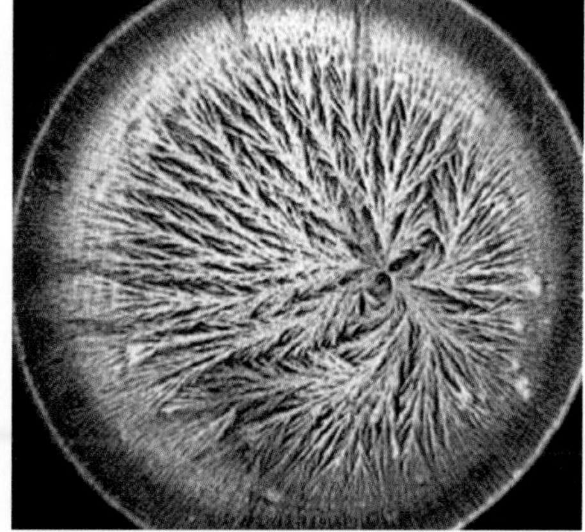

Left: Extract from conventionally grown beetroot treated with copper chloride. Right: Extract from holistically grown beetroot.[20]

Nutritional differences could be established quantitatively by simple means but it is more difficult to establish crystallisations such as on the previous page as proof of organic growing being superior. An observer can easily distinguish the clean intricate patterns of holistically grown produce from the broken and coarse patterns of old or conventionally grown produce. The analogy of a photograph of a jigsaw with pieces missing and slightly out of focus, compared to one that is complete and in clear focus, could be made to illustrate this difference. The consciousness naturally spots between the two patterns a teleological impression that appears to aspire towards an involuted and intricate order. The organic plant achieves this mysterious becoming better than the non-organic one. In Rudolf Steiner's agriculture lectures, a powerful emphasis is laid on increasing life and awareness to humans, animals, plants and even soil. A vitalist doctrine such as biodynamics does not sit well with the modern Western mind.

Today probably the highest expression of biodynamic farming can be found in the vineyards of France where the notion of *terroir*, an embrace of the total environment of the vine, produces distinctive wines at the higher end of the market.

Rudolf Steiner's methods are still going strong after nearly a hundred years and he has a formidable following worldwide. His belief was that conventional science was a study of dead processes because they did not acknowledge the existence of a life force. Gifted since childhood his experience of the world was at a considerable deviance to most people's day-to-day experiences that led him to develop a belief system called Anthroposophy where the normal here and now was interpenetrated by other dimensions. For someone from a different era who was a true genius and visionary in so many fields it is difficult to assess him in the same way you would others. Steiner was the kind of person who only comes along once in a generation, so although his beliefs have little basis in conventionally proven fact, they are still worthy of some consideration. In the area of organic growing, and particularly compost teas, his influence is unavoidable as all of the later approaches emphasising ecosystem processes have a strong resonance with his ideas.

Alan Chadwick, a former student of Steiner, created his own style of growing that had influences from biodynamics and was somewhat awkwardly named *Biodynamic/French Intensive*. In this approach he amalgamated many aspects of Steiner's spirituality with intensive techniques from early French horticulture. He also investigated and incorporated a number of ideas and techniques from ancient civilisations including Egypt, Mesopotamia, China, Greece and Rome.[22] His aim was not production, but a celebration of nature, where beauty was paramount. His gardens could be likened to the potager with a combination of vegetables, ornamentals and companion plants. A man of action and no significant written works, he often inspired the people he worked with. As with Steiner, Chadwick was strongly influenced by the phenomenology of Goethe, which sought a direct[23] and transformative connection with nature, which he attempted to unveil to his students/followers. From this philosophy stemmed a way of viewing problems in a garden as profoundly interconnected, which marks a firm line to conventional growing where problems are viewed in isolation.

> Nothing happens in living nature that is not in relation to the whole.[24]
> John Jeavons (2002)

In 1974 Jeavons took up the reins of Chadwick's approach and retained much of the Goethean flavour. Renaming it *Biointensive* growing[24] he rationalised the work of Chadwick and Steiner making it palatable for a wide American audience who were largely ignorant to organic practice.[24] Jeavons cut out much of the mysticism and astrology but retained planting by the Moon and pioneered special watering techniques that mimicked natural rainfall.[21] Biointensive farming has many finesses built upon traditional organic growing, which is suggested to have an incremental effect. As with Chadwick's approach, French style high intensity growing[25] with manure and deep digging was favoured with close planting on raised beds. Mechanisation was reduced still further than the simple traditional tools used in biodynamics. This approach marked a departure from Steiner's homeopathic style of herbal preparation[24] although some sprays were still used to discourage problem insects. Its practical significance in this account is that many of the techniques using water can be applied to foliar sprays and root drenches. Despite being incredibly productive and sustainable, it is criticised as being inappropriate for large-scale agriculture because of the lack of mechanisation and the high amounts of labour involved. This is a judgement using the criteria formed within models of conventional global economics where a few people produce food over vast acreages using powerful high impact machinery. If humans chose to return to the land to feed themselves and detached from almost complete reliance on the demand and supply chain of global economics, it could hold valuable possibilities. In any event many of the techniques can be experimented with and incorporated into standard growing practice and left to the individual grower to judge.

Natural Farming originated in the Philippines and takes a very intuitive approach incorporating Steiner's stirring procedures into their own compost 'Indigenous Microorganism' (IMO) tea preparations which advocates local experience and resources. The argument asserted is that encouraging indigenous beneficial microorganisms for the soil is desirable because they are best adapted to that area. It is fair to observe, in the light of recent evidence that once a group of microorganisms are established in an area they are more resistant to new colonies of organisms,[26] that it makes sense to use local stock. The teas combine fermentation and aeration with a variety of naturalistic procedures. The approach encourages the use of a variety of cultivars in crops and to develop a range of niches where different organisms can flourish. It would be very compatible with permaculture where a sensitivity to natural processes and structures is championed.

Many critics hold to the well established assumption that microorganisms are ubiquitous, which infers that because of their cosmopolitan distribution there is no local variation. Reviewing the information available, which is not extensive but robust enough to make a strong case, suggests that indigenous organisms are significant in local soil communities and that whilst there are a number of organisms that are very widespread there are also many that are unique to individual areas.[27]

The corporate answer to indigenous microorganisms is 'Effective Microorganisms' (EM), which can be exported and sold globally. Developed in Japan (1994) by Dr Higa[28] this combines a small group of aerobic and anaerobic organisms that have been isolated in a laboratory and are known to be beneficial for the breakdown of organic matter and disease suppression. This inoculant is applied to compost heaps, soil and plants and according to Dr Parr improves soil quality, crop growth and yield.[28] In the laboratory the strains can be developed to survive better than any other organism in controlled conditions, however the conditions that they will be introduced, in connection with organic crop growing are heterogeneous. A further criticism of this approach is rather related to the advantages ascribed to the indigenous approach on the previous page, that because these microorganisms are screened in an environment that is very much separate from normal conditions, they may not become well established or persist against well adapted and established indigenous organisms.

The *modern approach* to compost tea that advocates intensive aeration using a variety of electric bubblers was originally developed by the so called 'mother' of compost teas, Elaine Ingham.[29] She made a significant deviation from mainstream growing by emphasising soil health, fertility and food web resources rather than directly feeding crops, making a robust case for the value of the food web for plant health and productivity.[29] Ingham returned to the idea that a minute amount of material can have a powerful effect on fertility but she did not base this view on cosmic interactions as Steiner did, but on the notion of creating a chain reaction of burgeoning life. Stated another way in Jeavon's approach on the previous page is the view that life begets more life[24] and this is extended in Ingham's account. In many ways this approach re-connects with the approach of traditional farming before industrialisation, in conjunction with modern methods and technologies. Ingham is a recognised world authority on microbiology and she pioneered direct observation methods using microscopy to access soil communities. She is also another controversial figure. Her business interests often overshadow her scientific credentials and she has a tendency to exaggerate, presenting calculated guesses as facts with little solid evidence to back them up, which has caused her considerable embarrassment in one high profile case.[30] On a more positive note, her passionate support of organic growing and the role of compost tea has brought possible future alternatives to the attention of the mainstream.

Aerated tea has a strong industrial following in America with tea production operating on an enormous scale. Unfortunately, scientific studies regarding compost tea are often contradictory with some concerns that the procedures may incubate pathogens such as *E. coli*. This concern is countered by the assertion that *E. coli* does not do well in aerobic conditions and would be outcompeted.

A substantial problem with establishing the usefulness of compost tea is that there is no universally accepted standard for compost. Different studies used different materials and procedures for making compost for compost tea. All this disarray within research and tea production circles made findings difficult to replicate. Furthermore there was also a serious problem in researching compost teas because the microbiological processes are so intertwined and holistic that it is hard to investigate what is going on without compromising the process. Opponents of this approach assert that there is not enough scientific foundation[31] to warrant the enthusiasm with which consumers and companies have got involved. The problems with identifying what is happening with the complex interactions in the soil may point to a limitation with conventional scientific approaches. The general approach to research is usually strictly reductionistic, in essence treating nature as one would a piece of machinery, that you can explain how something works by separating and analysing the individual parts. Holistic theory is throwing up an important challenge to this orthodoxy with some significant contentions. A key observation from this position is that when you have a complex system, such as a watershed ecosystem or even a square centimetre of soil, emergent properties exist only when the different parts interact. It follows that if you attempt to dismember that whole, the results may not offer a full understanding of what is happening. For example, to understand the behaviour of social insects, it would be misguided to attempt to analyse hive behaviour by just observing a single bee.

We complete this review of approaches with Integrated Pest Management (IPM), which originated in America[32] and has since been exported worldwide. As indicated by the name, it's all about pests and as such it is rather the odd man out compared to the other approaches because its use is more partial. It could be seen as the mainstream response to flaws in its own system of pesticide use and advocates reform rather than revolution. This pragmatic approach attempts to absorb organic practices to use in conjunction with standard non-organic practice. As such it holds a useful role in plans to make a transition from conventional to more responsible and sustainable practice. Many of the preparations IPM advocates are acceptable within an organic growing system and in particular the use of botanical pesticides, which are described in more depth in Chapter Seven. Where IPM really comes into its own is in the sophistication with which it addresses uncertain problems in the field with observation, knowledge and timing.

A brief introduction to the food web

The food web is important to organic growers because it speeds up the breakdown of important nutrients and makes them available for plant use.[33] It also facilitates the interactions of beneficial organism support systems that can assist crop plants.[34] In normal agriculture these natural processes are supplanted by industrial chemicals and extreme human interference.

An awareness of the food web is useful because actions taken by growers to enhance crop growth and protect against plant damages, such as the use of various liquid preparations, create ripples within the natural community of organisms, which can have both positive and negative effects for crops and supporting organisms. Such an insight arms the grower with an increasingly deep knowledge of how crops interact and are affected by this web and ultimately his/her own interactions within it. Important in this approach is an awareness of the particular area that is intended for crop growth. In later chapters we will return to this matter in various aspects, but for the moment it would be worth noting down some of the organisms that are familiar in your local patch, whether they are conventionally viewed as pests, beneficials or of no apparent value to growing crops. Sometimes a little detective work is needed for tell-tale signs of activity.

In the classic and simplistic biologists' pyramid presented in an adapted form to the right, energy from the sun supports the growth of green plants through photosynthesis (producers), herbivores eat the green plants (first level consumer), second level consumers consume the herbivores, third level consumers consume the second level and forth level consumers (apex predators) eat the third level consumers. It was originally proposed by Lindeman in 1942.[35] This model gave a starting point for further consideration and provided a good illustration of many key factors at play. An important aspect of this kind of model is the illustration that size counts; larger organisms mostly predate smaller organisms. Furthermore, choices are made by organisms that tend to be the most energetically favourable, for instance a cat is not going to bother chasing a fly (except in play) when he/she can catch a mouse or small bird which is more of a meal. The biologists' assertion of competition, survival of the fittest being paramount, is emphasised, however from an ecological perspective many of the processes in nature arise from co-operation as well as competition, which is not acknowledged. For instance around 90% of all plants are dependent on mycorrhiza[36] for the supply of certain essential nutrients and observations of guilds and co-operative behaviour amongst numerous organisms is sufficient to suggest that competition between individual organisms is not as significant as the biologists assert.

Omnivory is an important confounding factor and the justification for putting mice and rats further up the pyramid than would be expected than if they were presented simply as herbivores on the first level. The more organisms that are added in the different sections the more is revealed of how difficult it is to generalise, lumping individual organisms together as herbivores or predators excluding the significance of omnivory and variances in the number of links in particular food chains. Organisms may dramatically change their eating preferences throughout their lifespan. For these reasons and many more, the concept of food *webs* was later introduced which gave a deeper insight into the interactions of specific organisms but made it more difficult to extrapolate general principles.

Continuing with our pyramid, if the organism dies of natural causes, decomposers return its remains to the soil and the whole sequence begins again.

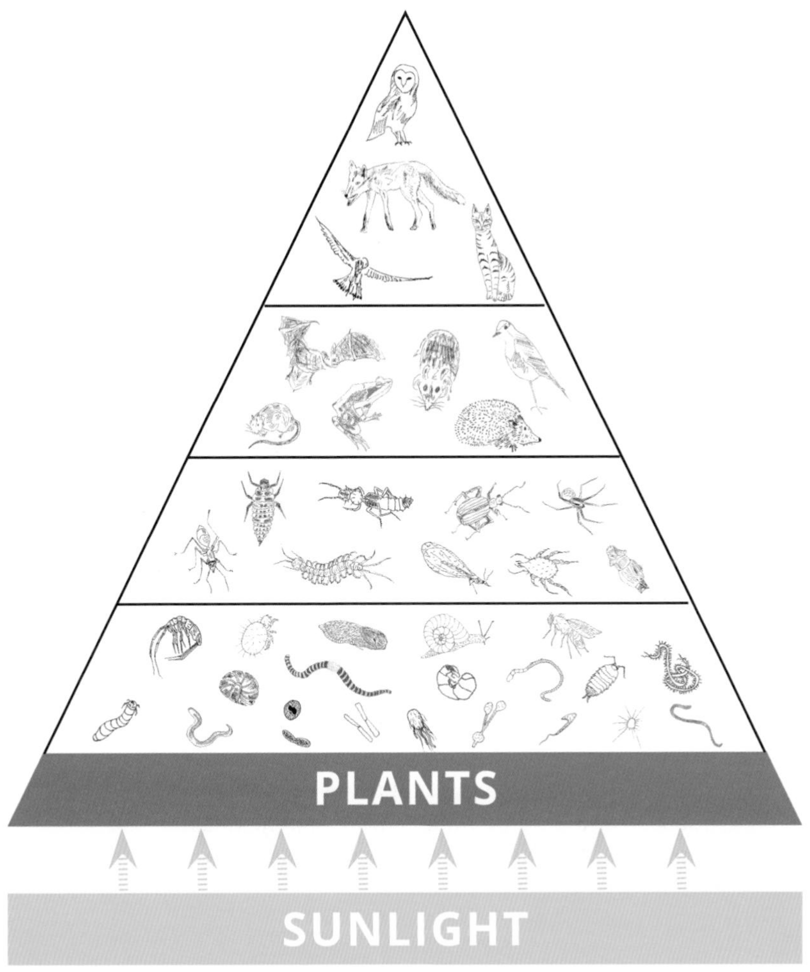

A terrestrial food pyramid for agroecosystems.

The emphasis in this model is this very noticeable and active group, and yet it accounts for only 10% of plant consumption.

The other 90% of plant material is returned to the soil without passing through this channel.[37] Another set of processes exist which are undeniable and of high significance to the grower which is dominated by the decomposers. They are not given the proper attention they deserve in models like this. Furthermore as an awareness of below-ground processes is growing, recent studies are highlighting another aspect of the food web that has been overlooked. Namely the influence of mycorrhiza that powerfully illustrates the idea of co-operation within the flow of energy and nutrients. Overleaf is an illustration of a food web that integrates these processes and brings us closer to what is actually happening in the field.

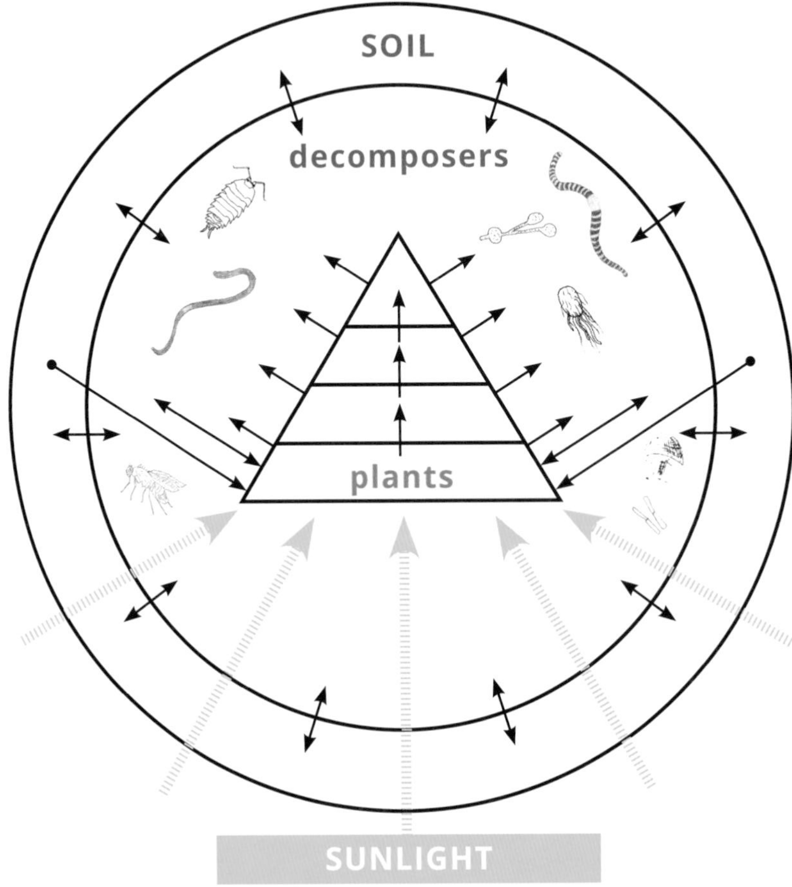

A terrestrial food web of predation and co-operation.

Note the double arrows attached to the plants that denote the symbiotic linkage between most plants, mycorrhiza and nitrogen fixing bacteria, which will be investigated in later chapters. In such a web, the classic food pyramid in the earlier figure still exists but is a subset of the more substantial decomposition processes where there is interplay on a number of levels. The single arrows pointing out of the pyramid indicate the death of that organism and attention of the decomposers, but there is also activity with the grazing of the decomposers, which is indicated by arrows returning to the pyramid.

An allotment plot, vegetable patch or a complete farm could be described as a small agroecosystem, that is an area controlled by humans in varying degrees to grow crops in varying degrees. If we consider the food web within this context, we find the top predators are feral cats, foxes and birds of prey. Their role is key to keeping down rodent levels and they act as powerful deterrents to crop raiding birds. Conversely, although rats and mice are undesirable in grow zones, they will undoubtedly attract major predators which create a deterrent effect towards other nuisance creatures such as pigeons that is possibly even more significant than actual kills. As an illustration of this point, it is well-known practice to employ professional falconers around council composting depots because of the profound scare factor it has on scavenging birds and in a more natural way the same process can operate on the growing site.

There is good evidence to support this top down view of biocontrol, but a caveat to this is that it is only considered most effective in 'homogenous' environments such as a farmer's field that only grows one crop. In the more heterogeneous territory of an allotment plot or permaculture farm, things are less clear and although it is hard to deny the influence of top predators it has been argued that their value is more varied.[38] For instance, a cat might not just predate rats and mice but also go for frogs (beneficial) which may help aphids (pest) to grow unchecked. In varied environments there are also varied resistances to different pests. So whilst one plant may become stricken, the same problem cannot spread unabated as may happen in a monoculture. Evidence from the bottom up school asserts, in heterogeneous environments rather than top down control of herbivores, that plant resistance is key and where there is excellent fertility and a good balance of nutrients many of the problems with crop pests and disease are held at bay.

Modern commentators on the current state of knowledge of food webs contend that the suppression of herbivores for the benefit of crops is essentially a synergy between top down and bottom up processes.[38] Growers must rely on their own knowledge and awareness of their individual locality with regard to the impact of beneficial predators and what provisions they supply. The importance of soil health and fertility on plant health, however, and the suppression of disease and resistance to herbivores is unequivocal.

Overleaf is a diagram of the interactions of an aspect of the food web that is of great importance to organic growers and brewers. The earthworm is a central player in the consideration of soil fertility and we will return to it in later chapters. They have many predators including birds, hedgehogs, and numerous rodents. Earthworms have been characterised as 'ecological engineers'[39] because they adapt their environment to improve the conditions for their survival which also happens to generate excellent conditions for growing plants. Their fertile casts in various combinations provide the raw material for many of the brewing approaches to follow. It is now common knowledge that microorganisms are present in heightened concentrations around earthworm middens and galleries.[40]

In proportions varying with their species, earthworms consume partially decomposed organic matter, soil, their own casts, fungi, bacteria and other meso and microfauna. They show clear preferences for certain microbes such as some fungi,[41] which are thought to provide an improved source of nutrients[42] that are available more quickly for consumption. An interesting aspect of this process is that whilst pathogens are destroyed,[43] certain microorganisms are not killed but activated by this passage.[44]

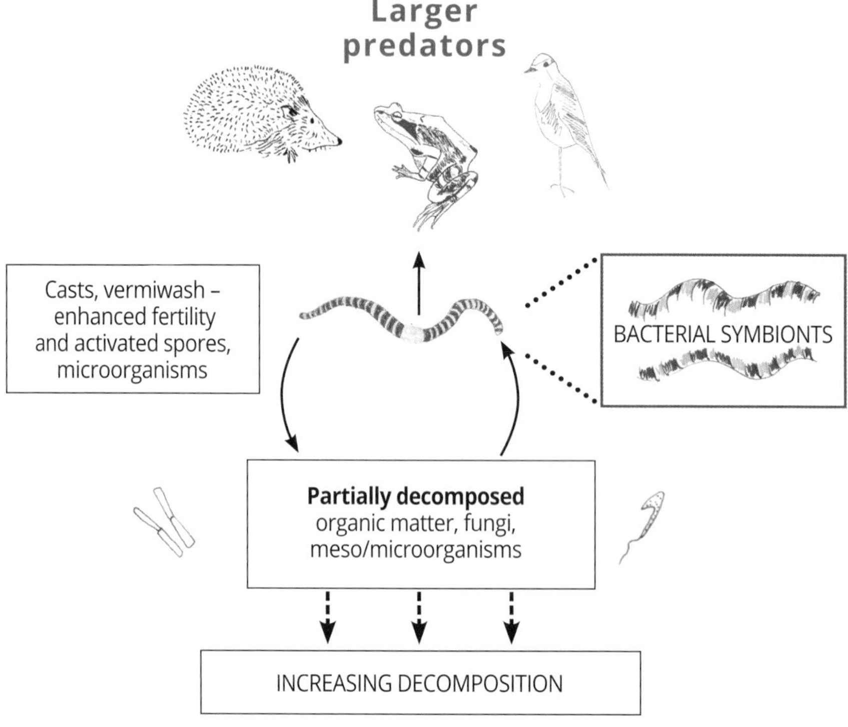

The food web for earthworms. The role of the bacterial symbionts in the earthworm's gut (nephridia) is still unclear but it is thought to be connected to protection from pathogens.[45]

Author's note

The chapters presented in this account of compost tea are very eclectic and engage with varied viewpoints. At the behest of the publishers the author has been asked to make a clearer division between what is science based and what is not. As a post-materialist[46] at heart it is not without some consternation that the author offers below some chapter-by-chapter sign-posting with regards to this matter.

In this chapter conventional views of the food web were reviewed, which have developed from considerable scientific thought. Flaws and oversights in these positions are considered from which point a new pattern for the food web was suggested (page 12). There is a great deal of empirical evidence for the food web. The tuned-up version of it, described earlier, is more relevant to the context of this book and is original to the author. It has not been subjected to rigorous scientific enquiry.

In Chapter Two the orthodox view of the essential elements is described, reviewed and critiqued. From the inconsistencies in the orthodox position on the essential elements additional concepts are suggested. From this point the author's own original view is presented along with a novel theory of co-operation between different organisms. The notion of holons (see Glossary), has been repurposed to be applied within this novel context. With regards to indicator plants the author would assert that the concept is scientifically sound but good literary sources are extremely difficult to come by. During initial research there was a vast amount of grey literature around indicator plants with many texts referring to texts with no scientific basis, which could reasonably be viewed as folklore, although it was not presented as such. To date no other text but this one exists as far as can be discerned which brings together so much useful information on indicator plants coupled with sound references.

Chapter Three draws from the established mainstream theory of nutrient cycles albeit from the perspective of a small-scale organic grower.

In Chapters Four and Chapter Five the organisms described are drawn from a background of natural history and scientific enquiry. Special emphasis is allocated to the behavioural traits of the various organisms within a context of growing crops. No mythical beasts are presented here.

In Chapter Six we use plant extracts and animal byproducts to feed plants. This in itself is not in any way controversial. The author has gone to great lengths to establish that certain essential or useful elements are present in the plants or other additions suggested and has taken great pains to draw from scientific studies. Of course the actual preparations are entirely original to the author and are derived from personal experimentation and practice with no large-scale science available to verify each individual preparation. The feeding suggestions at the end of the chapter are based on requirements of various food crops based on scientific and industrial studies of the plants at different stages in their development.

In Chapter Seven the use of plant extracts to defend plants is in no way controversial. Additionally although relatively new to contemporary science the use of plant extracts to trigger immune responses or to attract beneficial organisms is a part of established knowledge.

In Chapter Eight, conventional science based practice is in the centre ground. Using soil life in an enhanced, optimised way to speed up the composting process, break down seeds and disease is very much established mainstream practice. The way it deviates is that the welfare of living organisms, other than human, are given greater compassion and consideration.

The reference to Steiner's work regarding compost activators was included for the independent thinker to reflect upon and no inference was intended that it should be considered as being within science.

Chapter Nine. The use of aerated compost tea is a contested area within science. The arguments are finely balanced between it working and it being ineffective. Many studies are inconclusive. The concept of using beneficial organisms to improve plant growth is valid in theory, but obtaining proof is a more difficult proposition.

Chapter Ten. The topics in this chapter work from the fringes of science to outright abandonment. There is a great deal of activity in compost tea practice worldwide within this area.

The weight of evidence today suggests that the classical division between the observer and the observed has been refuted,[47] which leaves the whole basis of scientific enquiry on uncertain ground. If this discourse is not to become faith based, the real question now is not if consciousness affects matter but the scope of degree, and its implications. This is particularly cogent with regards to working with complex systems.[48] So whilst it is still a simple matter to suggest what is science or is not science as it is viewed in conventional mainstream practice, rather more difficult is suggesting what is effective or ineffective. When practices that work time and again are not accepted by the orthodoxy it poses a question to science itself and its boundaries.

The hard division noted above is one of the central tenets of contemporary science and depending on your philosophical persuasion can have profound implications for many of the practices that are considered non-rational and out of science. This topic is expanded upon further in Chapter 10. Furthermore there are many, many other reasons why certain practices are not endorsed by the orthodoxy and some of these reasons are difficult to dispute.

To help clarify matters the status of each main topic is dealt with very briefly below:

- In the introduction to Chapter Ten, a rationale for alternative practice is suggested that has a robust background in related evidence. The author would assert it is reasonable and within science although very unorthodox. Many of the implications of the above research described further in Chapter 10 simply have not been considered within this frame of reference before. A possible illustration of this point is the author's original outline of an approach to compost tea preparation suggested in connection with the butterfly effect which is logically reasonable if the evidence described is reviewed without prejudice, but is certainly very much outside of the conventional mainstream.

- Biocrystallisation mentioned in Chapter One and Chapter Ten is not in the mainstream of science but in a future where better systems of measurement are developed they may become so. The holistic water test could be considered a close cousin to biocrystallisation and is not endorsed by science partly because of Masaru Emoto's poor treatment of the topic and again partly because the results are difficult to assess scientifically beyond an intuitive level. Furthermore any theories noted in the text linked to vitalism, chi, vital energy and so forth are rejected by mainstream Western science because of a lack of evidence.

- The Natural Farming approach overall is not well researched and many of the practices such as those related to biodynamics are not endorsed by science. On the plus side the indigenous aspect of Natural Farming with regards to microorganisms has a strong background in research, which is discussed further in the text. The preparations are original to the author and do not have a research background other than anecdotal.

- Planting by the Moon has a good grounding in conventional research today. From a strictly scientific perspective the other cycles (with the exception of the Sun) have support on an energetic level but their effect on plants is generally purely speculative. There are notable exceptions in the text to this broad observation and the reader must bear in mind that simple physics dictates that water will be affected by these cycles whether in vivo or in vitro. In the context of the food web as a whole there is very clear evidence that many organisms are affected by these cycles.

- When we come to the *Tropical Zodiac* and *Sidereal Zodiac*, we are well into the twilight zone now. Applying these Zodiacs to growing is a non-rational practice because there is no scientific explanation for any positive results achieved. What is rather confounding is that when we come to the astrologically related Earth days and so forth, there is a reasonably good research background in studies that on certain astrologically derived days there is increased plant growth. The combinations of different planetary cycles in growing is highly speculative and an artefact of the author with no basis in conventional research.

In the interview section in the free pdf download <https://shop.permaculture.co.uk/compost-teas-the-interviews.html> the questions presented draw upon established science and also maintain an open stance towards alternative practice. The questions chosen initially reflected the interviewee's background and further questions developed during the course of the respective interviews.

Appendix One draws from established science and the history of science. An open stance towards alternative practice is maintained in congruity with the rest of the book.

Finally in Appendix Two Steiner's biodynamic preparations are described. The descriptions of processes and effects must not be viewed as within established science. The benefits of the amendments stated in the text are interpreted from his lectures and effects must not be viewed as within established science.

CHAPTER TWO

Essential Nutrients

Nothing happens in living nature that is not in relation to the whole.[1]

Knowledge of the essential elements is one of the cornerstones of crop production which we plan to harness in various preparations. The generally accepted definition of an essential element or more loosely, a nutrient, is described according to Arnon and Stout's three criteria (1943).[2] The first and most important is that for an element to be essential the plant will not be able to complete its life cycle without it. The second, which is rather encompassed by the first, is that an essential element's function cannot be replaced by a different element and finally the element must be directly involved in the plant's reproduction and growth (primary metabolism). This third criterion separates the roles played by some less than essential elements in secondary metabolism such as adapting to the environment or defending against enemies. A final point, which was not made explicit by these criteria, was that the element needed to be essential for most plants.

Currently 17 elements have been found that conform to Arnon and Stout's criteria[3] that are shown below. During certain periods over the last hundred years, authorities pronounced that all essential elements had been found only to be refuted with the announcement of a further element to add to the list. To date, the latest element to be added was nickel in 1987.[3]

There is still considerable debate over what is an essential element and matters are certainly not as clear cut as some texts would have us believe.

Essential Elements for Plants

Macronutrients[2] [> 0.1% of Plant Dry Mass]	Non-Mineral	Oxygen [O], Hydrogen [H], Carbon [C]
	Primary	Nitrogen [N], Phosphorus [P], Potassium [K]
	Secondary	Calcium [Ca], Sulphur [S], Magnesium [Mg]
Micronutrients[2] [< 0.01% of Plant Dry Mass]	Trace	Iron [Fe], Manganese [Mn], Copper [Cu], Boron [B], Zinc [Zn], Molybdenum [Mo], Chlorine [Cl], Nickel [Ni]

The primary nutrients and non-mineral elements are required in the largest quantities, then the secondary elements in lesser amounts. Following this the trace elements are required in minuscule quantities. The elements required in larger volumes do not make them more important than those required in smaller volumes. This relates to Sprengel's well-known Law of the Minimum, which was popularised by Liebig,[4] where the essential element that runs out first becomes the element most limiting to plant growth.

One of the methods scientists use to search for clues to essentiality is to analyse what is in plant tissue. The table overleaf is a breakdown of the amounts of the elements in corn tissue; although the figures vary from plant to plant, this gives a good indication of how the volumes are distributed. What is immediately striking about these volumes is the significance of the non-mineral elements, which highlight the importance of sunlight, moisture and well aerated soil for what are primarily carbon fixers.

Whilst food science was developing in past times it was very useful to have the criteria for essentiality, for instance when supplementing the nutrient deficient soils in Australia during the latter half of the 19th century,[10] but as time passed issues began to emerge. If just the essentials were added to a crop, they would survive, but there was no guarantee that the crop would do well, so Nicholas (1961)[11] added further elements to the list that were found to be just beneficial. For instance, sodium can perform some of the functions of potassium that would not be considered essential by Arnon and Stout's criteria but would be essential by this new definition. If an element was considered beneficial it had to have some positive effect on the plant in the right quantity. These additions made the theory of what was essential more in line with the approach of animal physiologists who held to the broader criterion that what was essential could just enhance growth and development.

Reviewing mineral element volumes overleaf in isolation can be misleading because plants accumulate certain elements that appear to have no benefit to them. Plant roots are known to be able to adsorb nutrients preferentially but it is also true that certain elements just happen to get adsorbed from their immediate proximity.[12]

There may be another explanation as to why plants adsorb disproportionate amounts of some chemicals for no direct reason. It could be an artefact of the way humans view plants in isolation from the system they exist in. In nature plants are thought to have grown and evolved in guilds, which can be described as beneficial groupings of plants. A classic, albeit human influenced example is 'The Three Sisters' where the corn provides support for the runner beans and the squash provides groundcover. Furthermore the possibilities for guilds do not remain solely in the plant kingdom; animals, bacteria and fungi can also be members.

Arthur Koestler has often been misrepresented[13] as suggesting that nature is a hierarchy of wholes. What he was really saying was more subtle and he used a new word 'holons'[14] to describe something that was whole and yet was intrinsically

Element Distribution in Corn

Element	% Dry Weight	Major Functions
Oxygen (O)	45	Essential component in plant structure. Major element in aerobic processes.
Carbon (C)	44	Essential component in plant structure. Present with hydrogen in all organic molecules.
Hydrogen (H)	6.3	Essential component in plant structure. Present with carbon in all organic molecules.
Nitrogen (N)	1.3	Nitrogen processes are a major factor in stem and leaf growth. Major component in proteins, vitamins, chlorophyll, hormones and enzymes.
Silicon (Si)	1.2	Proven non-essential in lab studies. May have an essential protective role in plants grown in harsher conditions.[5] Hardens plant tissues and protects plants against pests.
Potassium (K)	0.9	Necessary for making protein, sugars, starches, and carbohydrates. Important in plant metabolism. Improves cold hardiness and stem rigidity.
Calcium (Ca)	0.25	Essential component in plant cell walls. In acid soils increases availability of other nutrients, hence lime ($CaCo_3$). Sustains nodule activity in legumes.
Phosphorus (P)	0.16	Essential for flower and fruit formation. Needed for seed germination. Factors in nearly all aspects of growth and metabolism. Particularly useful in legumes because it aids nitrogen fixation.
Magnesium (Mg)	0.16	Important component in chlorophyll used in photosynthesis. Regulates uptake of other minerals.
Sulphur (S)	0.15	Important component of many proteins. Aids in chlorophyll synthesis and root growth. Increases nodule activity in legumes.
Chlorine (Cl)	0.15	Involved in a part of photosynthesis reacting with water.
Aluminum (Al)	0.11	Not essential but absorbed from the environment.
Sodium (Na)	0.03	A beneficial nutrient. Not essential to most plants[6] but essential to animals.
Iron (Fe)	0.009	Although not a component of chlorophyll it is essential in its synthesis.[7]
Manganese (Mn)	0.006	Speeds up photosynthesis, stimulates crop growth and essential for combining with proteins.[7]
Zinc (Zn)	0.003	Essential component of enzymes. A component in Carbon Anhydrase which is involved in an important aspect of photosynthesis in some plant groups.[7]
Boron (B)	0.001	Essential in seed production, regulates auxin (growth hormone), important in root growth, factors in nitrogen processes and is thought to have a direct effect on sugar synthesis.[7]
Copper (Cu)	0.0005	Is factored in numerous enzyme systems including the production of amino acids, proteins and root metabolism.[8]
Molybdenum (Mo)	0.0001	Essential to most organisms and to the nitrogen metabolism of plants. Essential for nitrogen fixation.[7]

Adapted from *Soil Testing and Plant Analysis* (1973) Leo M., Walsh D. and Beaton J.D. (Eds)[9] to include other organisms. Silicon, Aluminium and Sodium (in italics above) are not considered to be essential for plants.

part of a greater whole. Guilds can be seen rather like Koestler's holons, simultaneously pursuing their own survival whilst intrinsic within a broader cycle. A similar concept using different terminology is described in Bill Mollison's account of how a large area of land was remediated from a jungle of exotic weeds by someone setting up small 'nuclei' or clumps of companion plants which 'acted as one'[15] to recolonise the area expanding out from a secure stronghold.

The thesis suggested here is that plants function as guild members adsorbing the seemingly redundant nutrients to help the survival of the guild as a whole. Many of these chemicals such as silicon or selenium are essential to animals, which in some way they may repay in other services such as providing nutrients or seed dispersal. For instance, in winter, a hungry flock of pigeons may raid a brassica plant, replete with essential traces et cetera and then in turn the pigeons supply guano full of phosphorus for the next generation of seeds. If this thesis is true then a further interpretation of essential nutrients could be viable to encompass essentiality for dissimilar organisms 'acting as one' rather in the manner of a farm interacting as a single entity.

When considering the essential elements, we need to be aware of a further issue that has to be addressed if we are intending to eat our produce or indeed feed it to other animals. We need the crops to supply our own nutritional needs. There is considerable literature on food deficiencies in less technically developed countries but it should not be surprising that deficiencies are also emerging in 'developed' countries,[16] given the preference for highly processed foods and flawed agricultural practice. A topical example is a deficiency in selenium in British grown wheat[17] after companies switched from selenium rich American sources.

Below is a diagram that illustrates the established differences in essential elements between plants and animals. You will notice that the majority of plant

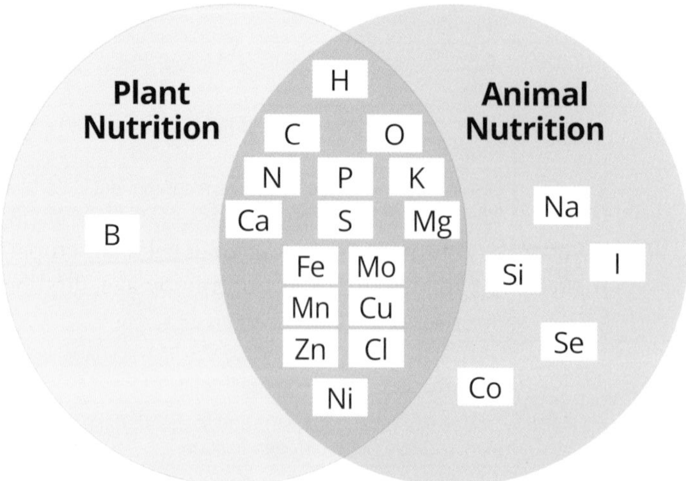

A Venn diagram of essential and beneficial elements in plant and animal nutrition.

and animal essential elements are the same. Currently sodium (Na), iodine (I), selenium (Se), silicon (Si) and cobalt (Co) are only considered to be essential to animals and beneficial to plants.

With regard to cobalt, another weakness is illustrated with the universal one -size-fits-all criterion for essentiality. It is essential but only to certain groups of nitrogen fixing microorganisms[18] needed by higher plants[19] to supply them vitamin B12, also essential to humans.[20]

Considering the current facts, further distinctions for essential elements need to be considered between higher plants, microorganisms and fungi, which emphasises beyond doubt the web of interdependence that we need to continue our existence.

Indicator plants, soil testing and beyond

The table on p.28 describes tell-tale signs of various deficiencies in your crop, however visual symptoms can be misleading and can only be used as an indication. Symptoms can sometimes be connected with other factors[21] such as acidity (which alters the solubility of various chemicals) and soil structure. Often deficiencies or excesses in one element can interfere with the uptake of other elements for example an excess of potassium can interfere with the uptake of magnesium and vice versa.[6]

A more reliable approach in some ways would be to send a sample to a lab for analysis but many small-scale growers may not be inclined to spend their limited budget on these services. Although such a test would give a solid indication of what is in the soil, there may be a disjunct between what exists in the sample and what is available to plants and other organisms. The lab test offers a classic example of an attempt to reduce an ecological phenomenon to chemistry, useful as an indication amongst others but potentially misleading. For instance, a result may indicate that there is plenty of phosphate, but without an indication of which forms in relation to soil fertility, could lead to poor choices in practice. An over emphasis on chemical analysis distances the grower from the soil and undermines the importance of soil health for sustainable fertility. In practical terms, there is no reason why a grower cannot do hard scientific tests and also explore less mainstream, more intuitive methods of gaining insight into the land. During actual practice however, the moment you begin to develop an empathy or more for your living surroundings, the less you want to microwave, dissolve, drown or suffocate live parts of it, so it can be rather a one way path after a certain point.

If these caveats are borne in mind, a few basic tests can add valuable gauges for consideration. Overleaf is an outline of the tests available to anyone that will be useful when consulting further chapters and selecting the most suitable preparations for your soil.

This section needs to be considered holistically in conjunction with the soil indicators described on pp.25-6. If you want to do more detailed tests, the United

States Department of Agriculture (USDA) website and 'Free Soil Test Kit' is a fantastic resource where a highly empowering approach is taken to describe step by step how quite a sophisticated soil test kit can be constructed, applied and interpreted at minimal expense.[22]

The basic soil pH test

These tests are widely available from garden centres and online. This is one of the most important checks you can do and is discussed further on p.26. An easy option is to obtain a handheld meter, which usually has multiple functions such as temperature and humidity included. Bear in mind that the pH meters only tend to work effectively where there is at least some moisture present.

Basic primary elements test (NPK)

This can give a handy rule of thumb indication but such tests can be misleading as previously discussed.

Secondary element tests

These are more difficult to find than the NPK tests and are more expensive. If you happen to be located in North America, some state agencies offer free testing for calcium, magnesium and other miscellaneous soil tests.[23] If you are spending over £40 ($50) on testing equipment, a professional soil test might be worth considering. Such tests can be re-appropriated from the pitch testing industry for comparable amounts.[24]

Soil compaction

Soil compaction is linked to bulk density (weight/volume). It can have a profound effect on the movement of nutrients, limit some soil organisms and obstruct plant roots ($>1.6g/cm^3$ is poor). We will return to this topic in Chapter Nine where soil improvement is addressed. If you have trouble getting your spade into the soil or have trouble penetrating the soil a few inches down in certain areas, you may have compaction. For your information, if you happen to like cracking walnuts with sledgehammers, there is a USDA test for it.[25] Furnished with these data, penetrometers (compaction testers) can be rather pricey and somewhat unnecessary. Experiential gauges can also be calibrated by comparing penetration in areas of soil frequented by heavy vehicles, playing fields and areas protected from footfall and rain.

When considering soil fertility and soil health, a lab test for soil carbon in its varied forms may also come in handy when deciding on amendments from Chapter Nine in particular.

Moving on from these tests the grower can take a walk around their land and identify what is growing. Season, climate and microclimate would also understandably factor heavily. Certain plants have a preference for certain soil conditions and invariably spring up when the conditions are right. Conversely some indicator plants such as bracken fern accumulate missing minerals in the same areas that are shown to be deficient.[26] The presence of certain plants gives us an indication of the soil's fertility, where there may be excesses or deficiencies, its acidity, and provides insights into a variety of other soil conditions. Below is a table of well-known indicator plants.

Indicator Plants

Indicator Plant	Soil Indication	Emergence
Bracken fern (*Pteridium aquilinum*)	Low K, Low P, acidic[26]	Spring[27]
Annual nettle (*Urtica urens*)	Acidic or low lime, tilled soil,[27] rich fertile soil[27]	Spring to Autumn[27]
Black bindweed (*Fallopia convolvulus*)	Compacted soil, poor drainage[26]	Spring[27]
Coltsfoot (*Tussilago farfara*)	Poor drainage, heavy soil[26]	Spring to Autumn[27]
Perennial sow thistle (*Sonchus arvensis*)	Poor drainage and alkaline soil[26]	Spring to Autumn[27]
Pineappleweed / Mayweed (*Matricaria discoidea*)	Hardpan[26]	Spring and Autumn[27]
Chickweed (*Stellaria media*)	Common in tilled soil[26]	All year[27]
Fat hen / Lamb's quarters (*Chenopodium album*)	Common in tilled soil[26]	Spring and Autumn[27]
Hedge mustard (*Sisymbrium officinale*)	Common in tilled soil[26]	Spring[27]
Shepherd's purse (*Capsella bursa-pastoris*)	Saline soils[26]	All year[27]
Dandelion (*Taraxacum officinale*)	Heavy clay soils[26]	Spring to Autumn[27]
Rhododendrons (and all other Ericaceous plants)	Acidic soil[28]	N/A

Some common sense is required here: people regularly put in plants that are poorly adapted to prevailing conditions and some plants keep going for a considerable time even when less favourable conditions are created. What we are looking for is an overall impression of plants that show some dominance in the area. If there are some uncultivated patches around, this can be very useful in this instance.

This is just a small illustration of what is possible in this area, which is often restricted by the availability of solid unequivocal information. When assessing an area, local knowledge may exist by word of mouth with very little put to paper and much finally comes down to the individual experience of the grower. Beyond this the grower possesses a special tool in their armoury that is often overlooked – their own awareness. The attentive grower should bear in mind that other cues are also important and should not restrict themselves to just looking at plants or doing soil tests, for example the presence of earthworms is a good indicator of soil fertility; but if they appear discoloured there may be a dangerous excess of some heavy metal.

All the various indicators can be put together like the pieces of a puzzle and point to answers that do not just rely on one set of facts, applying intuition and subtler features of the environment that are less amenable to established tests.

Acidity (pH) is possibly the single most important factor that affects mineral uptake in plants. It can lock up elements creating deficiencies or help release them for plant growth. The table overleaf shows the best pH for a selection of well-known crops. As an example of how you could use such a table, if you find that a certain part of your land is excellent for growing potatoes but a nearby cobnut is lacking in vigour, this could be an indication that your soil is on the acidic side. It is important to bear in mind that the pH scale is exponential so something with a pH of 5 is ten times more acidic than pH 6 so a difference of just one point can be very significant.

This process can be applied in reverse: if you look into what conditions favour a particular plant you can engineer the best conditions for it to grow. A classic example is to obtain an acidic soil for growing blueberries that prefer more acidic conditions. Consider a further example in a companion planting context; garlic (allium) grows well in the earlier and later periods of the growing season and contains a high proportion of sulphur. Clovers favour a higher content of sulphur to grasses,[33] so this suggests a favourable interaction for a tree guild. Garlic can thrive around a well-drained apple tree and supplies extra sulphur to the clover as its scapes break down which helps to resist the grass, in conditions that better favour the clover. When the clover plant dies it feeds the tree with its high nitrogen content. Applying this 'condition engineering' to compost teas if you have a field that is overly dominant with grass (which is known to compete and restrict tree growth), you can make a compost tea with a high sulphur content, apply it and sow some clover to compete with the grass. For legumes, in terms of ideal proportions[34] a ratio is thought to exist between nitrogen and sulphur of 8:1. This infers that as nitrogen feeds go up so must sulphur or it will run out and plant growth will be restricted.

There are two approaches to elevate specific nutrient levels once a deficiency is suspected or has been found. The first is to add it directly in the appropriate form. For non-organic gardeners, this would be the addition of a synthesised chemical. For the organic gardener either the addition of a chemical with organic status (e.g. certain rock minerals) or to find a plant that has a high proportion of that element as shown on p.29. When working with materials that contain minerals in heightened levels some caution is required because the quantities needed are

Favourable pH conditions for different crops

Family	Crop	pH
Solanaceae (Solanaceous)	Tomato[29]	5.5-7.0
	Potato[29]	5.0-6.0
	Chilli[29]	6.0-6.5
Alliaceae (Genus: *Allium*)	Leek[29]	6.5-7.5
	Garlic[29]	6.5-7.5
	Onion[29]	6.5-7.5
	Chives[29]	5.5-8.5
Umbelliferae/Apiaceae (Umbellifers)	Carrot[30]	6.0-6.75
Cucurbitaceae (Cucurbits)	Cucumber[30]	6.0-7.5
	Squash[30]	6.0-7.5
Polygonaceae	Rhubarb[31]	5.0-6.8
Brassicaceae (Brassicas)	Brussels sprouts[29]	6.5-8.0
	Cabbage[29]	6.5-8.0
	Radish[30]	6.0-7.5
	Swede[29]	6.5-7.5
Leguminosae (Legumes) (The more archaic and equivalent family grouping 'Fabaceae' is often used with broad beans)	Peas[31]	6.0-6.2
	Runner beans[31]	5.5-6.8
	French beans[30]	6.5-7.5
	Broad bean[30]	6.0-7.5
Amaranthaceae	Beetroot[29]	6.5-7.0
Asteraceae (Compositae)	Jerusalem artichoke[30]	6.75-7.5
	Lettuce[30]	6.0-7.5
Poaceae (Gramineae)	Sweet corn[29]	5.5-7.0
Rosaceae	Apple[30]	5.25-6.75
Grossulariaceae	Gooseberry[30]	5.25-6.0
Ericaceae (Ericaceous)	Blueberry[32]	4-5-5.5
Betulaceae	Cobnut[32]	7.5-8.0

so small it is easy to overdose, although with natural materials this concern is less serious because they are intrinsically more dilute.

The second option is to use a catch-all such as a feed that has a wide variety of essential nutrients. This is recommended because it is in a form that does not risk an overdose, it is a more holistic approach and it will also cover you for other deficiencies that are difficult to pinpoint directly.

Troubleshooting nutrient deficiencies

Deficiency	Visual Signs of Deficiency
Oxygen (O)	Growth below normal.
Carbon (C)	Plants become dwarfed.
Hydrogen (H)	Plants become dwarfed.
Nitrogen (N)	Retarded growth. Older leaves grow paler from light green to yellow then die.
Potassium (K)	Slower growth. Mottling, curling, spotty older leaves. Burnt effect on leaf edges. Weak stems and roots.
Calcium (Ca)	Root growth lessened. Terminal bud death and misshaped leaves.
Phosphorus (P)	Slower growth. Purple hues in veins of older leaves. Less fruits and seeds.
Magnesium (Mg)	Leaves droopy. Older leaves grow paler from light green to yellow then die.
Sulphur (S)	Yellow or pale green leaves. Lessened growth.
Chlorine (Cl)	Leaves droopy. Older leaves grow paler from light green to yellow and a few die.
Iron (Fe)	Older leaves grow paler from light green to yellow then die. Grasses take on yellow and green stripes.
Manganese (Mn)	Leaves turn white and drop. Major leaf veins more green than normal and leaves pale green.
Zinc (Zn)	Older leaves grow paler from light green to yellow then die. Misshapen roots. Bronzing or mottling of leaves.
Boron (B)	Essential for seed production.[5] Laterals die, terminal buds die. Thickened, curling leaves turn brittle. Many severe disorders from deficiency. No.1 in world for micronutrient deficiency.[6]
Nickel (Ni)	Minor deficiency, no visual symptoms. Higher deficiency, chlorosis along necrotic leaf tips. A component of plant urease. Essential for growth when urea is the sole source of nitrogen.[6]

Adapted from Williams (1992).[35]

Accumulator plants

Accumulator plants and elements that are present in heightened levels

Accumulator Plant	Elements Present in Heightened Levels
Yarrow (*Achillea millefolium*)	Phosphorus[36, 37] Potassium[36,37] Calcium[37] Magnesium[36,37] Copper[36,37]
Stinging nettle (*Urtica dioica*)	Nitrogen[38] Potassium[36] Calcium[36,38] Iron[36,38] Copper[36] Sulphur[38] Iron[38]
Kelp (order *Laminariales*)	Sodium[39] Iodine[39] Nitrogen[39] Magnesium[39] Calcium[39] Iron[39]
Comfrey (*Symphytum officinale*)	Nitrogen[36] Phosphorus[36] Potassium[36] Calcium[36] Iron[36] Manganese[36]
Garlic (*Allium sativum*)	Sulphur[40]
Indian mustard (*Brassica juncea*)	Manganese[41] Nickel[41] Copper[41] Selenium[41] Zinc[41]
Bracken fern (*Pteridium aquilinum*)	Potassium[39] Phosphorus[39] Manganese[39] Iron[39] Copper[39] Cobalt[39]
Horsetail (*Equisetum arvense*)	Silicon[36,42] Cobalt[36] Magnesium[36]
Clover (and other legumes) (*Trifolium* spp.)	Nitrogen [36] Sodium (White Clover)[7] Phosphorus[39] Molybdenum[43]
Sunflower (*Helianthus annuus*)	Manganese[41] Nickel[41] Copper[41] Potassium[36] Silicon[44]
Oak bark (*Quercus* spp.)	Calcium[39,45]
Dandelion (*Taraxacum officinale*)	Calcium [36,37] Iron[36] Copper[36,37] Potassium[37] Magnesium[37] Sodium[37] Cobalt[37] Phosphorus[37]

We finish this chapter with some photos of some of the indicator plants we have mentioned to aid you in identification. In the next chapter we take a look at how some of these elements cycle through the grower's fields and the broader environment which will help gain an insight towards improving soil fertility and enhancing crop growth.

Common Indicator Plants

First row: Perennial sow thistle, black bindweed, coltsfoot.[46]

Second row: Pineappleweed, chickweed, lamb's quarters.

Third row: Annual nettle, hedge mustard, shepherd's purse. Source (shepherd's purse and nettle) H. Zell.[46]

CHAPTER THREE

Nutrient Cycles for Growers

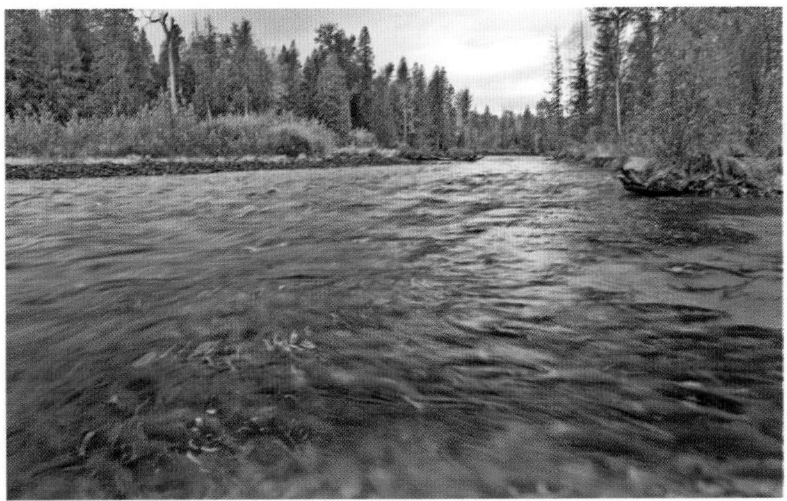

Sockeye salmon in Adams River, British Columbia. Photo: Katie Edmonds.[1]

The salmon from the great forests of British Columbia was chosen to begin this chapter to highlight the power and subtlety of natural systems. It is thought that without this fish, or possibly its passion, the forest would never have been.[2] As they swim upstream to their ancestral mating grounds they are predated, often by bears, and the remains replete with nutrients from the seas are discarded to decompose for other organisms.

I devote the following few pages to four of the most important cycles for growers – carbon (C), nitrogen (N), phosphorus (P) and potassium (K). They have been approached in a classical way with four diagrams showing the path of each of these elements, however the reality is a great deal more complex and interconnected.

With regards to carbon, recently there has been considerable research done relating to concerns over global warming, and the greenhouse effect that is intimately linked with the carbon cycle. Without wishing to detract from the gravity of these matters regarding the survival of humanity, the focus of this chapter is simply on developing an awareness of how the carbon cycle and the other important cycles relate to growing, its interaction with your growing area and how this can be used to enhance crops. Most importantly to tea brewers is the part of the carbon cycle related to the respiration of microorganisms, which we will look into further in this chapter and harness in a practical way in Chapters Eight, Nine and Ten. Regarding N, P and K these are the mineral elements that are needed in the largest quantities and tend to be the first to run out.

The carbon cycle

Carbon is the basis of life on Earth being central to photosynthesis and a structural component in many organisms due to its special bonding properties.

In an open environment the carbon dioxide given off in soil and plant respiration is added to the overall atmosphere and has less of an impact on proportions on a local level. But in closed spaces such as a greenhouse, variations can be tilted in the grower's favour. A well-known practice in large-scale horticulture is the use of carbon dioxide generators since increases in the proportion of carbon dioxide are known to enhance plant growth. Brought down to the small-scale, an interesting

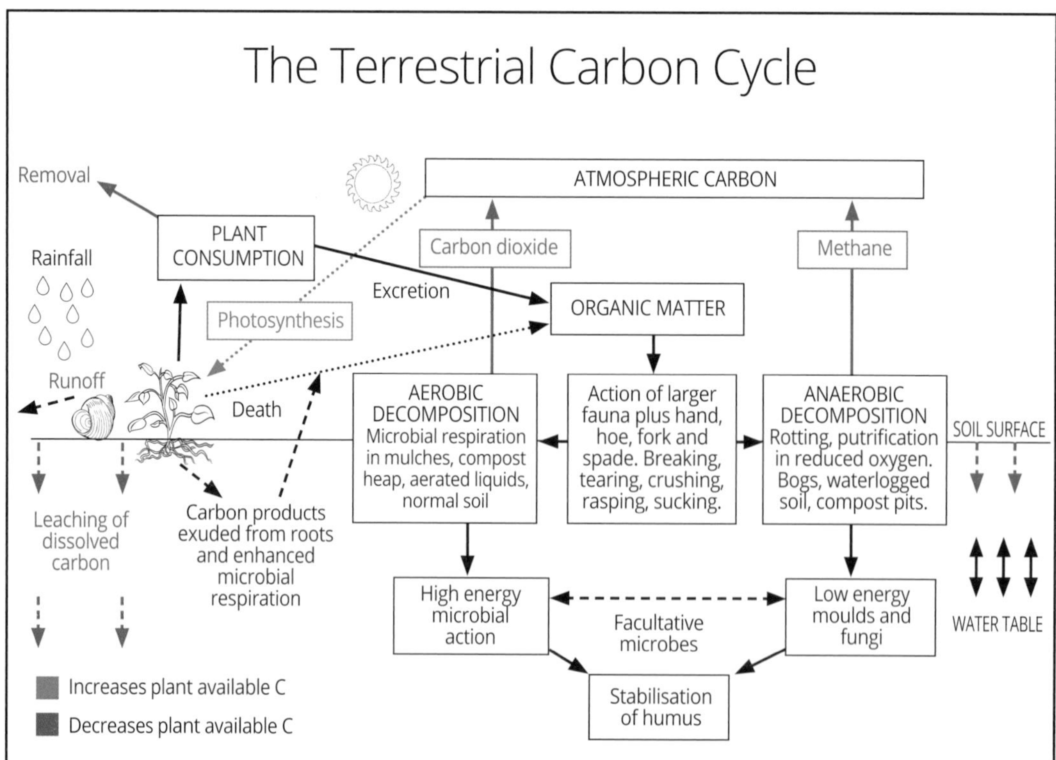

and naturalistic exercise in synergy could be the addition of a demijohn to the greenhouse, which is known to produce carbon dioxide during the wine making process. In Chapter Ten we do a partially fermented compost tea which would also be a novel addition to the greenhouse while it is being prepared, and after.

In the very partial view of the carbon cycle (left), plants fix carbon via the chlorophyll in their leaves using carbon dioxide and sunlight from the atmosphere to produce carbohydrates (energy) and oxygen. Within the plant this fuels activities that require energy and a significant amount is exuded through the plant's roots to fuel microbial activity. The plant is in no way a passive actor within the soil and at appropriate moments cuts off the supply of carbon to the soil microorganisms resulting in death, whereby the plant can claim nitrogen from the breakdown of that organism. The kind of carbohydrates produced by the plant are very simple sugars not unlike molasses. When making our aerobic tea we will use this similarity to bypass the normal plant processes to encourage reproduction of soil life on a massive scale.

Returning to the diagram, when the plant itself dies it follows the natural process of decomposition, which has one major branch. On the left side the aerobic processes take in oxygen from the atmosphere and produce carbon dioxide and on the right the anaerobic processes predominate, where alternatives to oxygen in the atmosphere are needed for their functioning. Although anaerobic reactions are usually presented as reactions not requiring oxygen, strictly speaking this is not the case since they still use oxygen but these organisms just do not draw it from the atmosphere. The products of anaerobic activity include carbon dioxide but also methane (contains carbon) and other foul smelling gases such as hydrogen sulphide.

The aerobic processes are faster and higher energy than the anaerobic processes and in the absence of oxygen things can switch to anaerobic very rapidly. Certain facultative microbes have the ability to adapt to both sets of conditions[3] as illustrated and during aerobic processes there is always some kind of anaerobic process going on in a diminished way.

In normal soil that is in a healthy condition for growing crops, aerobic processes are dominant whereas if soil is badly compacted or waterlogged things can turn anaerobic.

When organic matter breaks down there is a considerable amount of biological and chemical activity, which after a few months or years gradually slows down. Over considerably longer time spans this material continues to stabilise if left undisturbed.[3]

Take note of the shell on the left of the diagram; these little packages of calcium and carbon are often bemoaned by the grower and in extreme situations their residence can devastate crops. In a balanced organic system, they have a useful function as when the snail dies the calcium carbonate of its shell breaks down and the calcium is free to bond with hydrogen ions, reducing acidity.

The nitrogen cycle

A plant uses nitrogen in substantial quantities, more than any other mineral nutrient. A large quantity is taken up in the vegetative phase of plant growth and nitrogen plays an important role in photosynthesis. When it is deficient your plants may appear stunted or their fruit may ripen too early. A sign to look for is if your older leaves are pale green or yellowing starting at the tip of the leaf (see below), which then gradually spreads towards the centre.

Whilst nitrogen is highly abundant in the atmosphere, it is in a very stable form that is not generally directly usable by plants. It is difficult for plants to access and is considered the most limiting factor to crop growth that can rapidly become deficient in the soil. When it becomes available through nitrogen fixation or other processes, it is quickly snapped up by the many competing organisms or leached from the soil as often occurs with the use of non-organic fertiliser. With organic fertilisers such as blood and bone or hoof and horn, these could be considered slow release because the nitrogen is gradually delivered to the soil as the organic material breaks down and is less liable to leaching. An important reservoir of nitrogen that cannot be leached away is that held or 'immobilised' within living organisms. This live reservoir is of crucial significance to someone wanting to build up a vibrant food web and healthy productive crops.

The most important nitrogen fixing bacteria for growers are aerobic. Where the soil is waterlogged anaerobic processes kick in and other de-nitrifying bacteria release nitrogen from the soil to the atmosphere, which should be avoided at all costs.

A plant deficient in nitrogen from Epstein and Bloom (2004).[4]

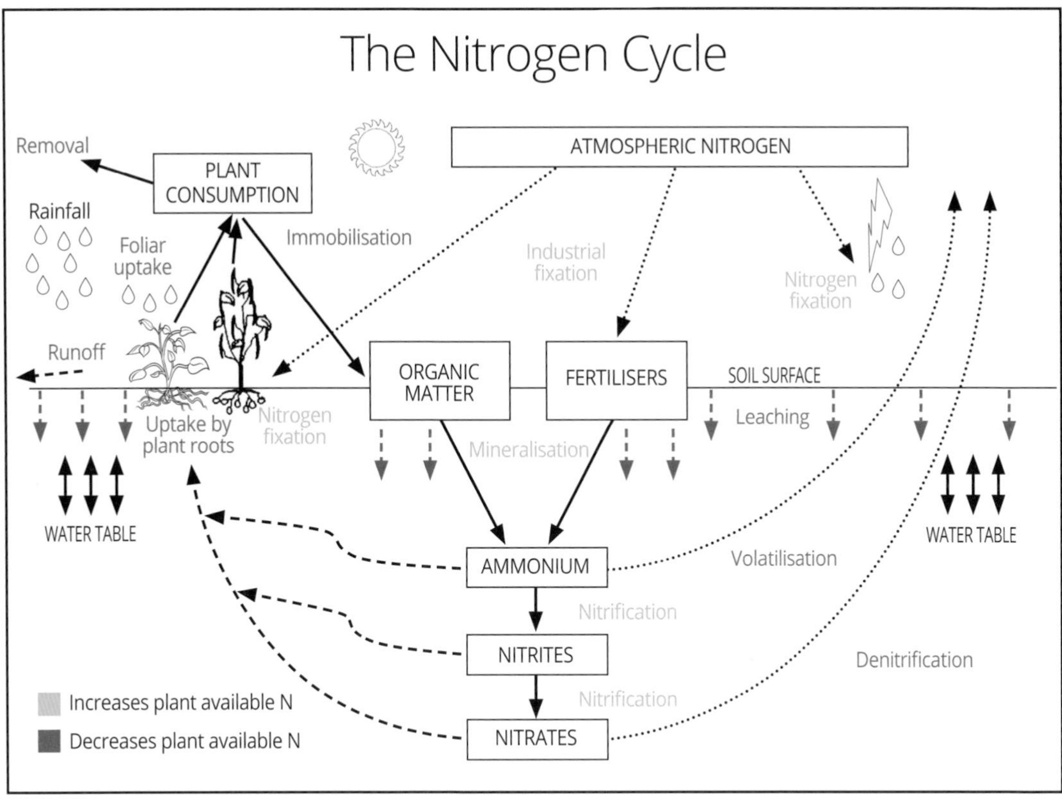

Above is a diagram of the nitrogen cycle which you participate in every time you eat a meal or excrete.

Nitrogen can come into your land fixed by symbiotic root nodules, by the input of manures, urine, not forgetting human urine, and from the decomposition of dead organisms. You may notice if you follow the arrows in the above diagram that there is no real endpoint; over time things just keep revolving. So while somewhere a nitrogen fixing *Nitrosomonas* is attached on a pea root nodule fixing atmospheric nitrogen, somewhere else nitrifying bacteria is converting ammoniacal nitrogen to nitrate. A key preoccupation for growers is to keep enough nitrogen available for uptake in the right forms at the right time for healthy plants, good crops and not to waste any. Nitrates in solution are readily absorbed by plant roots and to a lesser degree the foliage. Because of its high solubility it can be quickly consumed by competing plants or leech away. Approaches to getting more usable nitrogen to plants range from the synthetic chemical style of regular inundations of fertiliser, to controlled release pellets and the slower natural breakdown processes of organic matter favoured in organic growing. Legumes and other plants are used as green manure that fix nitrogen or absorb it in solution (centre left of diagram). Note the nodules on the roots of the plant to the right. These plants are part of a select group that form symbiotic relationships with fungi or bacteria to fix atmospheric nitrogen and hold a special place in the nitrogen cycle. We return to this topic in Chapter Four.

The skeletal outline on the previous page has profound implications for the grower and brewer of teas which we will now explore further. Shown in yellow are the processes that increase plant available nitrogen and in red those processes that decrease it. It starts with *mineralisation*, which is the microbially mediated process in the soil by which organic matter such as manure or dead plant material gets converted to a plant usable form. This process is less significant to the conventional grower because he/she bypasses it using industrially produced plant available forms involving the Haber process.

Mineralisation is essentially an aerobic process requiring oxygen and water. The first step is the conversion of ammonium in organic matter to nitrites by *Nitrosomonas* bacteria amongst others. *Nitrobacters* then take over converting the nitrites to nitrates which are very water soluble and become available in solution for uptake by plant roots. This is depicted by the plant on the far left of the diagram. Anyone wishing to focus on mineralisation in their tea making would be well advised to keep their brews between 20-35°C (68-95°F) as this is the temperature range where the nitrifying bacteria work best. Returning to the diagram, when the plant dies it will decompose and the whole process begins again.

The nitrogen cycle described so far is working in the grower's favour, however there is another aspect to this process. Where ground becomes waterlogged, anaerobic processes develop and *denitrification* kicks in. This is highly undesirable in areas intended to grow crops because nitrogen is lost to the atmosphere and all the effort involved in building up the best conditions is wasted.

Consider the bigger picture in conventional Western culture; food is harvested, eaten by humans and a substantial part is then excreted into the waste system and out to sea where denitrification takes place. So while nitrogen is never actually lost, it is lost to the field it was grown on and must be replaced in some way. In societies where less of a taboo was attached to human waste, it was returned by various means to grow more crops locally. For instance before industrialisation, China had a well-developed system of sustainable agriculture where 'night soil' was trenched, applying the ancient Indore composting method and used as fertiliser on a massive scale.[5] Whilst bearing in mind the potential of 'humanure' as a resource, care is advised because special precautions are required when composting it. Although this interesting topic is not central to this book a reference is included for further reading.[5]

> *Volatilisation* is the process by which nitrogen is lost to a field where ammonium in fertilisers and manures is converted to ammonia gas. It happens most when soil is too alkaline and is accelerated by windy and hot conditions. The muck spreader can be reassured that the manure will still retain plenty of organic nitrogen that is gradually released to the soil.

Leaching nitrates are drained from the soil by rainfall or changes in the height of the water table from below.

Immobilisation is where nitrogen becomes temporarily unavailable to plants when it is locked up within living organisms. This can be the time the organism takes to release nitrogen either through excretion or death and decomposition. Plants are known to influence the survival of different organisms in the soil, parcelling out or restricting microbial food through their roots. Many of the processes are not well understood but given the amount of nutrients released in this way it indicates that the processes are very important to the plants' biology. It is not necessarily undesirable to have nitrogen locked away in nearby organisms because it retains nitrogen that is potentially available at some point that might otherwise be lost to the immediate area.

The phosphorus cycle

Phosphorus is essential to many energy processes in plants[6] and is needed for the normal growth of stems and leaves.[7] Studies have found that there is a high proportion of P in seeds and the seedlings need plenty of phosphates in the surrounding soil to become established.[8] Hence phosphorus is very important in the early stages of a plant's life cycle. Where there is a deficiency your plant will have smaller leaves,[7] take longer to mature[9] and have a reduced yield.[7] Older plant leaves may show the first signs of deficiency by purpling as illustrated below.[7]

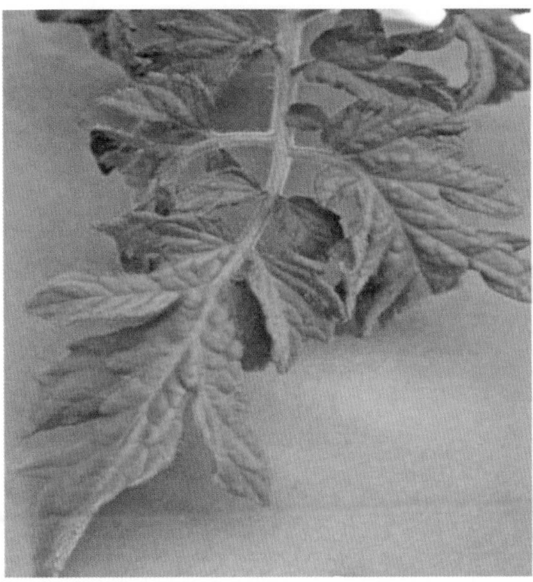

A plant deficient in phosphorus from Epstein and Bloom (2004).[4]

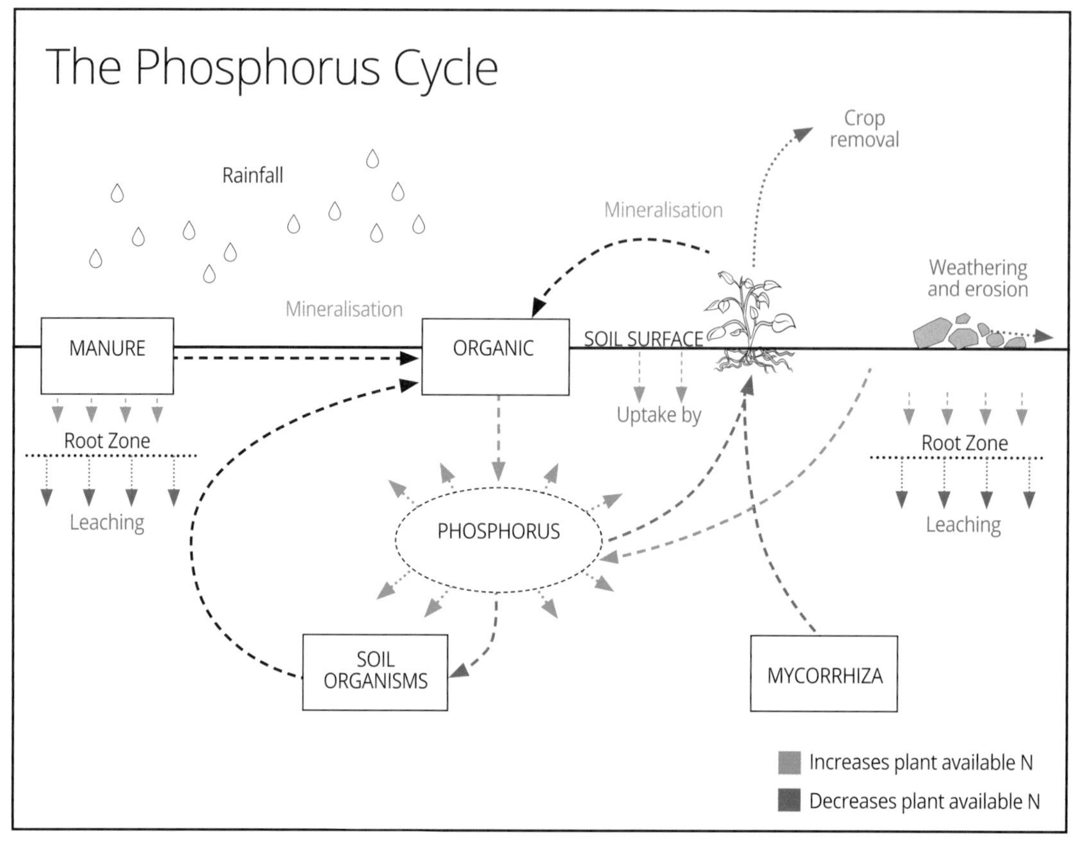

The cycling of phosphorus through your land, illustrated above, is very different to nitrogen because it is considerably less soluble and negligible within the atmosphere.[10]

Phosphorus has low mobility in the soil and its availability can be quite limited.[10] It is estimated that plant roots can only absorb phosphorus in soil water to a distance of ¼in.[11] This figure is diminished even further when the soil is dry. In cold soil phosphorus uptake is considerably reduced such as in spring,[12] so making phosphorus available to the plant early in the season at planting time can be very beneficial.

Phosphorus can easily be removed from the top layers of the soil by runoff and leaching.[13] Mycorrhiza, discussed further in Chapter Four, and some accumulator plants are very important to the location of phosphorus as their roots are known to sense and seek out sources[14] which can help bring back phosphorus from lower levels. When the accumulators shed their leaves in winter or die, the phosphorus is rescued from the lower layers and returned to the surface. Growers often overdo the additions of phosphorus and build vast reserves. If you have a pond that is choked up with algal blooms, this is a good indication that somewhere in your immediate vicinity considerably too much is being used and it is leaking in somehow.

Phosphorus is often the key growth limiting nutrient for algae[10] and if there is too much of it they tend to just keep growing.

The potassium cycle

Potassium is the third of our major nutrients necessary for all plant growth[7] and the right amount improves crop yield and quality.[15] It does not feature structurally in plants[7] but is essential to many metabolic processes.[15] Potassium also makes plants more resistant to disease and more hardy in winter.[16] Where potassium is deficient, a plant's growth is reduced: it can have a blue-green hue[17] and leaves can appear scorched as shown below. The older leaves are affected first[7] as with nitrogen and phosphorus deficiencies. Also as with nitrogen levels you can have too much plant available potassium in your soil, which can inhibit the proper growth of plants. In this regard potassium may interfere with the uptake of magnesium and calcium.[18]

Overleaf is a diagram of the potassium cycle. As with the other cycles described it is only a partial view and we have concentrated on the terrestrial processes because these are most relevant to growing crops. The key point to be aware of when exploring how potassium cycles through the soil, is that the processes are pH dependent. Potentially there is often considerable potassium in the soil except that it can often be in forms unavailable for plant uptake. When fertiliser, weathered minerals or organic matter are applied, only the potassium ions in solution leach into the ground and are immediately available for plant uptake.

A plant deficient in potassium from Epstein and Bloom (2004).[4]

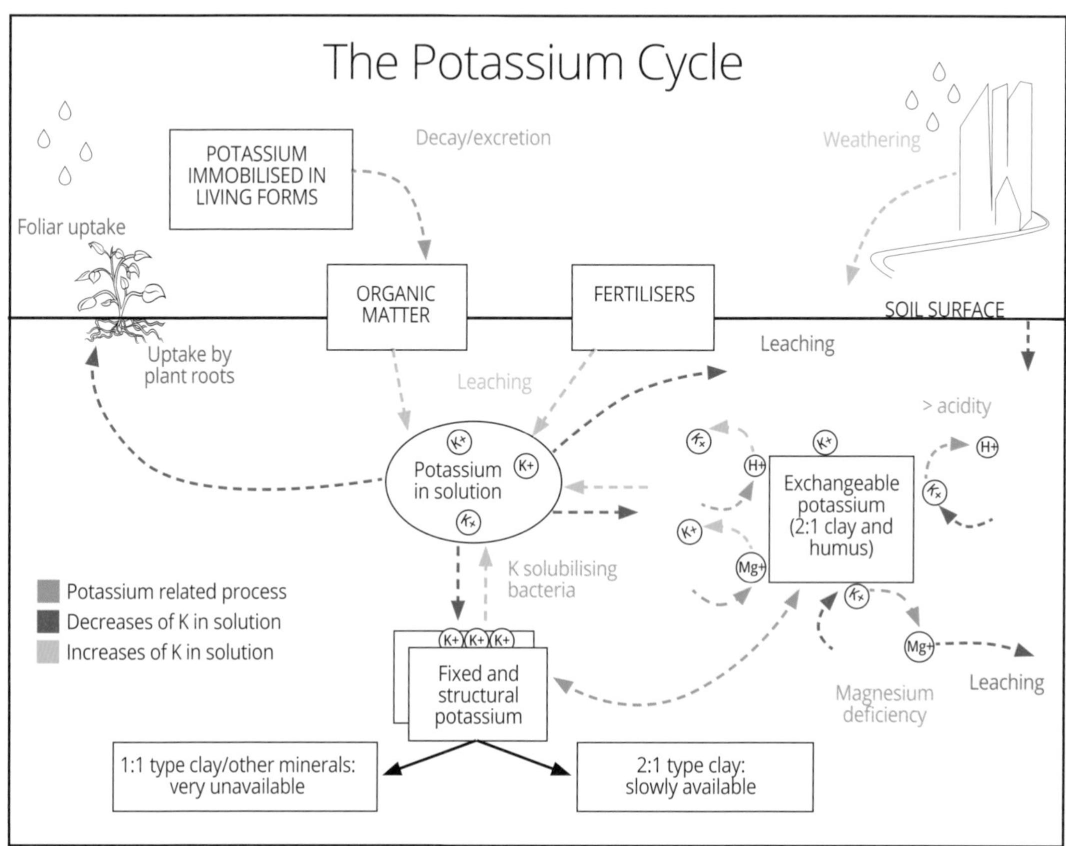

In the case of weathered minerals or ground quarried minerals the potassium is only released gradually. Potassium in solution is very reactive and what the plants do not immediately absorb through their roots is locked up due to other chemical processes or quickly leaches away. Because of this, small frequent applications are more effective than larger ones.

There are four forms of potassium in the cycle above – potassium in solution, exchangeable potassium, fixed and structural potassium in order of availability. Potassium ions in solution are the form that plants readily uptake via the roots and more partially through the leaves.

The clay fraction of the soil is very significant for potassium processes. The overall charge of clay is negative and this attracts positively charged potassium ions in solution. If you consider the exchangeable potassium in the diagram, potassium ions are adsorbed onto the surfaces of clay colloids at exchange sites, displacing other ions such as calcium and magnesium. Conversely where potassium is adsorbed to the surface of soil colloids or minerals, calcium or magnesium could displace potassium, which then may be leached from the soil or otherwise become unavailable.

If you have clay soil you can celebrate that you can expect to have plenty of potassium potentially available and then commiserate because it is not in a form that is easy for plants to use. With particular regard to potassium, an important concept to consider is that of Cation Exchange Capacity (CEC) which is defined as the capacity by which a material can retain cations (positively charged ions). It is abundant in soils with a high percentage of organic matter and clay.

With 2:1 type clay minerals such as montmorillonite or the well-known vermiculite, an octahedral layer of alumina is sandwiched between two layers of tetrahedral silica. As illustrated below, because there are only weak bonds between the layers, water can enter the matrix and liberate potassium in solution. The freeze-thaw process related to weathering is a powerful force for liberating potassium when it is considered that when water freezes it expands with a force that can blow rocks apart and burst water pipes.[10] This effect is noticeable in the well aerated soil of raised beds where the level of soil rises considerably in freezing conditions, leaving a fine and fluffy tilth afterwards.

With 1:1 type clays such as kaolinite there is one tetrahedral layer with a silica backbone bonded to one octahedral layer with an alumina backbone as illustrated in the simplified diagram below. This kind of clay has unfavourable properties that

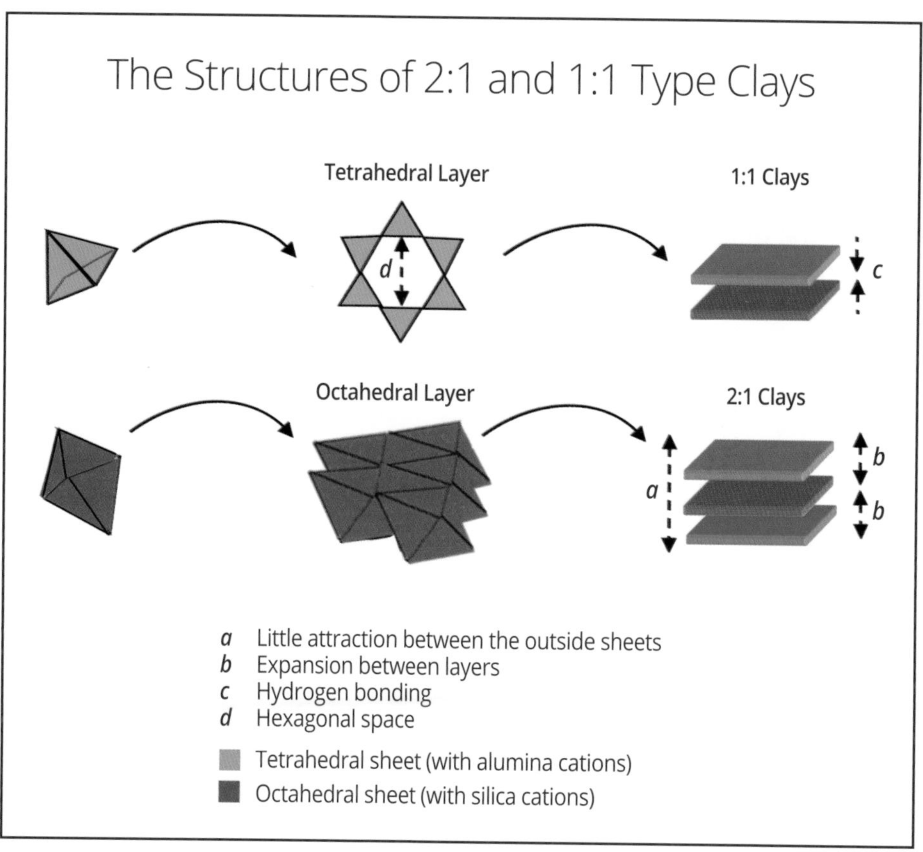

lock up potassium in its structure making it inaccessible. Because of the powerful hydrogen bonds linking the two layers no water can get in between, in contrast to the 2:1 type clays.

Considered from a purely abiotic standpoint, which is where many classical texts stop, you might conclude that when potassium gets into the layers of a 1:1 it is going to be useless for plants. The good news is that it has been established that biotic factors in the soil including many microorganisms and higher plants have the faculties to unlock these reserves.

Organic acids such as oxalic acid produced by organisms in the soil are thought to transform 1:1 clay into more expandable forms.[19] Rhubarb, which many growers have access to, grows in abundance and is a well-known source of oxalic acid which is used in its defences. This points to useful possibilities for people to prepare rhubarb teas to open heavy clay soils, although no studies are available to clarify this speculation. A suggested safe preparation is supplied in Chapter Seven, which in any event would still be an excellent natural pesticide if handled with care.

In a novel departure from conventional fertilisers, potassium solubilising bacteria have been discovered, 'KSB' for short. One such strain, *Bacillus mucilaginosus*, holds the potential to unlock the large reserves of potassium already present but unavailable. Although KSB research is still in its infancy, studies found KSB bacteria were shown to effectively increase plant growth, yield and nutrient uptake.[20]

In conventional high input systems where minerals are loaded in one end and lost through the other, the process is operating in a wider circle. What we are proposing is to draw this circle in, which is more sustainable, less costly and more synchronised with the cycles of nature that the grower can immediately influence. In such an approach nutrient cycling and the food web take on a new significance. In the next chapter we take a closer at the flora and fauna in intimate association with your crops below the ground.

CHAPTER FOUR

The Life in Your Soil

> All fertile areas of this planet have at least once passed through the bodies of earthworms.[1] Charles Darwin

The goal of any organic gardener is to encourage life to return to the soil after possibly decades of mismanagement and unsustainable practices. Gaining a greater insight into the lives of important organisms both visible and invisible to the naked eye can only help your stewarding of the soil and ultimately assist you in growing healthy disease resistant crops. Some organisms such as beneficial fungi and bacteria can be encouraged in the various ways described throughout this book. Others such as eelworms are troublesome and need to be discouraged. Rather than creating deserts by adding the poisons promoted by the agri-chemical industry, which devastate beneficial creatures and problem species alike, a healthy soil teaming with life is the best way to protect against disease.

In Chapter Three we looked at the most important nutrient cycles you need to consider as an organic grower. In this chapter we will look at some of the organisms involved in these cycles which interact, compete, form alliances, live and die in the vast web of life beneath your feet.

If you see a food web as a network of elastic bands between different types of organisms, what you want to do is to tweak the network in your favour in a way that creates excellent conditions for growing crops:

- Pests and predators in balance, not too many of anything
- Aerobic soil conditions
- Sufficient nutrients when they are needed by plants
- Enough moisture but not too much
- A dominant bacterial presence for most annual crops
- A dominant fungal presence for fruit trees.

Attempting to eradicate pest species is akin to stretching the bands of that food web past breaking point. If you succeed you have to deal with other consequences.

You might indirectly starve a pest's main predators and other pests may step up unchecked. It makes sense to know the kind of things that help beneficial species and create the right conditions for them to develop.

This chapter does not attempt to catalogue the myriad organisms in the soil that are useful to the grower, it only offers a snapshot of some of the main players. Many of the larger organisms will be very familiar, however a considerable proportion of soil life is microscopic and it is estimated that only 1% has been identified.[2] We begin with the largest of soil creatures first and gradually proceed to the smaller and then microscopic organisms.

Foxes, badgers, hedgehogs, mice, rats and voles all have potentially useful roles, circulating nutrients and aerating the soil. Rats hold very negative connotations for many people and it is true that they are vectors of disease, however the occasional rat or mouse on an allotment is a common sight and not really a cause for concern. Problems with rats tend to arise from the introduction of livestock and the discouragement of natural predators such as feral cats that are naturally attracted to such areas if they are left alone.

In compost heaps, rats are attracted to processed foods and will tend to take up residence in winter. This would mark an early phase of the composting process for that heap and it could be asserted that they have a natural role in speeding up the breakdown process. After this initial stage they will move on and their droppings will become sterilised by their passage through the earthworm.

Earthworms (family Lumbricidae)

The humble worm is one of the key architects of healthy soils. In UK soils there are upward of 26 indigenous species and another 20 species that have been introduced.[3] Overleaf is a photo of a compost worm (*Eisenia veneta*) at work in a local compost heap. It is part of an important group of litter dwelling (epigeic) worms that also includes the brandling (*Eisenia fetida*), also pictured. In identification keys[4] these worms can be differentiated from the others by the dark coloured stripy banding. Such worms reproduce quicker and produce more casts than soil dwelling worms such as the common earthworm (*Lumbricus terrestris*) and function best in compost heaps.[5] In later chapters we will be using vermicasts in a variety of ways to produce compost tea. Note the brighter band of the worms in the picture; this is called the saddle and is an indication of maturity that makes identification much easier.

Some people create their own mini worm farms because the worms are very vigorous feeders of home waste and the casts are so incredibly fertile. If you don't want to go that far you can encourage brandlings in a standard compost heap simply by keeping your heap dark, using a fork when you turn your pile and just supplying regular helpings of what you normally compost. Should your heap exceed one metre square the brandlings will retreat because of the increased heat, but will still be around on the cooler periphery. If your compost heap is one of the closed off types that are difficult for worms to leave, they may well cook. The brandlings can survive winter but become less active, so insulating your heap a little will help to keep things going. We continue with the composting theme in considerably more detail in Chapter Eight.

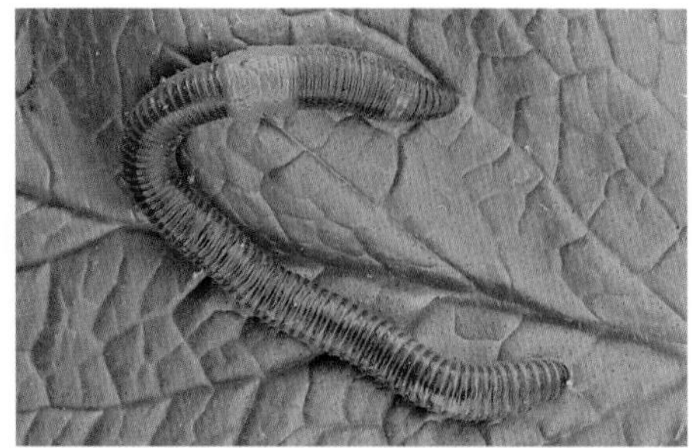

A mature compost worm (*Eisenia veneta*) in a local compost heap.

The common earthworm is one of the best known creatures in your garden and is an intrinsic part of a healthy soil's fauna. They play an important role in aerating the soil and produce fertile casts which facilitate aerobic microbes.[7] Earthworms also mix organic matter into the soil when they create their underground burrows.[8] The entrances can often be quite elaborately ornamented with organic matter and it has been shown that fertility around their dwellings is always considerably increased.[9]

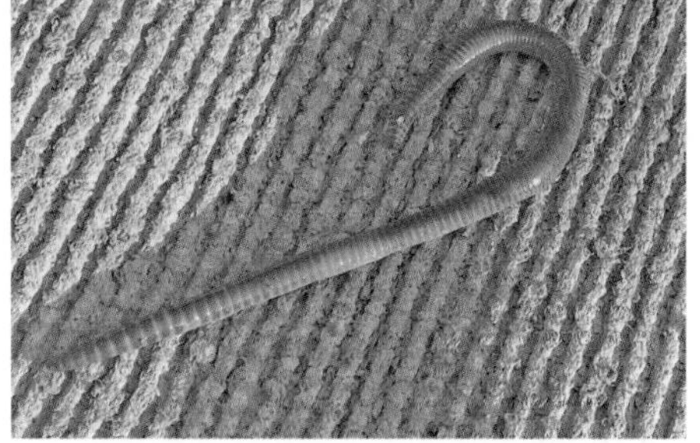

The brandling worm *Eisenia fetida*.[6]

For growers who have a respect and empathy for worms, digging is always an issue but sometimes, and especially in clay soils, it is necessary. To alleviate this concern, the judicious use of the fork can lessen the impact considerably with the spade only used for the finest of soils. The use of machinery such as the rototiller is justifiable for breaking up larger areas, but it is highly disruptive, bringing lower layers of the soil to the top which is unfavourable to cultivating the soil's fertility.

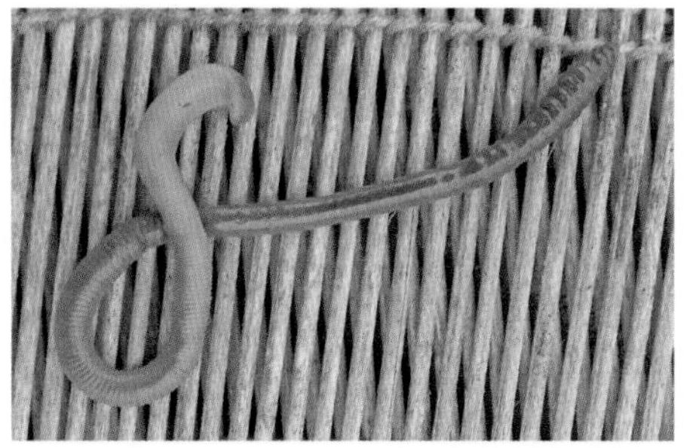

The common earthworm (*Lumbricus terrestris*). Note the tail (lower) flattens out into a paddle which is an important clue to identification.

Molluscs (class Gastropoda)

On the land, otherwise known as slugs and snails, these creatures cause problems for growers ravaging crops both above and below the soil. Despite this they do still have some more favourable functions for the soil. They move about on a trail of mucus which helps to condition the soil by making the soil crumbs stick together, they tear up organic matter and their casts contribute nutrient rich material. Furthermore snail shells contribute calcium carbonate to the soil which is the main component in gardener's lime. Gastropods can also help to break down woody material more quickly by digesting it with the cellulases they produce in their gut.[10]

They have many predators including birds, hedgehogs and occasionally other gastropods. Slugs and snails are known to be most active at night so if you can encourage hedgehogs into your plot on their nocturnal rounds you will be amply rewarded. A regular bribe of dog food may be just the right inducement but it might also attract rats.

You could make a habitat for a hedgehog to set up camp with a small box under an undisturbed pile of leaves or even a buried pallet stuffed with straw underneath your compost heap where the hedgehog will benefit from both warmth and food. For gardens it is crucially important that access points for hedgehogs are allowed for their nocturnal rounds.

Arthropods (phylum Arthropoda)

These are one of the most diverse groups of animals[11] that form the general category of 'bugs'. Strictly speaking they are invertebrates with jointed legs and segmented bodies. Familiar arthropod members include beetles, spiders, woodlice, mites, centipedes, millipedes and all other insects. Only a few of the most significant representatives have been included because the range of creatures in this group are simply too diverse to do them proper justice.

Beetles (order Coleoptera)

These are a very successful and diverse order of creatures even for arthropods. The ladybird (family Coccinellid) is the most celebrated member of this group and is well-known as a voracious predator of aphids, which we will return to in the next chapter.

The dor beetle. Photo by Käyttäjä Kompak (2005).[12]

In the soil, the dor beetle (*Geotrupes stercorarius*) is also held in high esteem because it industriously works manure deep within the soil.[13] If you lay mulches containing manure these beetles and others will almost immediately become more prominent on your land. Beetles such as the rove beetle (*Aleochara bilineata*) are beneficial predators and operate on the soil surface. The rove beetles are generalist predators, natural enemies of root maggots[14] especially on onion[15] and cabbage maggots.[16] Unfortunately they are also susceptible to pyrethrum, a botanical insecticide[17] discussed in Chapter Seven. They like to hibernate in grassy patches over winter but emerge a little late in spring to be useful as an early control. You can find them under rocks, in compost heaps[18] and inside corn husks.

Crustaceans (subphylum Crustacea)

From the grower's perspective the most important arthropod in this group is without doubt the woodlouse (*Oniscus asellus* is the most common). They can be a minor annoyance but are not in the same league as molluscs for crop damage, and woodlice have their uses. They are more interested in breaking down rotting wood than interfering with crops and are dependent on a moist environment. If you see them in your compost heap, be pleased because they are breaking down harder matter for smaller organisms and fungus mould. A fascinating behaviour of these creatures is that when they are in damp conditions they avoid the light but when conditions become too dry they move towards the light to find another moist place.[19] If you don't see any woodlice in your compost heap or it is full of ants this is an indication that it is much too dry.

Moving on from our brief tour of the larger soil creatures we proceed to the organisms that are at the brink of what can be seen by the human eye and then to creatures that can only be seen under a microscope, or in colonies. Many of these organisms function in the narrow film of water surrounding soil particles. An important zone to introduce here is that of the rhizosphere. This is the area in the soil closest to the roots and is where some of the real action takes place.

Arachnids (class Arachnida)

This group includes spiders, harvestmen, false scorpions and mites. We will return to spiders in the next chapter because they are also an important beneficial predator above the ground. Spiders most associated with the soil include purseweb spiders, trapdoor spiders and wolf spiders (family Lycosidae).

Wolf spiders (right) got their name because at one time they were thought to hunt in packs but their true method of hunting is to run down their prey at short range on their own.[21] The role of these spiders crosses over somewhat with the rove beetles on the ground predating whatever they can get hold of. These spiders are known to eat

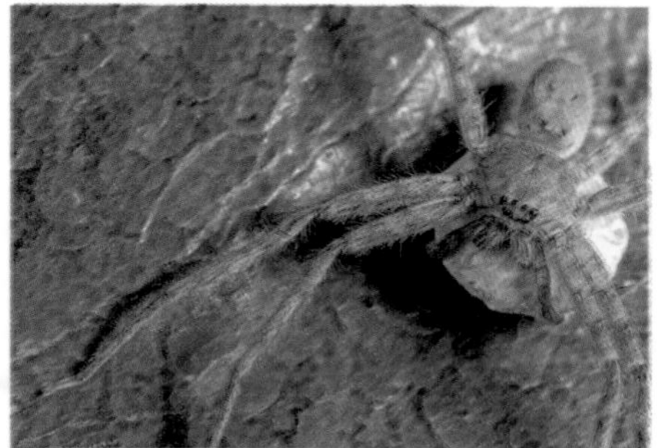

A female wolf spider with her egg sac. Photo ©Phichak Limprasutr – stock.adobe.com (2018).[20]

grasshoppers, crickets, aphids, scale insects, leaf hoppers, shield bugs, moths, butterflies, thrips, true flies and some beetles.[22] Their general preferences are insects 50-80% smaller than them[23] and sometimes they resort to cannibalism such as when their numbers are getting high.[24]

Wolf spiders run so quickly that it can be difficult to get a clear view of them, although they can be observed fleetingly with their conspicuous pale egg sac as captured on the previous page.

Rotifers (phylum Rotifera)

There has been a substantial amount of research conducted into earthworms, partly because they are relatively large and easy to study, however the rotifer is somewhat neglected despite being a valuable asset in the soil. They are aquatic and some are known to occupy the thin films of water around soil particles or in

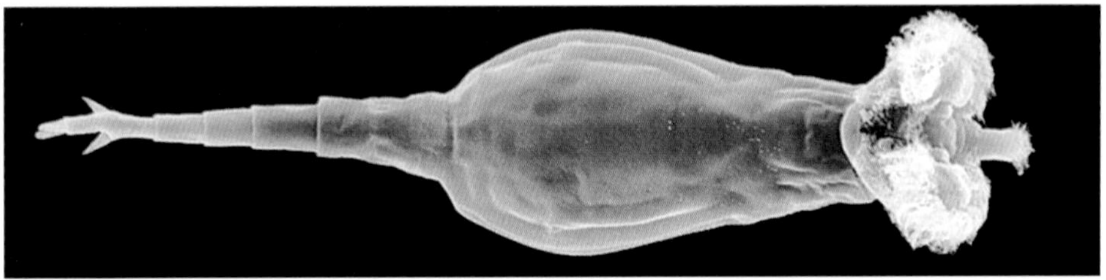

SEM (scanning electron micrograph) of Bdelloid rotifer.[25]

the compost heap.[26] The rotifer can grow up to 2mm in length so the larger ones can be seen unaided and should be of great interest to those following biodynamics since these creatures create their own small vortices to feed.[27] The vortex is thought to have a special significance for new life[28] and is a motif seen throughout nature from the birth of galaxies to the flow forms in a brook. The Bdelloid class of rotifer is exclusively female but other rotifers may have males and females or be asexual. In the photo above, note the rotary apparatus on the right that has an appearance of spinning when it's in action, propelling the rotifer and drawing food into her buccal cavity. Because of this facility the rotifer is an important creature for circulating air bubbles and nutrients. They are also helpful to the grower in times of low rainfall because some sessile types construct small tubes where they retain moisture.[29] They can be seen as minuscule packets of fertility and this has not gone unnoticed by the aquatics industry[30] where they are supplied to feed fish, corals and sponges that are a step up in the food chain. In the compost heap this would equate to nutritious food for beetles, centipedes, predatory mites, ants and springtails.[31] Interest in rotifers has even extended to home production kits available online, complete with air pumps.[32] In Chapter Nine we will be using a similar process to promote the growth of rotifers together with the many other organisms associated with healthy soils.

Protozoa (animal-like protists, kingdom Protista – a disputed taxon)

These are single-celled aquatic organisms ranging from between[33] 5µm and 500µm that can be found in the thin film of water around soil and in soil pores, and are found in considerable concentration in the rhizosphere. Studies indicate that protozoa have a powerful growth promoting influence on plants, grazing on fungi and bacteria[34] to return nutrients in plant available forms.[35] When there is an excess of nutrients that plants can't use, the presence of protozoa is linked to an increase in plant mass although the exact mechanism is unclear.[36]

There are three main groups: amoeba, flagellates and ciliates. Some amoeba are testate having shells of silica or chitin, but the naked amoeba are the most important of all the protozoa in the soil, because they have an amorphous form and can alter their shape to get to places that are inaccessible to the others.[37] They move around by means of pseudopodia or 'false feet'.

To get an idea of scale, if you picture a bacteria around the size of a ball bearing, an amoeba would be the size of a football. Amoebas' method of predation is to engulf their prey, creating a food vacuole that is then digested.

Ciliates are the largest protozoa and move around quite smoothly spinning along in the shape of a sine wave, by rippling the numerous rings of hair-like cilia on the outside of their bodies.[33] They do well in anaerobic conditions and have a preference for consuming anaerobic microorganisms. When there are a disproportionate amount of ciliates in the soil, or indeed in compost teas, they are a good indicator of anaerobic conditions.[39] Unless you are the owner of a rice paddy, anaerobic dominance is unfavourable for crops.

Flagellates are the smallest of the protozoa and feed on bacteria. They are so named because of their whiplike flagella that propel them along.[33] If you ever see one under a microscope their movement appears quite erratic.

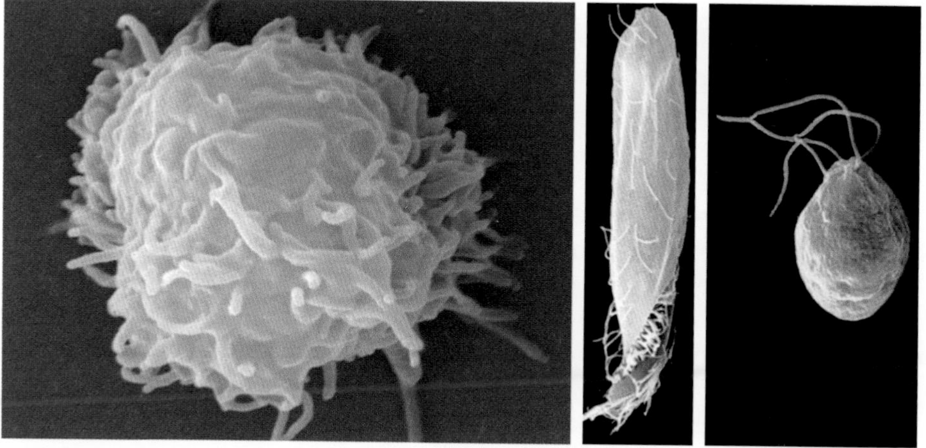

From left to right: An amoeba;[38] *Metopus hasei*, a ciliate common to poorly aerated soils; *Polytomella* sp., a common soil flagellate. SEM photos by W. Foissner.[40]

Nematodes (phylum Nematoda)

Nematodes, otherwise known as roundworms, are some of the most widespread of the soil's inhabitants and ubiquitous in most soils, active wherever there is moisture. These small blind worms grow between 0.3mm and 3mm in length[41] so with difficulty some of the larger ones can be seen unaided.

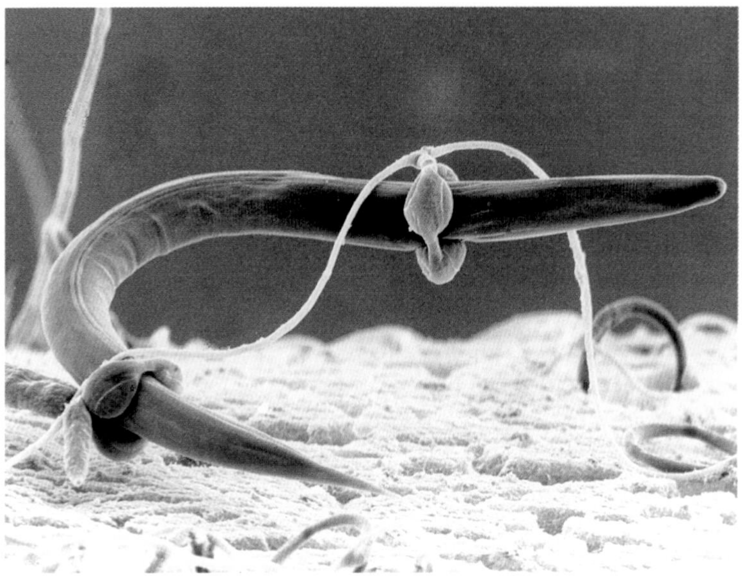

SEM of a nematode (*Caenorhabditis elegans*) trapped by predatory fungus (*Arthrobotrys anchonia*) on a tomato plant root.[42]

Finding the best way to categorise nematodes has caused taxonomists many sleepless nights. One practical approach, which is suitable for our purposes, is to differentiate them primarily via mouth parts. The ones that are a problem to growers are the plant parasitic nematodes with needle stylets illustrated overleaf, which are used to pierce growing vegetables such as root knot nematodes, dagger nematodes, needle nematodes and root-lesion nematodes. These small worms can often be vectors of serious crop diseases[43] and in certain phases of their life cycle can be highly mobile in the soil, travelling through soil water and even catching lifts on, or in, larger organisms.

The sedentary nematodes such as *Meloidogyne* spp. are considered the worst of pests[45] and for temperate climates this would be *Meloidogyne hapla*; they spend a greater part of their lives inside plants and seriously disrupt the plants' normal growth.[46] The nematodes do not seek to kill the plant as they need the plant alive whilst diverting moisture and nutrients. The female nematode consumes considerably larger amounts of resources than the male. Where the nematode population becomes too high inside the plant they have a chemical signalling mechanism that triggers a response that causes the females to change sex.

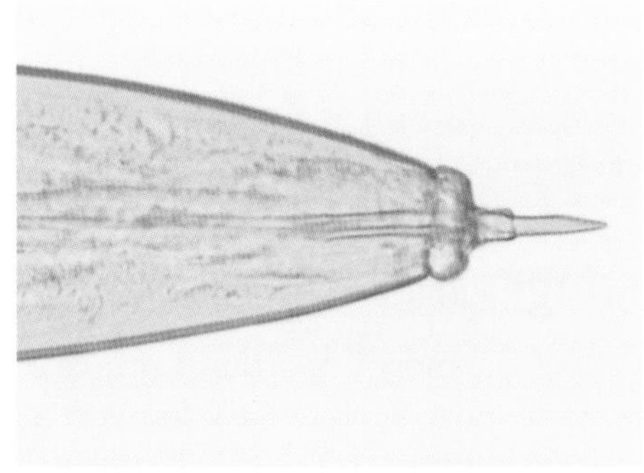

SEM of a Nematode with stylet. Tarjan *et al.*[44]

Once sedentary nematodes become established on a plot they are virtually impossible to eradicate and some chemically harsh methods such as methyl bromide (MBr) are being phased out because of their environmental impact and toxicity to humans. Rather than eradication, growers, both organic and otherwise, have been looking to manage the problem using rotations, organic amendments and incorporating plants that nematodes avoid. Research studies found with particular reference to nematodes that encouraging beneficial organisms by adding organic amendments had a holistic beneficial effect which kept the problem nematodes from causing disruptions to crop plants.[47] The food web nurturing teas in Chapters Nine and Ten would be useful in this respect as is our section on marigold tea in Chapter Seven. An important tactic that targets nematode eggs is the application of high chitin amendments (e.g. exoskeletons of crustaceans: woodlice, crabs, lobsters) that attract organisms that break down the nematode egg shells.[48]

The nematode has many enemies but one stands out as possibly the best for those interested in permaculture and edible landscapes. The delicious and carnivorous fungi *Pleurotus ostreatus*, overleaf, is a valuable ally to humans, not least for the esteemed oyster mushrooms they can produce. It has been the topic of research to control nematode numbers[49] and suggests interesting possibilities for polyculture growers who can use the spores as a potent addition to compost teas. In nature they are known to grow from the sides of trees as illustrated but mycologists have also found that they grow abundantly in high carbon substrates such as straw, which can be mulched around crops, triggering valuable synergies for gourmets and crop protection.

The pest nematodes attract a disproportionate amount of attention since the greater proportion of nematodes either have no impact on crops or are distinctly beneficial. Many important beneficial nematodes are being investigated to predate pest nematodes such as those in the Steinernematidae and Heterorhabditidae families. They go by the popular label of entomopathogenic nematodes (EPNs) and predate a wide range of soil organisms including black vine weevil, currant borer moth,[50] fungus gnats, other weevils, scarabs, cutworms, webworms, billbugs, mole crickets, termites, peach tree borer and carpenter worm moths.[51]

A further realisation is their importance in nutrient cycling where they operate on a number of trophic levels making nutrients more available to plants more quickly. Their grazing limits bacterial numbers, which has the effect of stimulating the bacteria into greater activity, increasing decomposition of organic matter and bringing immobilised nutrients back into motion. Their movement through the soil introduces microbes to fresh substrates and they provide aeration as they move through the soil pores.[52]

Fungi (kingdom Fungi)

One of the most important features of certain fungi for the grower is that some types of fungi, such as Basidiomycetes, form mutually beneficial growths around and within plant roots called mycorrhiza. Although this symbiotic alliance has been studied as early as 1885, its importance has not been sufficiently acknowledged by the growing community and yet it is essential to the proper growth of over 90% of plants.[54] Study after study has shown that the 'infection' of plant roots by mycorrhiza has a profoundly beneficial effect on the plant, with roots growing far more prolifically than plants grown with no mycorrhiza. The host plant makes a trade with the fungi in carbon rich exudates, which the fungi thrive on, in exchange for essential plant nutrients which are supplied by the fungi.[55] In popular horticulture the symbiotic associations with legumes are often cited with regard to fixing nitrogen in crops but many texts attribute even greater importance to the value of mycorrhiza for transporting immobile essential nutrients such as phosphorus,[56] copper and zinc.[57] Furthermore, mycorrhiza can bestow a

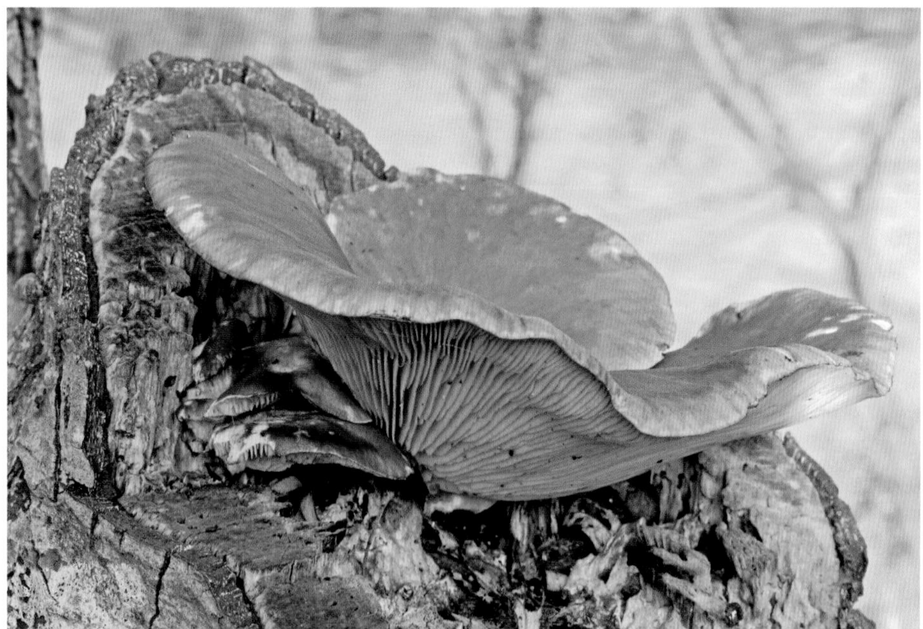

The fruiting bodies of predatory fungi *Pleurotus ostreatus* growing on the side of a tree. Photo by Vaughan Fleming.[53]

number of other benefits such as added protection to the plant from pollutants in the soil,[58] enhanced disease resistance[59] and resistance to high temperatures.[60] An interesting facet of mycorrhiza is that the same fungi does not restrict itself to one plant[61] which may have useful implications for horticulture in enhancing the flows of different nutrients.

The SEM of a nematode trapped by fungal hyphae (p.50) is an illustration of the way certain fungi can suppress pest numbers. These beneficial fungi act as sentries around plant roots in exchange for exudates. When a nematode is caught in this way a mechanism is triggered and in the space of a few seconds a lancelike part of the fungal hyphae penetrates the prey[62] and uses digestive fluids to break down and absorb the valuable nutrients within the nematode. Other methods of capture in fungi include the generation of minuscule adhesive knobs that work like fly paper[63] whilst some fungi produce nets to trap prey[64] and could reasonably be called the fishermen of the soil.

Mycorrhiza fall into two broad categories, with ectomycorrhiza fungi that grow around plant roots and endomycorrhiza that penetrate within. There are two main types of ectomycorrihiza but these are generally considered to be associated with non-agricultural plants,[66] however there are a number of exceptions and this view relates to conventional agriculture; for the alternative grower there are greater possibilities to explore. Of the ectomycorrhiza there are Basidiomycetes and Ascomycetes, which include those fungi with the fruiting bodies we

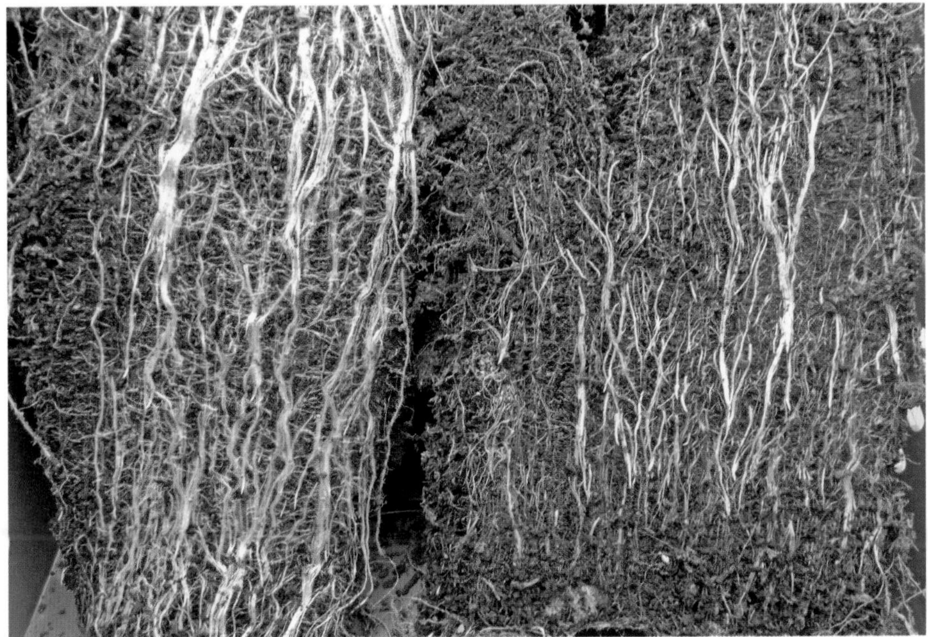

Mycorrhiza surrounding a root.[65] In this photo the mycorrhiza is paler with blunter tips apparent.

use as edible mushrooms such as the oyster mushrooms on p.52. Below ground the fungi form a kind of jacket structure, illustrated on the previous page.

There are five main types of endomycorrhiza[67] but there are only two which are of direct significance to horticulture and these are Arbuscular Mycorrhiza (AM) and Ericoid Mycorrhiza (ERM). Overall the AM fungi are by far the most important to horticulture because they are the most cosmopolitan both in terms of the terrestrial conditions they can survive in[68] and by the range of plants they infect.[69] Unlike the ERM fungi, AM fungi cannot survive without a host and their structures are much finer, forming a 'Hartig Net' at the interface of the root where all the nutrient exchanges take place.[70] Such a structure becomes so deeply integrated between plant and fungi that it's considered to function as a single unit.

'Arbuscule' literally means 'little tree' which reflects its structure within the plants they colonise. This special structure has self-similarity as it can be viewed from a number of different magnifications but still appear the same. The significance of this kind of structure in a variety of natural phenomena is unclear but it is known to be more dominant in areas of great species diversity.[71] Whatever forces shape these structures in the microcosm also appear to be those that shape the great rivers and oaks which hold the same motif. Comparing the forces at work between the microcosm and the macrocosm there is a considerable disparity in the type of known forces at work and yet the structures maintain their self-similarity.

The ERM form beneficial associations with plants in the Ericaceous family,[72] which includes blueberries and cranberries (*Vaccinium* genus). These are plants that specialise in growing in highly acidic and harsh conditions. Blueberries are very dependent on their mycorrhizal hosts with the plant not functioning properly without this association. Some types of these plants have poorly developed root hairs[73] which are important for absorbing nutrients. Instead this role is taken over by ERM which facilitates greatly enhanced performance.[74] In conventional horticulture this association is known to be restricted by fertiliser use, soil compaction and intensive mineral use[75] which in many ways replaces it.

In nursery plants, infection by ERM is often poor,[76] which suggests possibilities for importing inoculants from wild to cultivated stock. The best place to hunt for material infected with ERM is in a peat bog; a handful is all that would be needed to apply the techniques in Chapter Ten.

Find a vigorous looking plant, preferably of the same family as you are growing. You might collect soil from around a wild 'low bush' blueberry *Vaccinium angustifolium*, growing vigorously in some heath, which could work well for a tea preparation for your more cultured 'high bush' blueberry *V. corymbosum*. Failing that, soil from around a wild cranberry, bilberry, huckleberry, azalea or rhododendron would be acceptable.

Bacteria (domain Bacteria)

Bacteria are arguably the smallest of living organisms with sizes between[77] 0.5µm and 5µm and are the most numerous in the compost heap.[78] Soil bacteria are rod shaped and some possess primitive flagella similar to protozoa. There are two types, autotrophs that mainly use sunlight as an energy source and heterotrophs that mainly use carbon dioxide. The most important type in the soil is the heterotroph.

Bacteria have a key role in cycling major nutrients and the natural world we see today could not function in the way it does without them. They also have valuable roles in remediating soil contaminated by pesticides and herbicides which they break down over time.[79]

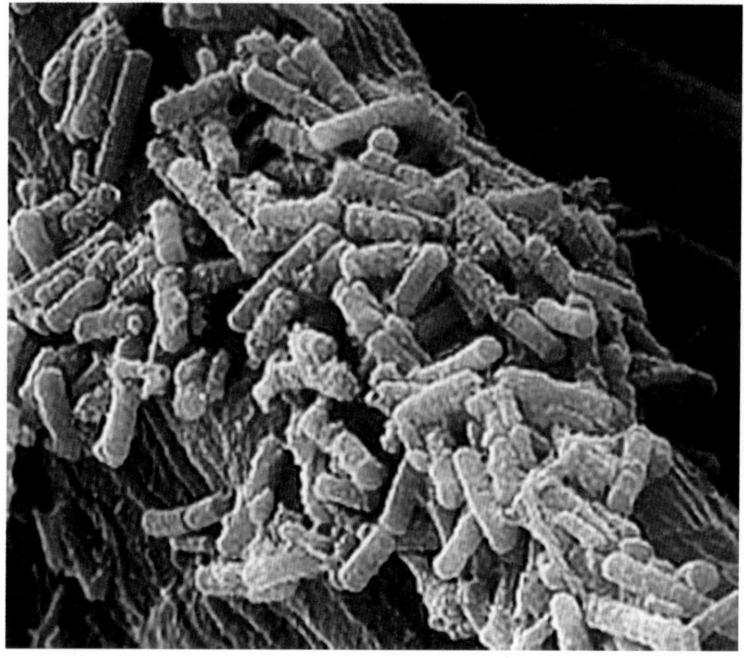

SEM of thermophilic bacteria, *Bacillus stearothermophilus* by Tim Sandle.[80]

The bacteria in the above photo are important in the composting process at higher temperatures. In composting, bacteria are known to operate in succession with one type of bacteria taking over from another when the conditions are right. The role of bacteria in the compost heap will be looked into further in Chapter Eight.

Of particular interest to the grower are the bacteria involved in the nitrogen cycle such as the rhizobium nodules pictured overleaf. When the nodules are

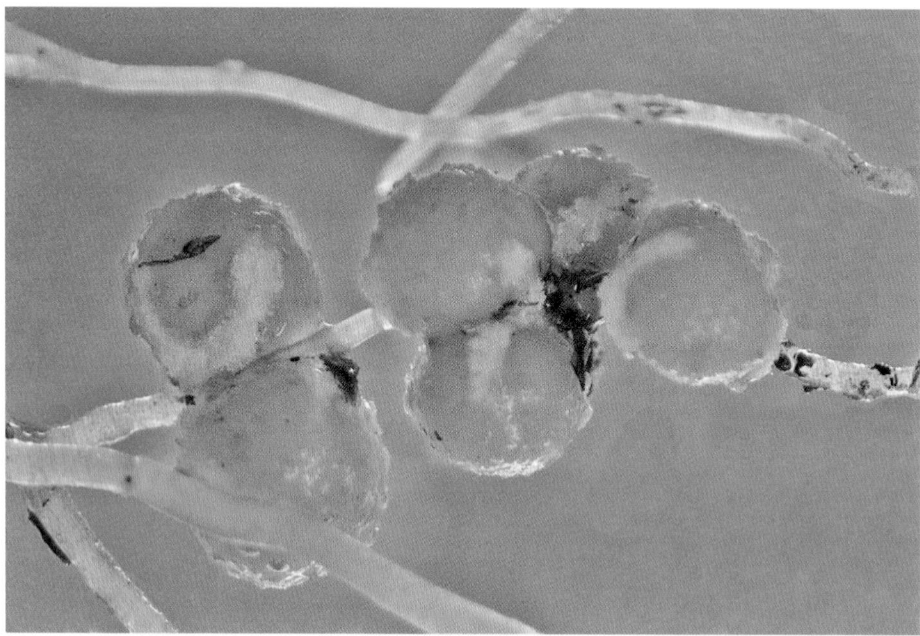

A plant root showing nitrogen fixing nodules. Photo by Tomomi Suwa.[81]

effectively fixing nitrogen they turn from white to a light pink. Other important bacteria in the nitrogen cycle are *azotobacter*, *pseudomonas* and *clostridium*.

Actinomycetes are a significant group of bacteria for composters that are sometimes mistaken for fungi because of their mycelia-like threads.[82] They are significant because they are effective at breaking down hard carboniferous material such as lignin in woody material or chitin in the exoskeletons of small insects.[83] One type comes into action during hot composting[84] and another type works in the later curing stages which look like a thin pale web spread across the heap.[85] Because of its fragile structure it functions best when it is undisturbed by turning and is responsible for the wholesome earthy smell of fresh compost.[86]

Another important category of bacteria is 'Plant Growth Promoting Rhizobacteria' or PGPR for short, which is the topic of contemporary research in sustainable agriculture. These diverse bacteria have valuable traits in their interactions with crop plants such as maintaining crop health, suppressing pathogens, making plants more resistant to pests and provide growth stimulation by the production of plant hormones which release a variety of minerals for plant uptake.[87]

One such PGPR bacteria is depicted overleaf and is considered somewhat elite for its traits and capabilities. A further application of the vortex is shown in the structures of this recently discovered social microorganism, *Paenibacillus vortex*. It is widespread in the soil and is considered to have great potential as an ally to plant roots.[89] Working as a coordinated group these advanced organisms

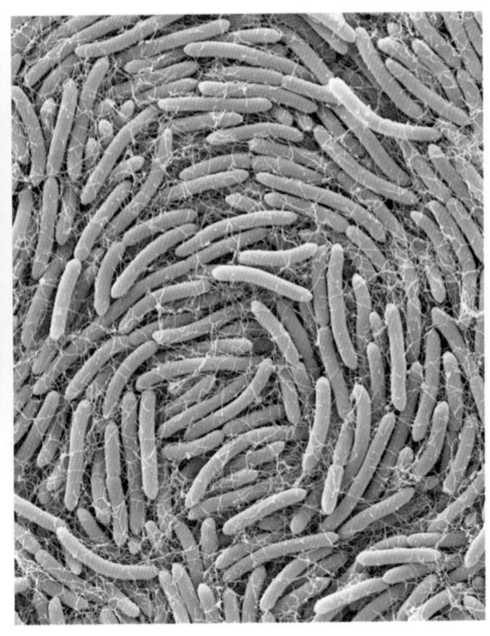

Above is a whole colony of *Paenibacillus vortex* from an agar plate. The dots are the epicentres of the vortex. A close-up of the centre one is shown by the SEM on the right.[88]

can swarm, form alliances with plant roots, search for food and exhibit traits of multicellular organisms. Their complex interactions bestow massive advantages over small organisms that just act on their own. The vortex comes in as a way of movement in rolling forms of billions of bacteria interlinked which can transform depending on the type of surface.[90]

Whilst nurturing the creatures of the soil is of crucial importance for an organic grower, another world exists in the world above the soil, which is intimately linked. In the next chapter we will take a closer look at this.

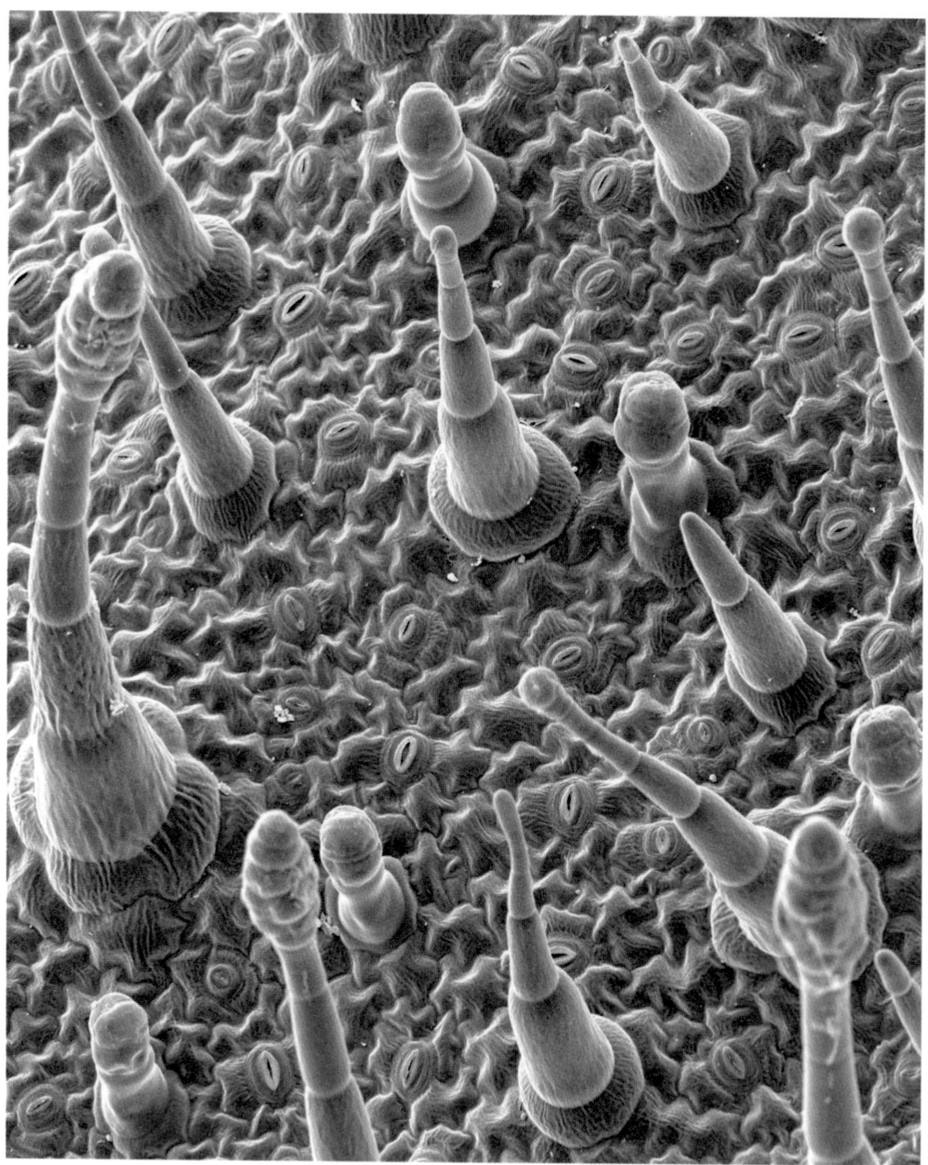

'Under the microscope, aerial plant leaves resemble eerie landscapes, with deep gorges, tall peaks and gaping pits that riddle the waxy surface.' Johan Leveau[1]

CHAPTER FIVE

The Life in Your Leaves

Many organisms in the food web perform valuable horticultural services on your land. This can be pollinating plants, assisting in breaking down organic matter,[2] predating problem species, contributing useful substances to your soil or suppressing disease.[3] Sometimes matters are not entirely clear cut with certain creatures performing useful services and on other occasions being a nuisance, such as certain birds predating snails but also raiding strawberries. It is down to your own ingenuity and hopefully books like this to try to ameliorate the disadvantages and exploit the advantages.

In a practical book of this nature, there is no way we can even begin to give a thorough treatment of all the organisms that operate above the soil that are important to horticulture. Therefore to avoid any digression, we have selected a few that have a significant impact on crops that you can effect either positively by encouraging your food web as in Chapters Eight, Nine and Ten or disrupt with the biological sprays in Chapter Seven. Certain organisms mentioned in this chapter spend various phases of their life cycles in the soil and have been included because their greatest impact is above the ground. The life cycle of a beneficial can be intimately linked to the pest it attacks so an understanding of the relationship between pest and predator is important when planning biological measures of control. It is a well-known practice to tolerate a small number of pests, simply to keep your beneficials going, so if you are too successful at control you may conversely be reducing the long-term pest resistance on your land. For organic crops what a grower does not do is almost as important as what they do do. As home preparation of natural insecticides is considered in Chapter Seven, some of the examples in this chapter illustrate many of the potential consequences of indiscriminate use of sprays whatever the origin.

An important distinction to consider when reviewing beneficials and pests is whether they are soft bodied, because this is a significant factor as to how much they will be disrupted by the various sprays. A deeper understanding of important beneficials may avert disaster in this area. For example, bees are affected by pyrethrum but if you know they go back to their hives after dark, you can spray when they are not around. It is regrettable that when using sprays it's inevitable that certain beneficials are going to be hit and it's up to the individual gardener what line they draw.

An area of particular importance to us is the phyllosphere, that thin film of life around and to some degree inside the aerial parts of plants which can be understood in a similar way above ground as the rhizosphere is within the soil. Later on in this chapter I will introduce some of the microscopic organisms of the phyllosphere. Many beneficial bacteria play important roles above the ground attacking pests, harmful moulds and forming beneficial alliances with plants. Unfortunately, there is conclusive evidence that many of the fungicides currently on sale cause serious damage to fungi and protozoa communities although bacterial communities can return very quickly.[4] It is a cause for concern that such communities may lack the indigenous nuances they possessed previously as aggressive 'global' bacteria will have a greater opportunity to gain a foothold.

Firstly we will take a look at some of the larger friends and foes.

Slugs and snails (class Gastropoda)

Slugs and snails appear in Chapter Four and they get a further mention here because they are such important pests in foliage. They favour moist conditions and prefer to do their ravaging in the dark. One of the best times to find them is a couple of hours after sunset. Evidence that they are active in your plot includes silvery trails of slime. They can be a particular problem for organic growers because many of the organic practices make favoured habitats for them. For instance, during the day they will seek out sheltered damp conditions such as in heavy mulches or under tarpaulins. In the spring, snails can often be spotted laying their eggs in greenhouse plant pots. The good news is that only a few species are a real problem.[6] They are susceptible to various tactics and a preparation is included in Chapter Seven which is based on the ancient practices of monasteries and is the subject of contemporary research at Newcastle University. The use of agri-chemical controls against gastropods has serious drawbacks since they are often poisonous to children, pets, beneficials[7] and are generally not something you want around food crops.

In the close-up overleaf, the rasplike radulae of *Helix aspersa*, which approximates to a tongue in humans, indicates its potential to damage crops or speed the breakdown of organic material on the compost heap.

Mites and ticks (subclass Acari)

Some can be seen by the human eye and grow up to two millimetres. The most significant creature in this group to growers is the infamous red spider mite, *Tetranychus urticae*. Its arch nemesis *Phytoseiulus persimilis* is pictured overleaf and is available today commercially as an effective biological control.

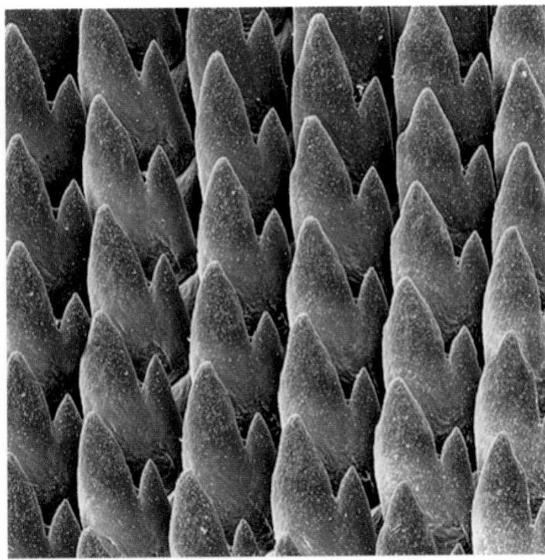

Left: *Helix aspersa*, with slime trail and droppings.
Right: A close-up of the rasplike radulae of *H. aspersa*.[5]

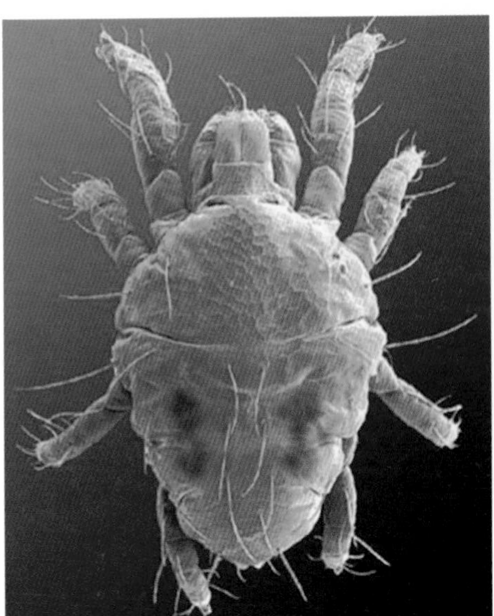

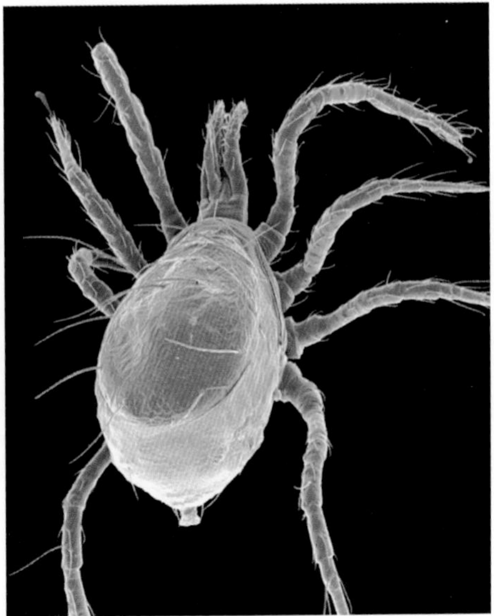

Left: The red spider mite.[8]
Right: Its predator *Phytoseiulus persimilis*.[9]

The red spider mite spends one phase of its life in the soil before moving to the leaves. As an example of how the balance between important organisms can be upset by crude measures of pest control, an incident of increase in red spider mites was traced back to the spraying of DDT (Dichlorodiphenyltrichloroethane). This substance, which was eventually banned because of its horrendous effects on the food chain, was applied to fruit trees as a general pesticide to kill codling moths. Unfortunately it also killed another significant predator of the red spider mite, *Panonychus ulmi*, and escalating the problem further the eggs of the red spider mite are DDT resistant.[10] Whilst in a balanced system this mite would have been of minor significance,[10] in the artificial environment created, this combination of factors was enough to promote the red mite spider to major pest status as it multiplied unchecked.

Aphids, whitefly and scale insects (suborder Sternorrhyncha)

These are major pests and as a whole are one of the most troubling to large-scale agriculture, because presentation can be everything if orders from supermarkets can be lost over a minor blemish. Potentially they can be vectors of many diseases and the honeydew they secrete on leaf surfaces can support the development of harmful moulds.

Regarding aphids, an interesting but troublesome feature is that certain ant civilisations farm aphids, thereby transferring disease to fresh plants. Some ants feed the aphids to certain grubs that the ants prize as a food source. They favour these grubs so much that if their colony is relocating they will choose these grubs over their own young.

One of the things that make the Sternorrhyncha such successful pests is that some species can reproduce without mating. When a greenfly gives birth, a further greenfly may already be growing in the offspring. This process is a form of parthenogenesis.[11] They can reproduce 'normally' when the conditions are right but this feature can give them a jump start when building an infestation. A single greenfly can generate a whole colony and when that area gets overpopulated a biological trigger is released and they can develop wings and migrate to the next patch.

Additionally, aphids are more of a problem because of the conventional grower's use of synthetic fertilisers. This practice makes nitrates available in excess to plants creating soft growth which these insects prefer. According to Garden Organic, aphids deposit double the number of eggs on plants grown with chemical fertilisers when compared to organically grown plants.[12] Organic plant foods are more slow release and the plants are less vulnerable to aphid attack because they are more resistant. It must be added that the industry is aware of this problem and certain pellets have been designed to restrict the speed of nutrient release.

Normally for someone growing organic and diverse crops on a small scale, Sternorrhyncha are merely a minor nuisance, if occasional imperfections in produce are overlooked. In any event, they are soft bodied and a number of home preparations will work against them that are described in Chapter Seven.

Sternorrhyncha are often described casually in terms of colour such as whitefly, greenfly, black et cetera, but this masks a taxonomically complex and diverse group. Overleaf as examples, we begin to describe a few of the most common variants, but realistically when you see a greenfly as pictured it may be one of

Left: A species of greenfly.[13]
Above: The vociferous ladybird beetle larva whose emergence is often linked to aphid life cycles.[14]

hundreds of similar or even identical looking species.

One greenfly that is not widely reported is an exception in this brief account of beneficials and pests. The nettle aphid (*Microlophium carnosum*) is not a pest and is not directly a beneficial. As its name suggests its habitat is the stinging nettle patch and they are not a threat to crops. Being autoecious they spend their entire life cycle on the same species of plant. What is interesting about nettle aphids for the organic grower is that they can be used as a valuable early food source for aphid eating beneficials such as hoverflies, lacewings and other pest predators. The lady beetle is known to lay its eggs in proportion to its prey so having a high quantity of *M. carnosum* has got to be a good thing. The aphid multiplies rapidly in April and reaches its peak in June.[15] Having this aphid resident lets your beneficials build up in number for later in the season when they can move to consuming pest species.[16] Of course being soft bodied this aphid can be affected by chemical sprays and many of the natural sprays mentioned further on.

If you cut back your nettles in May you might starve your beneficials of their early food source. If you want to harvest your nettles for making tea, the best time is mid-June as this will evict your beneficials from the patch at a time when you want them to go for the real targets. If you cut later in July you may be too late to let your beneficials have their maximum impact.[17]

Whiteflies are significant pests in the Aphidoidea superfamily. They are around 1.5mm long, have two pairs of wings folded over one another front and back and mini antennae on their heads.

Trialeurodes vaporariorum (family Aleyrodidae) is the most notable whitefly in greenhouses as it does best in warm conditions[20] and is a pest of solanaceous plants, cucurbits and legumes.[20] The behaviour of whitefly is often a strong indication to their presence as when they are around in considerable numbers they fly off at the slightest interference. After you have seen them a few times

Whitefly (*Trialeurodes vaporariorum*).[18]

you will get a sense of their traits and can recognise them even without seeing them directly.

In the greenhouse, the predatory wasp *Encarsia formosa* is a specialist in parasitising the Aleyrodidae[21] only.[22] It kills its prey by feeding on its hemolymph (whitefly equivalent to blood) and by injecting eggs inside the whitefly's body (oviposition) during their immature stages. The mature wasp is only around a millimetre long so they are not the most noticeable of predators. A useful indicator of this parasitic wasp's presence is the noticeably blackened whitefly pupae on the underside of leaves which can be distinguished from unaffected pupae. This special parasitoid is somewhat mysterious outside of the greenhouse as it has only been studied rigorously in controlled conditions.[23]

Another important whitefly, *Aleyrodes proletella*, which is not an aphid but a true bug, can be found on brassica and if rotation is not employed this fly can build up to excessive levels. It can be distinguished from other 'whitefly' by the dark flecks on its wings. Because whiteflies are soft bodied, sprays can have an effect, but when established they are difficult to control. Over a season whitefly may run through numerous generations increasing exponentially each time. The beauty of using natural predators is that they appear to have evolved to take full advantage of their prey's life cycle. With our predatory wasp, *Encarsia formosa*, its life cycle is three times faster which infers, at least theoretically, that it has the capacity to multiply at greater exponential rates than its prey to hammer down pest levels. In her lifetime, our parasitoid may oviposit up to 165 eggs[24] and might kill another 100 hosts from her feeding,[25]

The parasitic wasp *Encarsia formosa*.[19]

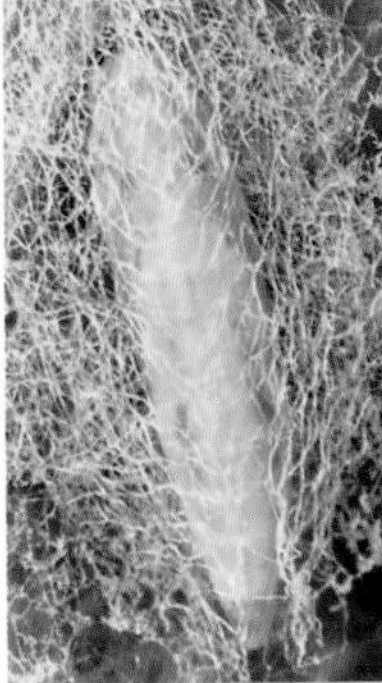

Above: One of the Diamondback Moth's key predators the Diadegma wasp ovipositing an egg.[28]

Far left: The Diamondback Moth larva.[29]

Left: DBM pupa.[30]

not bad for such a tiny wasp. Despite these impressive statistics our parasitoid is a little lackadaisical in her predation, getting bogged down in honeydew, craggy surfaces and wandering around leaves somewhat indiscriminately. They tend to be most effective when whitefly have not reached excessive numbers where other big hitting beneficials move in, such as lacewing and lady beetles. Earlier we briefly touched on aphid reproduction and it is interesting to note that our parasitic wasp can also become parthenogenic when infected with the bacteria *Wolbachia*. There are few studies to refer to here but it is conceivable that such an infection may benefit the grower and facilitate *Encarsia formosa* to keep up with the aphids when they hit more considerable numbers.

There has been an interesting development with a close relative of *Encarsia formosa* that has recently been found breaking through its pupa in Kent by Dr Andrew Polaszek of the National History Museum. It was originally thought to be a wasp just new to the UK[26] but it turned out that it was a completely new species[27] which at the time of writing is yet to be named. What is particularly interesting about this discovery was that it occurred outdoors. Its significance to horticulture is yet to be researched but if this predatory wasp is as effective outside as *Encarsia formosa* is in the greenhouse, then this could be very useful to organic growers. Whether this was a freak event or this wasp has naturalised in the UK is also unknown. It is possible that the wasp has been around for a long time and because of its size has just gone unnoticed. Further information is eagerly awaited.

Moths and butterflies (order Lepidoptera)

Since moths are part of the same order as butterflies, anything that kills moths will generally also kill butterflies. Some moth species can be valuable pollinators[31] and it must also be borne in mind that not all moths are agricultural pests.[32] In terms of British Lepidoptera, most are not pests.[33]

Regarding crop damage, the most significant phase of the moth's life cycle is the larval stage and for fruit the adults of some moths can cause serious commercial damage,[31] although much of this is merely cosmetic and could reasonably be dismissed.

Important lepidopteran pests include the butterfly larvae of the Pieridae family, which are brassica pests and moth larvae of family Tortricidae, which are tree pests. The codling moth (*Cydia pomonella*) is a fruit tree pest[33] and is the most important member of this last group. The family Sphingidae of the Sphinx moth is an important pest family for solanaceous crops (tomato, chilli et cetera).[34]

As a part of attempts to detach from toxic pesticides, research in many countries has been directed towards developing natural attractants to trap crop damaging moths. A very promising process originating from indigenous approaches has been the use of fermented oak sawdust that has been shown to attract most moths.[32]

The diamondback moth (*Plutella xylostella*) or DBM for short is one of the most important pests of brassica crops worldwide.[35] It also has a special claim to infamy for being the first pest in the field to develop a resistance to DDT[35] and the biological pesticide Bt toxin.[36] It is not native to temperate climates but is highly mobile and in warm conditions can fly in on jet streams to wreak havoc on crops worldwide. The adult female can lay between 200 and 400 eggs[37] between the veins of leaves[38] mainly on the upper surface.[37] These white eggs incubate for up to eight days when they hatch into the green caterpillar pictured on the previous page. They then get to work on the underside of leaves,[38] so anyone wanting to use the biological sprays in Chapter Seven would be wise to check there when spraying. Evidence of their presence is windowlike holes in cole crops. Over around 15 days the caterpillar will go through four instars and grow to 12mm, finally forming a pupa[39] and the adult moth emerges approximately nine days[40] after. If the pupa has been parasitised by our friend the predatory wasp, the resident inside will appear dark as compared to the normal DBM process of turning from pale green initially to white.[41]

Because this moth has such a quick turnover in its life cycle they have the potential to go through up to 20 consecutive life cycles accumulating each time, although in temperate climates this tends to be around four cycles[42] before the winter halts it. As with other pests, chemical attempts to control these are rather unsuccessful, tending to damage important predators while the DBM develops resistance.

The DBM has a plant enemy, wintercress (*Barbarea vulgaris*), since it is attracted to this plant but any eggs it deposits fail to survive.[43] It cannot survive cold winters and has its own very special parasitic wasp, *Diadegma semiclausum*, which can decimate up to 75% of DBM populations in the field.[44] There are other parasitic wasps that have DBM as hosts but *D. semiclausum* is by far the best that is known.[45] This small wasp is 6-10mm long and soft bodied. In many ways it is like *E. formosa* in the way it injects its eggs into the pupa. DBM is its key host but it is possible that it has an alternate host for the winter like its close relative *D. fenestralis* which overwinters on hawthorn.

Our wasp may struggle to keep up the diamondback's population explosion and sometimes does not live up to its potential. One explanation for this is that the DBM can access food stuffs that are richer in nitrogen in highly fertilised fields than the equivalent diet of the wasp, which is another good argument against excessive use of fertiliser.

The cabbage white butterfly (*Mamestra brassicae*) is a key pest of brassica crops and many other crops besides. The adults can be seen between May and September and are attracted to nectar, sugar and light. The caterpillars of *M. brassicae* like to hide underneath leaves by day close to the soil[47] as do many other caterpillars. They can overwinter as either pupae or larvae. It is predated by the lesser known parasitic wasp *Cotesia marginiventris* which predates the larvae and is considered a good pest locator. In a real sense it could be considered as one

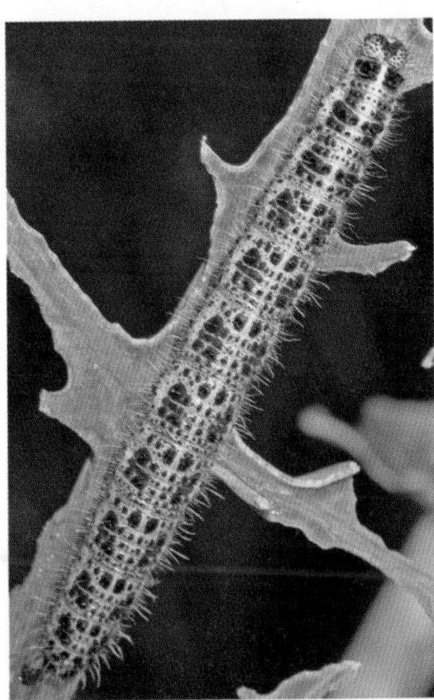

Left: Cabbage white larva on a brassica leaf by Soebe (2005).
Right: An adult cabbage white moth with pupa by Entomart (2009).[46]

of the plant's emergency services. Recent studies have established that the wasp responds to the volatile alert chemicals emitted by the plant to summon this beneficial to predate on active pests.[48]

Green or common lacewings, family Chrysopidae, are important generalist predators. They form part of the larger order of Neuroptera or net winged insects. A major predator of aphids, the larvae or aphid lions, will also devour a broad range of other organisms[53] including insect eggs, insect larvae, aphid honeydew,[54] caterpillars[55] and the red spider mite.[56] In terms of pest predation, lacewing are most significant in their larval phase[57] which lasts approximately three weeks[58] and can grow up to 10mm in their final instar.[59] They resemble miniature alligators rather like ladybird larvae but less colourful. Our aphid lions are ferocious and extremely cannibalistic if they run out of their usual prey.[60]

By conservative estimates they may consume up to 200 aphids a week[61] before they pupate and turn into the adult lacewing pictured below. During the final stage of their development you may find their cocoons slipped discreetly in various shady nooks. The adults are 12-20mm long,[62] soft bodied, have long antennae and shining golden eyes. They feed mainly on honeydew, nectar and pollen.[57] Female lacewings can lay around 200 eggs in their lifetime.[62] The eggs are oval, just under 1mm long and hang beneath leaves[54] on silken stalks[55] as below, which protects

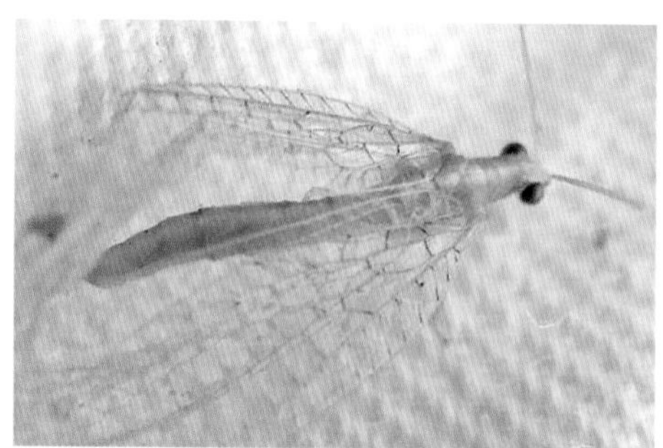

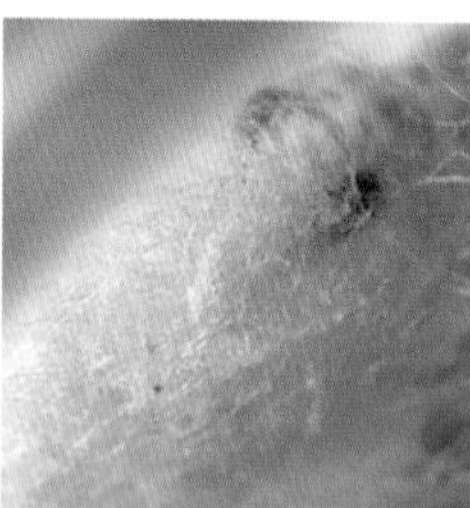

Above, a green lacewing[49] and its ferocious larva,[50] pupa (left) and, bottom, lacewing eggs on stalks.[52]

the hatchlings from other predators[62] and cannibalisation.[63] When first placed on the bristly stalks or pedicles the eggs are green and gradually turn grey before hatching.[64]

The phyllosphere and compost teas

The phyllosphere is the surface of the aerial parts of plants and the interior[65] according to some authorities. Its inhabitants are microorganisms called epiphytes and include bacteria, fungi, archaea and protists. Larger organisms do interact with this zone and this area should in no way be considered as separate from the rest of the food web.

Of particular interest to the grower are those organisms that can suppress crop diseases, combat pests and interact with plants to conserve and supply nutrients. How well a grower can influence these beneficials is a topic of considerable dispute.

Organisms that live in the phyllosphere are often quite distinct from those found in the rhizosphere, which is understandable when you consider how different the conditions are. Many experts in this field characterise the phyllosphere as a harsh environment for its inhabitants with considerable variations in moisture, sunlight, nutrients and temperature as compared to the rhizosphere.[66] Not only are there wide variations in conditions across the phyllosphere but conditions can alter rapidly. Some recent studies have mitigated this bleak picture with the discovery that the residual water on leaf surfaces is wetter than was previously thought[67] which offers greater potential for certain microorganisms to survive in dry conditions as long as they can access this narrow film. Surprisingly it was found that this film varies very little in terms of water content. For the brewer of compost teas this adds further credibility to assertions that foliar sprays can have a positive long lasting effect on crops since with a stable albeit minuscule supply of water, organisms have more of a chance of becoming established.

A further useful finding that could have important implications for the foliar tea brewer has been that leaves at different stages of growth have different kinds of bacteria in different volumes.[65] Young leaves have the most bacteria which are most closely related to populations of bacteria in the rhizosphere.[65] As leaves mature, the type of bacteria become more dissimilar.[65] Intuitively this would suggest the best timing for spraying an aerated compost tea, which will be dominant with rhizosphere bacteria, would be in the early stages of plant growth and such a spray may be developed from soil bacteria. It must be stressed however that whilst in the light of these findings such teas should have good chances of success, the formulations are pioneering and highly speculative. In this instance we are very much past the point where confidence can be backed by numerous related trials.

Most microorganisms like damp conditions[68] and establish more easily in the shady corners available in the phyllosphere where they are safer from UV radiation.[69] So someone wanting to attempt to enhance or initiate a colony of beneficials with a foliar spray could choose to spray in the evening, pre-dusk, and make sure the leaves are damp before spraying. Foggy and still, but not raining are good conditions for spraying.

As previously discussed many bacteria and fungi are beneficial to crops. Some are naturally occurring and could feature in 'Indigenous Microorganisms' approaches as in Chapter Ten and others have been isolated by researchers attempting the 'Effective Microorganisms' approach. Antagonists are beneficial fungi that attack crop diseases and entomopathogenic fungi target insect pests. A key concern with these latter beneficials is to what degree they attack non-target insects. Control measures using these fungi have been suggested as possible components in IPM as an alternative to wide spectrum pesticides.[70] A handful of the most important fungi are described below which survive use in foliar sprays.

Beauveria bassiana is an entomopathogenic fungus found naturally in soil and on plants. It needs warm, humid weather to spread and infects a wide range of pest species such as aphids, thrips, corn borers, Colorado potato beetle and Mexican bean beetle. Recent studies indicate that in low dosages it was relatively harmless to lady beetles, lacewing[71] and honey bees.[72] Infected pest larvae turn from white to grey. *B. bassiana* occurs naturally in the phyllosphere of many plants including nettles and it is suggested that insects are important vectors of this beneficial fungi.[73]

Verticillium lecanii is an important fungal predator of aphids. It is used in greenhouses to control whitefly, aphids, thrips and mealybugs and is well known for its usage in sprays.[74] This fungus does best in humid and cooler temperatures which makes it a good candidate for temperate climates.

Trichoderma harzianum is a very widespread fungus found in agricultural soils and decaying wood. It is known to locate and grow towards pathogenic fungi, which it parasitises and hence it is a useful ally to protect crops from fungal diseases.[75] In the compost heap it is useful for breaking down recalcitrant cellulose, reducing pathogens and raising the ratio of nitrogen to carbon.[76]

Paecilomyces fumosoroseus is a generalist fungus[77] that has a broad host range including the diamondback moth[78] and a range of mites.

We leave our brief tour of the food web now and begin the more practical part of this book with the traditional approach to making compost teas. A number of variations and innovations on this theme have also been introduced.

CHAPTER SIX

Feed Preparations

The Haws watering can in action, designed to simulate natural rainfall.

In this chapter we get down to the actual detail of preparing feeds by the traditional steeping method introduced in Chapter One. A number of techniques will also be introduced such as extraction methods for comfrey and the steps you need to take to prepare kelp meal. We will be applying some important resources that are world renowned for their fertility enhancing qualities. We will also be incorporating some plants that have fallen out of the attention of the mainstream but have a history of usage, such as yarrow. Some of the plants you may have come across are seen as weeds, but have surprising and useful roles for the home brewer.

As discussed in Chapter Two, plants uptake substantial amounts of nitrogen (N), potassium (K) and phosphorus (P) so it is understandable that they form a large proportion of both conventional fertilisers and organic feeds. Secondary calcium (Ca), sulphur (S) and magnesium (Mg) trace elements are also important as previously outlined, but are only needed in smaller volumes. Many of the plants described in this chapter have multiple benefits for the organic grower such as refuges and food stores for beneficial insects, which will be noted as we go along.

An important point is that a resource may not hold all the necessary nutrients to give your hungry tomato or pumpkin a boost, so with this in mind some preparations use combinations of plants and composts.

If you follow good organic practice for a basic level of soil fertility, many of the concerns with nutrients and their availability have been addressed in a holistic way that bypasses the necessity of tinkering with the finer details. In Chapter Two we introduced how to assess what may be lacking in your soil and in Chapter Three we looked at the cycles of some of the most important nutrients. In this chapter we hope to replenish many of the nutrients removed by harvesting and other causes with various preparations to furnish your plants with what they need, but before we proceed, we need to consider some issues with chlorine.

Chlorine issues

This only applies to people using tap water; if you are using rainwater, which is recommended, you can discount this section entirely.

Chlorine is used in domestic water supplies to kill pathogens within pipes. The amount of chlorine in drinking water varies and for public health must be between 0.5 ppm and 2 ppm.[1] Laboratory studies indicate that even at these levels it can disrupt or kill some beneficials. For instance it was found that rotifers (beneficial) were stunned at 2 ppm of chlorine.[2] A small amount of chlorine in your feeds will not be disastrous and it is an essential element to most organisms in trace amounts, but why have damaging amounts there when it can be easily removed?

The simplest solution is to pour out your water, leave it open overnight and agitate it in sunlight as the UV rays will help the chlorine to disperse. The 'aquarium method' of adding various potions to your water to disperse the chlorine is not recommended because it introduces chemicals to your brew with uncertain effects and is something of an overkill.

A more challenging problem related to chlorine is chloramine. This substance does not disperse when exposed to air like chlorine. There are two methods, one costly and one free, but not quite so effective. The costly answer is to pass your water through a carbon filter. The process is as simple as buying the product and pouring it through your filter. Some may consider the use of carbon filters as overly elaborate and unnecessary because the chloramines work by attaching themselves to organic matter,[3] so if chloramines come into contact with vermicasts or compost this substance will be removed by a small sacrifice of living organic matter. Hence the more cost effective method would be to mix in some organic matter then strain the solution. For the purist this is not acceptable; they want the best possible environment for their microherds to thrive in, but realistically no one should lose sleep over their preparation being contaminated by chloramines.

Comfrey (*Symphytum officinale*) in flower.

Comfrey tea

An important variety is Russian comfrey or Bocking 14, which unlike the wild plant does not set seed. Comfrey (*Symphytum officinale*) is a perennial and grows up to 1.5m. It is rich in phosphorus and potassium so it is very good for fruiting plants such as tomatoes, cucumbers and peppers. Comfrey is also a useful source of manganese, calcium, iron and cobalt.[4]

It is one of the most popular support plants for organic gardeners and with good reason. Comfrey makes an excellent forage for bees and can be used as a top dressing and as a compost activator. It also makes an excellent barrier against encroaching weeds and its size offers possibilities as a wind shield. Comfrey is known as a dynamic accumulator because it can draw nutrients from deep within the soil.

Instructions

Steep 1kg of comfrey leaves in a sack with 10 litres of water. Cover but do not seal, stirring every couple of days. Leave for 10 days. Dilute 7 parts water to 1 part feed for foliar applications. For root feeds it can be used undiluted using a container sunk into the ground.

Diagram of set-up apparatus for comfrey extract

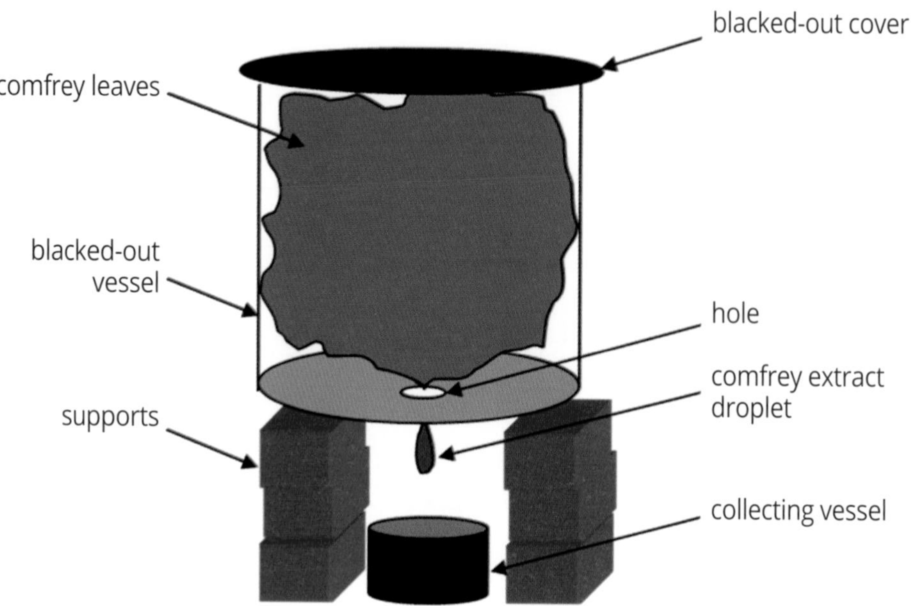

Concentrated comfrey extract

As an alternative to the steeping method an extract can also be prepared which does not require any water. It can be stored for up to six months and does not smell as pungently as the steeped comfrey. You simply need a container with a small hole in the base and a collecting vessel as shown in the diagram above. It is important to cover the vessel to protect it from UV light or it will compromise the breakdown process. It has been known to use an old bath with a carpet thrown over for this method.

> To get things started, stuff comfrey leaves into the hole and then add more leaves and press them down well. Place a collecting vessel below the hole. Within a few days the dark comfrey extract will start to drip into the vessel.
>
> Leave to drip into the collecting vessel for 14 days when it can be sealed. For foliar application dilute 10 parts water to 1 part feed. For root feeds it can be used 2 parts water to 1 part feed.

Stinging nettle tea

For the organic grower the stinging nettle (*Urtica dioica*) is much underrated as the living plant is an excellent support plant for beneficial insects.[4] This perennial is used by ladybirds for laying their eggs and is also popular with butterflies. Nettles make a nutritional general purpose feed with high nitrogen, potassium, calcium, magnesium, sulphur and iron, but low phosphorus.[5] Nettle tea is useful on its own but is improved further when mixed with something high in phosphorus such as comfrey.

Anecdotal reports suggest that nettle can enhance the growth of neighbouring plants, intensify their essential oil content and thereby increase the plants' aromatic/insect repellent qualities.[6]

Instructions

With gloves (!) and shears, tear up 2kg of leaves. As they are very fibrous this will increase the surface area and help them break down. Place in a hessian sack with 10 litres of water. Cover loosely and stir every couple of days for 10 days. Use after three weeks.

For foliar application dilute 6 parts water to 1 part feed. For root feeds it can be used 3 parts water to 1 part feed.

Stinging nettle (*Urtica dioica*).

Comfrey and nettle tea

With this tea you get the best of both worlds with a good balance of macro and micronutrients. The nettles bump up the nitrogen levels and the comfrey bumps up the phosphorus – this mix is highly recommended as a general purpose feed. An additional advantage is that in the UK comfrey and nettles are very abundant.

Ingredients

0.5kg nettles
0.5kg comfrey

Instructions

It is somewhat redundant to suggest that when picking nettles wear gloves or you will get stung. Also use gloves for the comfrey because although the fine hairs don't have the same jolt as nettles they still irritate the skin. Chop up the nettles as in the solo nettle preparation.

Put the 0.5kg of nettles in a sack with 10 litres of water and cover loosely. Stir every couple of days for 10 days. Pick 0.5kg of fresh comfrey, tear it up roughly and put it in the sack with the nettles. Stir every two days for another seven days. For foliar application dilute 6 parts water to 1 part feed. For root feeds it can be use 3 parts water to 1 part feed.

Clover tea (*Trifolium* spp.)

Clover is well-known as a green manure crop, but its high nitrogen and phosphorus content[7] also makes it an excellent candidate for compost teas. There are a number of varieties of clover each of which have special features. Crimson clover (*Trifolium incarnatum*) is best for ground cover, whilst white clover (*Trifolium repens*) is low growing and is favoured for intercropping with root vegetables. It is very tough and resistant to being trampled on and it is superior to red clover as a bee forage because of earlier nectar flows.

Yellow clover (*Melilotus officinalis*) is vigorous, very hardy, deep rooted and in the past was favoured as a green manure.[8] Along with alfalfa[9] it was highly regarded for breaking up and conditioning clay soils.[10] With the advent of cheap fertilisers its resilience began to be seen as a nuisance and it fell out of general usage.[8] Most significantly it is an important molybdenum accumulator[11] along with the nodules of other nitrogen fixing plants[12] which can be used to good effect in the next recipe.

Instructions

Steep 1kg clover in a sack with 10 litres of water. Cover but don't seal and stir once a day for 10 days. The residue can be used as a mulch or put on

White clover (*Trifolium repens*) from a local plot.

the compost heap. For root feeds dilute 5 parts water to 1 part tea. For foliar feeds 3 parts water to 1 part feed.

Legume boost

For legumes it is important not to give them too much nitrogen, especially nitrates, because it inhibits the nitrogen fixing processes and encourages favourable conditions for invasion by grasses.[13] Phosphorus is often the major limiting nutrient in legumes[14] so yarrow, pictured overleaf, has been chosen as a base for this preparation because it accumulates it in high quantities.[4] This plant is also an excellent habitat for beneficial predators and grows readily from seed with minimal maintenance. Magnesium, a secondary nutrient, also gets used up quicker in legumes[15] and again yarrow is a good plant source.

This mix is partly a supplement for nitrogen fixing bacteria that supply nitrogen to legumes to enhance their growth. Garlic is a sulphur accumulator and is added to stimulate nodule activity. Nickel, cobalt and molybdenum in trace quantities are very important to nitrogen fixing processes[16] and plants are included in the recipe overleaf that have them in heightened quantities.[4] If you consult Chapter Two, a list is included of accumulator plants, so if you do not happen to have one of the plants below there may be a possible replacement. It would be best to have a number of different sources included so your crop plant can absorb nutrients in the form that suits it best, at the appropriate time. The addition of garlic has the dual effect of pest repellent qualities (dealt with more specifically in Chapter Seven) and increased sulphur which is particularly important to legumes.

Yarrow self-seeded in an allotment in north-east NE England. The dense foliage is an excellent haven for ladybirds.

Eggshells have been incorporated because legumes also need higher amounts of calcium[17] and this also raises the pH slightly.

Ingredients for root drench

12 finely ground eggshells (calcium)
1kg yarrow (phosphorus, calcium, magnesium)
0.5kg dandelion, not in seed (phosphorus, calcium, cobalt)
20 crushed garlic cloves, equivalent to two heads (sulphur)
100g yellow clover roots, or other nodulated legume roots (molybdenum)
1 pinch sunflower seeds, ground finely with eggshells (cobalt, nickel)
0.5kg kelp meal using preparation method on p.81 or organic supplier (wide range of trace elements)

Instructions

Finely chop dandelion, clover roots and yarrow then steep in 4 litres of water, cover, stir every day for a week, and strain. On the last day grind up eggshells and sunflower seed to a fine powder then mix in crushed garlic to make into a paste. Stir in 1 litre of water to create a suspension. Prior to application, mix this in thoroughly with the strained yarrow, clover and dandelion liquid then apply. Use the residue around your legumes as a mulch.

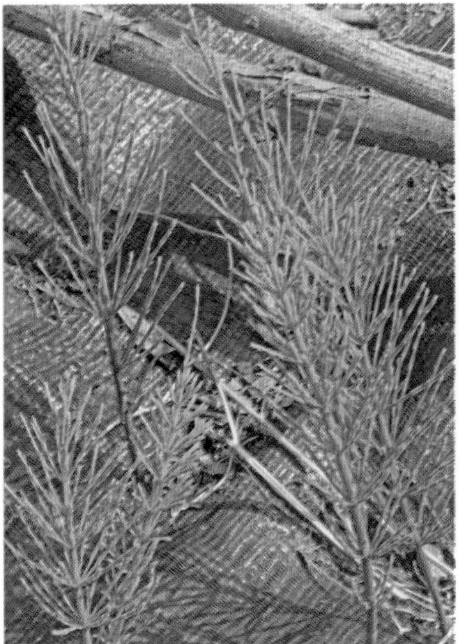

A close up of marestail (left) and a rough patch of marestail growing in a neglected spinach patch.

Root and shoot stimulator

As in the legume boost, yarrow is again used as a base because of its high phosphorus content that is the major limiting nutrient important to root growth along with the complex interactions with other nutrients to various degrees.

The other key ingredient for this mix is *Equisetum arvense*, otherwise known as marestail/horsetail and is the scourge of growers in the UK. It is a recalcitrant and ancient perennial weed that is virtually impossible to eradicate. Practice suggests that soil cultivation, improving drainage and depriving this plant of light and open soil can eventually succeed. In early spring it grows spore producing stems that look similar to asparagus and then later in the season it develops the familiar marestail form above.

The roots go down to at least two metres in the ground and being another dynamic accumulator[18] it draws valuable resources from these levels. The good news is that it has many uses for the organic gardener. It is high in silica,[18] cobalt and magnesium.[4] Its value for roots lies in the silica which falls into the beneficial category as introduced in Chapter Two. From laboratory studies silica is considered non-essential, however in the outside world it plays an important role in hardening plant tissue and protecting plants against pests and harsh conditions[19] which we will return to in the next chapter. It also has a valuable further role enhancing the soil's fertility through improved soil structure and maintaining nutrients in plant available forms.[20] Despite these advantages the sight of marestail on plots is rarely celebrated.

A number of studies have shown that plants grown in soil with optimum amounts of active silica compounds, such as the compounds in marestail, substantially increased the mass and volume of their roots as compared to plants without this addition.[21] We are less concerned with the inert forms of silica found in some minerals because it can be present in the soil but is not very accessible.

The addition of properly made compost as described in Chapter Eight adds a biologically live element to this mix, supplying many important nutrients, soil conditioners and growth promoting substances to encourage root growth.

Ingredients

0.25kg *Equisetum*
0.5kg yarrow
0.5kg homemade compost (for foliar spray, 1 tbsp vegetable oil as wetting agent)

Instructions for root drench

Simmer 0.5kg of fresh *Equisetum* in 10 litres of water for three hours to release the silicon into the water[22] and add this to a suitable vessel. Add fresh yarrow to the vessel with another 10 litres of water. Cover but don't seal and leave to steep for 10 days, stirring once every couple of days. Dilute 7 parts water to 1 part feed. Mix up with fresh compost at the last moment, strain and apply. The residue can be used as a mulch or added to the compost heap.

Instructions for foliar spray

Same as above except immediately prior to application mix in a small amount of vegetable oil (15ml) as a wetting agent.

Kelp tea

Kelp algae (order Laminariales) is a large type of seaweed of around 30 genera. It is high in potassium,[23] excels in 60 trace elements[24] and is a good source of iodine[25] which is important for human and ruminant nutrition, but the principal benefit of kelp is its possession of special plant growth hormones which are well known to science.[26] It is hypothesised[27] that the combined effect of the dearth of trace elements and plant hormones have a synergistic effect which stimulates the plant. In a study with beans[28] it was found that a foliar application of kelp improved the yield of beans by as much as 24% with considerable gains in numerous other crops including potatoes, beets, tomatoes, peppers and sweet corn.[29] It also gives a better resistance to frost, seed germination and uptake of nutrients. Better resistance to insects and fungal diseases have also been reliably reported.[30]

If your levels of N and P are well supplied by your soil you will be pleasantly

Kelp (order Laminariales) found on the UK coast.

surprised at how well your plants grow after applications. In some mixes such as the 'multi-nutrient tonic' overleaf, further additions are made to bump up the primary nutrients.

Making kelp meal is an important intermediary step to using kelp in teas although it is not something you would want to do in your home. It acts as a beacon to many different creatures especially certain flies as its aroma is somewhat pungent, but its effect on crops is renowned.

Kelp is very abundant on UK coastlines and large quantities are washed up yearly. It is possible to buy kelp in manicured packages, but why not take a trip to the seaside and pick your own! You might have to contact the local council in that region to see where the unpolluted areas are and if it is permitted.

Meal preparation

The advantage of preparing a meal is that it can be stored and used when you need it. Homemade meal is far superior to shop-bought meal because the live microbes have got much more chance of survival and less have been lost during processing. The meal is simple to make if you follow the instructions to the letter but there are many ways you can go wrong making seemingly innocuous changes. Spread out the seaweed on a pallet to dry making sure the pallet wood is clean and untreated. If it is a sunny day it is good to leave it in the open but do not allow rainwater to get to it. Overnight it is necessary to set up a raised cover. An old tent open at the sides works fine for this, or lay it in your shed. Do not place a cover in direct contact with the seaweed as this blanches the seaweed and may cause it to cook, ruining the whole process. Do not stack the seaweed as we do not want it to rot, we just want let the air get to it to preserve it. This process works very well for other seaweeds also, such as bladderwrack, another excellent growth enhancer.

Liquid feed preparation

Mix 0.33kg of fresh kelp or kelp meal with 10 litres of water. If you are doing this in summer you can use a dark coloured container to absorb the sun's rays, however the weather does need to be warm. Leave this covered but not sealed, in the sun for three days, mixing twice a day. If the weather is too inclement you can heat the mix on very low heat for six hours although the smells will not make you popular. The heat should be lukewarm between 18-25°C (64-77°F); at no point allow the mix to get too hot as this will disrupt or kill the microbial life.

Whichever method you use, the brew should then be strained and applied directly. What is left in the straining can be used as a mulch or put on the compost heap. For foliar feeds, a tablespoon of fish oil can be added as a surfactant and then dilute 5 parts water to 1 part feed. For root applications dilute 3 parts water to 1 part feed.

The multi-nutrient tonic

This special preparation contains a wide range of elements and the added power of natural seaweed hormones. It offers the range of nutrients you get in some highly priced non-organic fertilisers and does not suffer from the drawback of being stored in polythene for long periods to undergo partial anaerobic decay. The process is a little more involved than many of the other preparations but the extra effort is well worth it.

Ingredients

0.33kg nettles
0.33kg comfrey
0.33kg kelp

Instructions

Fill your container with 10 litres of water. Put 0.33kg of nettles loosely in a sack and suspend in the water. The nettles are added first because they take longer to break down. Stir every couple of days for five days then add 0.33kg of freshly picked comfrey to the sack. Stir every two days for another five days. After this period, begin your kelp feed preparation as described in the 'kelp tea' section. Finally mix it all up in a barrel or water butt. Apply 3 parts water to 1 part feed for roots. For foliar feeds add 1 tablespoon of fish oil and dilute 5 parts water to 1 part feed; apply immediately.

Vermicast tea (steeping method)

Vermicasts are higher in minerals and growth promoting substances than normal compost[31] and considered by some to be a panacea for agriculture since the casts enhance plant growth to an extent comparable to synthetic fertilisers with none of the drawbacks.[32] It has also been cited as possessing numerous disease suppression and pest repellent qualities.[33] Because of the inoculating effect of being passed through the gut of a worm these casts have a very low level of pathogens[34] and are considered completely safe for direct usage although because of their concentration they need to be used sparingly or diluted, such as below.

The casts can be generated using worm bins with kitchen waste, specially prepared beds, bought from worm farms or collected by hand.

Instructions

Mix 0.33kg of matured worm compost in a sack and place in a bucket with 10 litres of lukewarm (approx. 25°C / 77°F) water. Lift and replace the sack a number of times, stir and agitate the water with something with smooth sides such as a wooden cooking spoon. Do this for 15-20 minutes. (Some worm bins come complete with a tap at the bottom to drain off leachate; if you have this available don't use it for making teas using this method as it is not mature enough.) Use immediately. For foliar application dilute 5 parts water to 1 part feed. Root applications 3 parts water to 1 part feed.

General manure tea

Manure from pig, sheep, poultry, horse, cow, rabbit or guinea pig is fine, but it is important to know where you are getting it from because you do not want to start off with something that is full of agri-chemicals. Avoid manure from dogs, cats, foxes and humans because of disease issues. There are special ways of dealing with these manures but it is beyond the scope of this book. Another problem with dealing with manure is that it is of variable strength based on the type of creature, their diet and the manure's maturity. For this preparation, a manure that is well rotted is recommended. The instructions below err on the side of caution because if the mix is too strong it can burn leaves and roots.

Instructions

Place 0.5kg of manure in a sack and place sack in a bucket with 10 litres of water. Cover and stir every two days for 10 days. For foliar application dilute 5 parts water to 1 part feed.

Root applications 3 parts water to 1 part feed.

We complete this chapter with a table for how you might apply these preparations with specific crops at different stages of their life cycle.

Compost tea amendments linked to life cycles of various popular crops

Crop	Frequency	Stages in life cycle				Comments
Potato	Every 2-3 weeks	Sprout Development 0-14 Days	Planting/ Vegetative Growth 15-40 Days	Tuber Initiation and Bulking 41-80 Days	Maturation[36] 81-130 Days	Nutrient intake highest at tuber bulking stage[35] especially potassium and nitrogen[35]
			G	Alternating H and I	Alternating E and G	
Tomato	Every 4 weeks then every week after fruit set	Germination and Growing On 0-15 Days	Transplant and Flowering 16-30 Days	Fruit Set and Fruit Development 31-61 Days	Fruit Ripening and Harvest 61-210 Days	Phosphorus, calcium and magnesium needed at a constant rate[37]
			B	A then H	Alternating G and I	Higher potassium at flowering then higher nitrogen[37]
Cucumber	Every 2 weeks then every week during fruiting	Germination and Vegetative Growth 0-7 Days	Transplant/ Vegetative Growth 8-55 Days	Flowering and Fruit set 56-85 Days	Fruit Swell and Harvest Period 86-111 Days	Unusual for a crop because needs more potassium than nitrogen[38]
			G Then H	Alternating A and C	Alternating F and C	
Sweet corn	Every 2 weeks then every week when cobs swelling	Germination and Growing on 0-14 Days	Transplant/ Vegetative Growth 15-60 Days	Tasselling and Pollination 61-70 Days	Cob Swell and Harvest 71-100 Days	Nitrogen and potassium the most growth limiting primary nutrients over phosphorus[39] which is most needed in earlier stages[40]
		F	H then B	Alternating I and H	Alternating I and H	

Crop	Frequency	Stages in life cycle				Comments
Peas	Every 2-3 weeks	Germination N Fixing Nodulation and Growing On 0-14 Days	Planting and Vegetative Growth 15-40 Days	Flowering 41-50 Days	Flowering/ Pod swell/ Harvest/ Flowering 51-70 Days	Low nitrogen and feeding the nitrogen fixing bacteria also. Special attention during flowering.
		D	E	A	A	
Runner beans	Every 3-4 weeks	Germination N Fixing Nodulation and Growing On. 0-12 Days	Transplant and Vegetative Growth 13-44 Days	Flowering 45-90 Days	Flowering/ Pod swell/ Harvest/ Flowering 70-90 Days	This legume likes a little nitrogen to get started
		G	D	E then F as Foliar spray	A	
Brussels spouts	Every 4-6 weeks	Germination and Growing On 0-28 Days	Transplant and Vegetative Growth 29-168 Days	Vegetative Growth and Button Formation 169-179 Days	Button Formation/ Harvesting Period 180-300 Days	Likes balanced amounts of nitrogen and calcium but not too much
		G	E then C	H	Alternating G and H	
Garlic	Every 4 weeks	Planting Cloves e.g. December. Bulb Swell 0-59 Days	Scape/Leaf Growth (early spring) 60-116 Days	Bulb Swell and Scape Growth 117-229 Days	Scape Yellowing/ Harvest 230-244 Days	Higher nitrogen good early spring but then lessen for bulb growth[41]
		E	B then G	E	---	
Chilli	Every 3 weeks then weekly after flowering to onset of ripening, then every 2 weeks	Germination and Growing On 0-49 Days	Transplant / Greenhouse and Vegetative Growth 50- 84 Days	Flowering and Fruit Set and Fruit Growth 85-79 Days	Fruit Ripening and Harvest Period 80-120 Days	Needs highest amount of nutrients,[42] especially potassium[43] from flowering to the onset of ripening
		G	H	G	H	

Key
A-Comfrey tea. B-Stinging nettle tea. C-Comfrey and nettle. D-Legume boost. E-Root and shoot stimulator. F-Kelp tea. G-Multi-nutrient tonic. H-Vermicast tea. I-General manure tea.

Here ends the recipes using the traditional approach to compost teas that has been used for thousands of years in various forms and is still in use today worldwide. In the next chapter, we will look at the vexed topic of botanical pesticides for tea brewers that organic growers can use if they are desperate enough. We will also dabble in synergy and introduce a novel preparation.

CHAPTER SEVEN

Teas for Protection and Control

The use of blasters in conventional agriculture to apply insecticides, fungicides and herbicides.[1]

Pesticides have come a long way in the last few decades from the reckless attitudes towards nature and all the environmental wreckage that entailed, towards a grudging acceptance to working in more harmony. Many of the synthetic pesticides hawked by the global agri-business trade have been outlawed and no safe synthetic alternative currently exists. Because of this, greater attention has again been directed to botanicals, namely plant extracts, that tend to be less destructive and linger less in the environment.[2] A home brewed botanical pesticide can have a novel advantage over something industrially produced, because a substance from natural sources possesses a greater range of active compounds which are harder for a pest to evolve a resistance to than a single lab produced chemical.

Today the organic grower works spade by spade with the chemical grower, so many of the problems that beset the chemical grower are transferred to the organic grower's fields. The chemical grower's plan is high risk in that they grow single crops of the same cultivar. The preoccupation with detrimental regimes of pest control can be seen as an aspect of this scene. If the particular conditions are suitable for a pest it has more potential for growth in a field that has only a single crop and in principle the same applies to crop diseases. The grower is constantly waging a losing battle as pests gradually develop a resistance to the current brand of pesticides. From the point of view of agri-business this situation is ideal; it could be said that they are waging a winning battle with another business opportunity every time the current round of pesticides become redundant or are found to be environmentally unsound.

If the organic grower lived in a world where there was no chemical grower there might be no need for this chapter. The idea of attempting to surgically remove a pest is against the philosophy of organic growing where a balance is struck between a diversity of organisms. The organic grower would have fewer concerns about pests and crop failure because the ecology of the crops is inherently more resilient. The grower will have a different set of concerns, such as issues over productivity and challenging the false perception that all unblemished, uniform produce is healthy. If something goes wrong with one crop they will have crops of other cultivars that have not been stricken. If a particular pest attempts to ravage a crop it will be in less devastating numbers than a pest that has found ideal conditions in a vast single crop. In organic growing, natural enemies linked to a particular pest are nurtured and any pest that is starting to get out of control will soon be brought back in check with an influx of predators. For instance, the life cycle of the ladybird is intimately linked to the emergence of aphids with ladybirds seeking out areas where aphids may be prevalent such as in a nettle patch or a stand of yarrow. If all the wild patches have been removed to squeeze in a few more crops then another safety net has effectively been removed.

A further interesting example of a natural balancing mechanism is the way the wolf spider will expand their numbers where there is abundant prey and then regulate their own population by cannibalism when they run out of food.

Unfortunately even if you are a dedicated organic grower you still have to deal with factors outside your control in the best way you can. You may be near a farmer's field for instance or your site may be next to a grower that has built up a whitefly problem from growing brassicas on the same spot for decades and has unwittingly killed all the whiteflies' natural predators. The philosophy of tackling the symptoms rather than the causes can lead the grower down an often unnecessary cycle of destruction with spiralling pest numbers and costs. Desperate situations sometimes require desperate measures and many of the recipes in this chapter should only be used in an absolute emergency. To mitigate this issue the ingredients and preparations chosen are those that have minimal impact on beneficials and still leave the plant edible to humans.

I concentrate on preparations that have an excellent track record as botanical pesticides, have ingredients that are easy to get hold of and can be cultivated locally. The pests and diseases that may beset a plant are too numerous to mention in a single chapter so I will only note some of the most important ones, where there is a good indication that the preparations here can have an impact. Towards the end of the chapter I describe two teas that initiate a 'Systemic Acquired Resistance' (SAR)[3] which is an intrinsic natural process that exists in an enormous range of plants. Basically if a plant is attacked, a defence mechanism is triggered which operates within the whole plant. These teas initiate these responses without the attack.

When investigating pesticides you will come across the terms target and non-target. This simply means that some preparations target a specific problem taxon, for instance spider mites, and some are less specific, being non-target and hitting a broader spectrum of organisms. You will also come across pesticides that are systemic and non-systemic. All this means is that systemic pesticides go into the plant and non-systemic pesticides stay on the surface.

The concept of half-lives is sometimes used with regards to how persistent a pesticide is after it has been applied. If a particular pesticide has a half-life of three days its effectiveness will halve every three days. So if you have 100% effectiveness on the moment of application, your pesticide will be 50% effective after three days, 25% after six days and so on. The point of using half-lives is that products can linger a very long time in negligible quantities and this kind of information can make more sense when judging a product's safety. It's unfortunate that with regards to applications of pesticides to the soil, half-lives can be misleading and should only be treated as a rough estimation because their pathways rarely follow such a predictable pattern as there are a considerable number of confounding factors such as weather conditions, runoff and the amount of organic matter in the soil.

The foliar preparations presented here use liquid soap combined with water as described below. The reason for this primarily is that the soap solution acts as a wetting agent which helps the preparation stick to the surface of the plant considerably better than just water alone, which makes it stay longer on the plant and enhances the impact.

A note of caution must be added for this chapter that although almost all the preparations here are derived from natural locally available sources, it does not necessarily mean that they are harmless so some degree of care is needed, the extent of which will be noted in each case.

Making liquid insecticidal soap with soapwort

(*Saponaria officinalis*)

Soap sprays have an insecticidal effect on their own with nematocidal properties and are indicated to affect the breathing apparatus of insects. A further feature of soap sprays is that they can help protect the plant against infection by wiping clean the feeding apparatus of certain pests when they penetrate the plants with their mouth parts. Depending on your level of engagement with self-sufficiency you may want to produce your own liquid soap. A method for one such solution is described here. Alternatively you can go to your local pharmacy and obtain an enema soap which works fine. Under no circumstances use washing up liquid because the chemicals it contains are too harsh and definitely not a product for organic growers. Finally, if you like bees, it is best during flowering time to spray in the evening when the bees have gone back to their hives.

Soapwort (*Saponaria officinalis*) growing wild in North East England.

Soapwort was one of the staples of the cottage garden and is popular today for its beauty and heady fragrance. It was brought over to the UK during the Middle Ages by Franciscan and Dominican monks from Northern Europe as 'a gift of God intended to keep them clean'.[4] Later the plant was taken to America by the Puritans. Its main active compounds are saponins which are mildly toxic. Today soapwort is important industrially for making halva and the head on beer. It is a vigorous perennial herb and grows up to a metre high. For our purposes the soapy compounds of saponin makes a useful wetting agent as a base for other preparations in this chapter. The added benefit is that saponin compounds have excellent insecticidal properties in their own right;[5] in a recent study aphids and caterpillars were killed or disrupted by saponins.[6]

It is very important to use soft water for this preparation; hard water interferes with the process, therefore rainwater is ideal. When preparing liquid soap for up to 10 litre sprays you only need a standard sized cooking pan because the final solution is to be diluted 1 part liquid soap to up to 5 parts water/pesticide solution.

Pulverise the soapwort roots and finely chop the whole plant, roots and all. Half fill your pan (2.5 litres capacity) with the whole soapwort plant and fill up with water. Bring to a boil then simmer for 30 minutes or until the solution becomes viscous. Allow to cool and then strain using panty hose. You are now in the possession of a very useful and versatile liquid.

Marigold tea (*Tagetes* spp.)

French: *Tagetes patula*
African: *Tagetes minuta*
Mexican: *Tagetes erecta*
Mexican mint: *Tagetes lucida*

Using marigold (French, African or Mexican) extracts has had considerable attention from the scientific community.[7] There is no doubt that the marigold of the Tagetes genus is an important group of plants for organic growers in whichever species or form it's used.[8] Although the pot marigold (*Calendula officinalis*) shares its name with the Tagetes marigolds mentioned above, it should otherwise be thought of as a totally different plant as it is of a different genus. The pot marigold does have pest repellent properties but has also been indicated to be a food plant for the larvae of certain pest species such as the cabbage moth,[9] which is fine if you are using it on a trap crop but not so good as a natural pesticide. For this reason a preparation for this plant has not been included.

There has been intense interest amongst researchers of biological pesticides in their search for a pest specific, cheap, environmentally friendly alternative to

Tagetes erecta.

many of the synthetic pesticides that have been banned recently. The attraction of the *Tagetes* marigolds in particular is that they tick all the boxes and have a number of active compounds with excellent fungicidal, insecticidal[10] and nematocidal properties.[11] Nematodes are a very important pest and it is estimated that they reduce crop yield by 12% worldwide.[12] In Chapter Four we described a variety of different nematodes. Below is a photo of the effects of one of the more harmful types, the root knot nematode (*Meloidogyne hapla*). A symptom of this infestation is the development of galls on the roots of crops which are quite dissimilar from the beneficial symbiotic associations between roots and mycorrhiza described in Chapter Two. The galls depicted below are formed by nematodes distorting the growth of the plant to create a suitable environment for them to continue their life cycles. Whilst they bear a superficial resemblance to the rhizobium nodules in Chapter Four they appear considerably more grotesque than the neat jackets of rhizobium which sublimely vary their pink hues with the fixation processes.

Root knot nematode infestation on parsley roots. Ruabete T.K. (2005), Plant Protection Service.[13]

Preparation

Take 1kg of fresh marigold flowers or seeds, pound up with a pestle and mortar and leave to steep for 48 hours. Filter and dilute 1:2 with 5 litres of dechlorinated water then mix in 150ml (10 tbsp) of soft soap. Since the active part of marigolds break down in daylight, it would be prudent not to prepare your mix in direct sunlight and you will need a lightproof container to store it.

Chrysanthemum tea (*Chrysanthemum cinerariifolium*)

This tea needs to be handled with a certain amount of respect. Its most important active compounds are pyrethrins which are found in the flowers.[14] Although perhaps surprisingly it is of low toxicity to humans, it is highly toxic to all insects and will disrupt or kill beneficials and pests alike.[15] Pyrethrum is known to knock down or kill many common pests such as aphids, cabbage white, Colorado potato beetle, flea beetle, gall midge, grasshoppers and thrips.[16] It is also harmful to fish and to a lesser extent mammals.[15] Its action is to attack the nervous system, and when not present in high enough quantities to kill it, acts as a repellent. One useful feature of pyrethrins, unlike most other mainstream insecticides, is that generally insects have developed less of a resistance.[17] This could be attributed to pyrethrum having more than one active compound.[18]

Pyrethrins break down due to exposure to light, which has both advantages and disadvantages. As such it has a low persistence when sprayed onto crops,

Pyrethrum (*Chrysanthemum cinerariifolium*). In 'Kenya's Pyrethrum Industry 2005'.[14]

making it very safe for food growing. The drawback is that after a few hours of sunlight the pest disrupting effects are negligible.[15] It bonds quickly with the soil and hence does not travel far when applied.[19] Pyrethrins are considered one of the safest insecticides and mild compared to synthetic pesticides. Chemists have made a synthetic equivalent to pyrethrins, called pyrethroid, which is more lethal and has more stability on application.[20]

An interesting development was the discovery of 'synergists'; these are substances that may not have an insecticidal effect on their own but when combined they dramatically increase the effectiveness of a particular insecticide.[21] With regard to pyrethrum, they can slow its breakdown in the environment or interfere with the organisms' ability to deal with the poison. Another useful feature of some synergists to factor in is that they prevent insects developing a resistance[22] and may have the potential to be more pest specific. For the non-organic grower who uses pyrethrin, various chemicals were found which acted as synergists[15] such as piperonyl butoxide. An exciting progression in the research of natural botanical insecticides was the recent novel discovery of dill seed oil (*Anethum sowa*) as a synergist of pyrethrin,[22] which is natural and suitable for organic use. A combined recipe is included below. Other natural pyrethrum synergists include nutmeg, parsley, sassafras and safrole oils.[23] I have included a selection of recipes with other synergists in the following pages for you to try but please bear in mind that the use of synergists in biocides is so recent that there are few corroborating studies, so anything you do with them will be somewhat pioneering.

A further possibility with regards to companion planting, that is beyond the scope of this book, is whether there is an enhanced repellent effect when dill is planted with pyrethrum as compared to planting pyrethrum alone. This remains one of many open questions that are yet to be explored and reported upon.

For this recipe, you need to find the right kind of chrysanthemum (*C. cinerariifolium*) and to maximise the amount of pyrethrin, get your timing right. They can be grown outside in temperate climates and have been celebrated in the potager for hundreds of years, however they do best in the greenhouse in more exposed areas. When harvesting your pyrethrum, the most pyrethrin content is in the flowers and seeds. The flower head is actually numerous flowers together. If you look closely at the yellow part (calathid) of the flower you will see a number of concentric circles, containing disk florets. When four or five of these circles begin to open this is the optimum time to harvest.[24]

Preparation

Pound up 75g of dried pyrethrum flowers and seeds in a pestle and mortar, mix into 1 litre of soap solution. Perform this process in a shady place somewhere out of direct sunlight and artificial light. Dilute with 4 litres of water, strain and use immediately. For maximum effect the best time of day for application is evening, just as it is getting dark.

Pyrethrum and dill tea

(*C. cinerariifolium* and *Anethum graveolens*)

Studies indicate[25] that the best ratio between pyrethrum and dill for the synergistic effect to be most effective is 1:3.6 which this preparation takes on board. Emulsion of dill oil and powdered pyrethrum in water is proposed which needs to be thoroughly shaken up before use.

Take 19g of dill seed, grind and steep in 0.5 litre of vegetable oil for 48 hours and strain through a fine cloth; stockings are ideal. Grind up 56g of dried pyrethrum and mix with 1 litre of soap solution. Dilute with 3.5 litres of water and use immediately. Use as you would the pyrethrum spray on the previous page.

Chilli tea (*Capsicum* spp.)

The main active compound in chilli is capsaicin, which accounts for its pungency in cooking. As an insecticide it acts as an anti-feedant, stomach poison and repellent.[26] Kenyan studies found that pests were reduced by 50% in cabbage fields from the use of chilli spray and suggested a time period of two weeks between spray applications to prevent build-up.[27]

Capsaicin is a powerful substance in natural pest control which is one of the most persistent in this chapter. Kenyan studies report good results against aphids and grasshoppers. It must be viewed as a broad spectrum insecticide which will disrupt or damage any insect it is in close proximity to. So if you are desperate to use something on your crops to attack pests, watch where you are aiming to minimise disruption to beneficials.

As anyone who cooks with chillies will know, they burn. Wear gloves if you are concerned and keep them away from your skin and eyes. Because the strength of chillies varies broadly between different varieties and actually within the same varieties also, it is wise to just try your solution on a few leaves and over dilute at first

Birdseye chilli growing in a northern English greenhouse.

because if the solution is too strong it may cause problems. Birdseye chillies were chosen for this tea because they are so easily available, of a suitable strength and are also one of the least temperamental to cultivate. Other varieties can be used but you need to work out how hot they are and dilute accordingly. A table is shown below giving the amount of chillies you need for an equivalent strength of 100g of birdseye from a selection of well-known chillies. The higher and lower figures indicate how much the strength of chillies can vary within the same variety.

Chilli comparison chart

	Lower (SHU)	Higher (SHU)	Average	Weight of chillies (g)
Scotch bonnet	100,000	350,000	225,000	33
Birdseye	50,000	100,000	75,000	100
Cayenne	30,000	50,000	40,000	188
Jalapeño	2,500	8,000	5,250	1,429
Poblano	500	2,500	1,500	5,000
Bell pepper	0	0	0	-

The measure used to gauge the strength or pungency of chillies works on an exponential scale called Scoville Heat Units (i.e. 1, 10, 100, 1,000 SHUs etc.) and goes from 0-15 million! Birdseyes are around the centre of the scale at 50-100,000 Scovilles. If you have Jalapeños, best just put them on your pizza because at 2,500-8,000 Scovilles you will need a lot of them to be effective over the same area. If you have some Scotch bonnets however, at 100,000-350,000 Scovilles, be on the safe side and just put in 30g at first. If you can find the Scoville rating for a particular chilli you can make a quick 'back of an envelope' calculation for the amount of chillies you will need. If you plug your Scoville Heat Units (SHU) into the formula below you will get the weight of chillies you need to use to make 5 litres of natural pesticide (equivalent to using 100g of Birdseye chilli).

$$\text{weight (g) of chilli} = 100 / (\text{SHU}/75{,}000)$$

Preparation

Select 100g of fresh birdseye chilli, making sure the chillies appear healthy and are not showing any sign of disease such as fungus or rotting. Pound the chillies in a pestle and mortar. Put the pounded chillies in a bottle and shake vigorously with 5 litres of lukewarm (approx. 25°C / 77°F) soapy water. Filter the water through a cloth to remove the chillies, before spraying. Chemicals in the soap will also contribute to the extraction of capsaicin. Use with caution.

Garlic tea (*Allium sativum*)

According to one of the foremost authorities on organic growing, Garden Organic, garlic is effective against diamondback moths, Colorado potato beetle, pulse beetle, whitefly, wireworm, false codling moth, imported cabbage worm, khapra beetle, mice, termites, moles, Mexican bean beetle, mites and peach tree borers.[28] Also researchers at Newcastle University found its extract to be effective against slugs and snails.[29] It possesses fungicidal, bacterial, nematocidal and repellent properties, and as such is very much in the broad spectrum category.[28]

Garlic contains sulphur which is used with other ingredients in the pesticides industry and was one of the first insecticides known to humans. There is solid science behind the usefulness of garlic as an insecticide as it has numerous active compounds which attack pests;[30] in particular the compound alliin, which produces allicin vapour when bruised or crushed. When activated it has a half-life of two and a half days in juiced form before the insecticidal effect wears off.[31] The vapour has powerful repellent properties and is a good indication as to why garlic is a popular companion plant around trees. With regard to harvesting crops and timing there is still a residual aroma associated with the garlic for up to a month after application.

Preparation

Pulverise and mash 150g of fresh garlic and place in water for three hours. Tease out the liquid by straining through stockings or blend to a fine paste and thoroughly mix with soap solution to 5 litres. It may need mixing up again prior to application. Use within two days but ideally straight away.

Garlic, chilli and onion tea
(*Allium sativum*, *Capsicum* spp. and *Allium cepa*)

Practice suggests that combining natural biocides can have a complementary effect greater than that of the individual ingredients. This tea is general purpose and used against leaf eating pests.

Chop up 40g of garlic, 35g of onion and mix in 1 tbsp of chilli powder. Add soap solution to 5 litres, strain and apply.

Marigold, chilli and onion tea

This is another well-known combination which can be used against red spider mites.

Chop up 50g of ripe chilli pods, 50g of marigold leaves and 50g of onions, steep in 2 litres of water for three days then strain. Add 3 litres of soft soap solution before spraying.

Marestail tea (*Equisetum arvense*)

In Chapter Six we used marestail to stimulate root growth, and in this preparation we use marestail for its protective features. Its high silicon content can be used to good effect as a foliar plant 'hardener' to strengthen the plants' resistance to insect attack. There is considerable interest in mainstream commercial research towards the use of silicates in disease suppression.[32]

Equisetum has fungicidal properties and attacks many of those fungal pathogens in the blanket term 'damping-off'.[33] It is also a botanical insecticide[34] and anecdotal reports suggest that nettles may be a synergist of marestail, enhancing its pesticidal qualities, but no scientific studies have been found in this area. In biodynamic circles it is used in a well-known remedy against damping-off[35] and fungal diseases in general[36] that is sprayed onto seeds and seedlings to protect them (see Appendix Two).

Preparation for anti-fungal spray

The shoots of *Equisetum* are best picked in early summer because they break down easier.[37] Heat 0.5kg of *Equisetum* in 2 litres of water for 20 minutes but do not boil. Dilute with another 8 litres, cover but don't seal and leave to steep for 10 days, stirring once every couple of days. Strain and use. For foliar application use 5 parts water to one part soap solution. The residue can be used as a mulch or put in the compost heap as an activator.

Delux wheast

Also spelt as whest, it is a contraction of whey and yeast. This preparation takes a rather different approach from the others in this chapter. Plants are well known today for their ability to recruit beneficial insects via a variety of chemical cues.[38] This refinement of the traditional agricultural brew uses the same principle. It is crude by comparison to natural processes, but effective. If considering the purchase of wheast from suppliers, some care should be taken after recent scandals regarding milk products and melamine contamination.[39]

Wheast is known to attract lacewings, ladybirds and syrphid flies[40] to pollinate and help defend plants from attack by pests. The yeast contains living organisms so care must be taken to make sure they are in a suitable state for use. Keep in the fridge until you want to use them, mix up the ingredients in ambient conditions and use a wooden spoon.

Paint

0.5 litre of lukewarm water
500g pot of plain organic yoghurt (at room temperature)
28.35g (1oz) fresh baker's yeast
4g of organic honey (approx. 3 tbsp)

Mix up the ingredients to create a thin paste. Apply immediately to leaves with a brush.

Spray

Dilute with a further 2 litres of water to prevent clogging.

If you want to finesse this recipe, mix in the pollen that is preferred by the beneficial you want to attract. For green lacewing this includes pollen of leek, cabbage, carrot and alfalfa, plus wild forage from families Chenopodiaceae, Asteraceae, Ranunculaceae, Gramineae and Aristolochiaceae.[41] Their favourite appears to be *Brassica napus* (rapeseed and swede) although there are strong indications their preference is mainly based on availability.[41]

Willow water (*Salix* spp.)

This preparation is special in many ways and is very versatile as a dip, drench or foliar spray. Most trees and shrubs in the *Salix* genus have remarkable rooting qualities. This feature is attributed in large part to an important family of natural growth hormones called auxin that can be applied to other plants with our preparation. We discussed SAR (Systemic Acquired Resistance) in the chapter introduction and this is one of the many complex mechanisms that are attributed to the enhanced rooting phenomena with these auxins.

Of particular interest to our requirements is indole-3-butyric acid (IBA) found in the greatest concentrations in the growing tips of *Salix* spp. and salicylic acid (SA)

The weeping willow likes its feet wet and is often found near water. As an ornamental it can be spotted in gardens, 'the best time to take a cutting is when no one else is looking'.[42]

found in the bark. It is known to be moderately soluble in boiling water (also acetone, alcohol and ester), which is sufficient for our requirements.[43] SA has been found to increase the number of flowers, plants' size, earliness of flowering and yield which is attributed to improved root length and density.[44] SA has important functions in most plants and although the processes are not entirely understood it is known to be involved in signalling[45] and in particular the SAR response. If a plant is chewed on or under attack from some pathogen, SA helps to trigger the plant's defences. Making an amendment of SA initiates a whole plant response ahead of any attack. A remarkable aspect of this response is that in many plants it can also act as an alarm to nearby plants because of an ester that exudes into the air. This ester, methyl salicylate, well known in wintergreens (*Gaultheri* spp.), repels herbivores and in some instances recruits beneficials.[46]

IBA is thought to be the most important hormone for rooting. It is also effective in protecting plants against stress of various kinds and improving yield.[47]

To extract IBA many of the commercial rooting dips use alcohol, which is phytotoxic. This cocktail is then diluted so it does not damage the plant. It is possible to do this without a chemistry set using vodka but it is not recommended for the reasons already noted. IBA is often seen as insoluble in water which accounts for the alcohol extraction methods, however looking deeper it emerges that it is poorly soluble at 250mg/l,[48] which is sufficient for the small proportions needed to induce rooting and the related SAR responses. Synergists of this process include gibberellins,[49] other growth hormones, soil bacteria and carbohydrates.[47] These other constituents are found in concentration in vermicompost[50] so a small addition of some freshly harvested, cast-rich, live compost as described in the following chapters, is recommended to finish off the mix.

Equipment

Clippers
Heat resistant glass bowl approx. 2 litres capacity

Ingredients

1 litre rainwater or dechlorinated water (see Chapter Six)
250g willow tips
50g of home compost (see Chapter Eight onwards)

1 Find a willow tree – you can refer to a key but the weeping willow is very distinctive and the photos on the previous page should be all you need. Willows are known to grow close to water and are popular as an ornamental tree in gardens.

2 Take around 100 cuttings of approximately 15-20cm from new growth on the tips; this should yield around 250g of cuttings. This stage should not take more than 15 minutes.

3 Remove the leaves and cut into centimetre-long pieces.

4 Use a blender or a pestle and mortar and make a pulp; this increases the surface area further and speeds the following process. If you choose to use a pestle and mortar, which can be useful off grid, a small amount of fine horticultural sand can aid the process.

5 Measure and boil 1 litre of water and pour into your bowl then add the willow pulp.

6 Stir every four hours for 16 hours. You now have enough of this solution to propagate a small forest.

Rooting dip

The active chemicals in this preparation work in very small quantities which are still the topic of contemporary debate. In a recent study it was established that compared to a control of normal water, the willow water delivered gains in rooting of 10-20%.[51] If you have a number of cuttings it is best to decant the willow water into different containers of 2.5cm depth so the strength of the solution in the original bowl does not get depleted. Often the diameter of the cutting can give you an idea of how many cuttings you can do. The volumes on the previous should easily propagate 200 tomatoes and 50 tree cuttings.

The remnants can be used as SAR inducing mulch or composted. Generally tree cuttings take longer to root than herbaceous plants. It is worth doing some research on the particular plant you want to propagate, as there is a great deal of variation in rooting speed within the different plant species. It is also important to do the procedures you might normally do when taking cuttings such as reducing foliage and keeping them out of the sun till the roots develop. As a general rule of thumb leave woody cuttings in the willow water for six hours and herbaceous cuttings three hours. It is also important to get the cutting in the dip as quickly as possible, ideally as you cut them to prevent air being drawn into the stem which would interfere with the rooting process.[51] Heating mats and humidity domes are optimal extras for these activities depending on the level of difficulty experienced with certain plants.

Mulch and root drench

The same as from the previous page without the blending step (4), this will make the solution less viscous and some of the active chemical will be retained in the willow pieces. Use the liquid on the soil by your plants and lay the solid pieces on top.

Foliar spray

The same as the rooting dip above except in step (5), mix in 800ml of boiling water followed by 200ml of soapwort solution (see previous section) wetting agent an hour afterwards.

Rhubarb leaf tea (*Rheum rhabarbarum*)

The making of rhubarb tea takes a different approach from the other teas and is used with a different purpose. It is an established practice in organic growing to place rhubarb leaves over areas where you want to reduce the vigour of weeds and this tea is simply an extension of this process. Rhubarb produces oxalic acid and other compounds as part of the plant's natural defence mechanisms as it has proven insecticidal qualities.[52] These compounds can be concentrated by boiling off excess water using a technique well known to chefs for intensifying flavours. It must be stressed that rhubarb leaves are toxic and it would not be a good idea to use the preparation vessel for any culinary purpose afterwards. A further warning would be to make sure the area is well ventilated for this process and not to inhale or stand over the brew too much. With this in mind, a favourable, albeit unscientific, way of preparation would be onsite by a campfire using a worn out pan placed on some bricks. Leaves from mature rhubarb are best because the oxalic acid content is known to be higher in older plants.[53]

This mix is effective for killing or weakening perennial weeds and could be applied to a trap crop not intended for consumption. Rhubarb leaf spray is also known to be effective against sucking insects[54] but it is not recommended to use as a spray on food crops because of its toxicity. Regrettably this has not stopped it from being used as a pesticide by some in the mainstream farming community. In a recent U.S. survey significant residues have been found in coffee, tea and tomatoes.[55] Excessive consumption of food with high oxalic acid content can interfere with calcium absorption[56] and contribute to kidney stones together with other health problems related to the calcium salts that the oxalates bond with.[57]

In another study the undersides of cucumber leaves sprayed with rhubarb leaf spray created a systemic resistance in that plant to anthracnose[58] and it could be extrapolated that this resistance might also apply to other unstudied fungal diseases. Oxalic acid was found to be the active compound, but because it is unclear how long the oxalic acid stays on, or indeed how much is absorbed into the plant, its use as a fungicide on food crops is also not recommended. On the plus side, it is conceivable given the SAR response as it occurs with this novel application, that this could initiate SAR responses in nearby untreated crop plants via the esters from the trap crop, which could help to generate a protective zone to discourage pests in conjunction with the other holistic methods discussed.

Preparation

3kg of rhubarb leaves (approximately half a wheelbarrow full)
10 litres of rainwater or dechlorinated water

Making sure gloves are worn, chop the rhubarb leaves finely. Place the leaves in a jamming pan with 10 litres of boiling water and simmer until the solution is reduced by a half (5 litres). Strain the mix, combine with 0.5 litre of soap solution and apply as a foliar spray when cool.

Other important plants for pesticide teas

This book has concentrated on plants that are suitable for use by organic growers, that are easily available and easy to grow, however a number of valuable plants currently being used worldwide do not fall within these criteria and are still quite important. It would be remiss if a selection of them were not at least given a mention. One of the most important is neem (*Azadirachta indica*), which possesses a variety of potent insecticidal properties, has a minimal impact on beneficials and is non-toxic to warm blooded mammals.[59] It is particularly effective on cabbage pests[60] including the diamondback moth which we looked at in Chapter Five. Its drawbacks for a temperate climate are that it is a tropical plant and grows to around 20m, which may be tricky to cultivate in a normal greenhouse.

It does have a relative, *Melia azedarach,* that can grow in temperate climates but the effects are considerably weaker.[61]

Various tropical and sub-tropical plants containing rotenone such as *Derris elliptica* and *Tephrosia vogelii*[61] also deserve a mention in this section, however they do not have organic status because of toxicity issues and are not recommended for that reason. Rotenone is extremely toxic to fish, toxic to beneficials and moderately toxic to humans.[62]

Zimmu is a new arrival, at least to Western horticulture, and grows naturally on the hills of Himachal Pradesh in India, so it is not the easiest of plants to get hold of and there is little information as to how well it grows at lower altitudes. It is a cross between *Allium cepa* and *Allium sativum* as it is presented in some journals, sounds rather unapproachable until you realise that this is simply the Latin name for the common onion and garlic which doesn't sound so impressive. What is impressive though, is its outstanding anti-fungal properties which were tested to good effect on early tomato blight[63] along with a number of other plant diseases.[64]

We leave the topic of pesticidal teas now and in the next chapter proceed to preparing the most important ingredient for making aerated teas and one that is at the heart of every organic grower's plot.

Left: Fresh uncured home compost. Right: Cured home compost.

CHAPTER EIGHT

Compost for Aerated Compost Tea

Composting is a process that occurs normally in nature. When something dies it naturally breaks down, undergoing a biological process involving a vast range of organisms that eventually return it to the soil. The composter aims to create the best conditions for this process according to the time available, the amount of effort, the materials to hand and its intended use. In this chapter we will describe a system of hot composting that the home composter can use for making compost tea. We will also describe a cheat version that takes less effort but is not strictly correct.

The simple act of stacking a pile of organic material allows greater insulation than if it was just spread out on the ground which allows the heap to increase in temperature and microbial activity to intensify. The composter may enhance this insulation by covering the heap in some way. The balance of ingredients in terms of carbon (C) and nitrogen (N) plus the degree of aeration and moisture are key elements in this process. There are very easy and intuitive ways of doing composting which we will be looking at, and after preparing a few heaps you tend to get a sense of what works. The best balance between carbon and nitrogen for composting[1] is 25-30:1. You could picture this as bread and butter for microbes. The bread is the carbon rich material for energy and butter is the nitrogen for protein. Composting will still work if the balance is not quite right but this is the ideal. Overleaf is a chart of the different balances in a variety of common natural compounds that can be used in your heap. The aim is to balance out the high carbon materials with the high nitrogen; these are often loosely referred to as browns and greens. To save any possible confusion it should be added that many texts including this refer to manures as 'greens' due to the high nitrogen content, even though they are usually brown in colour.

Composting stages

The composting process has three distinct stages. The mesophilic stage operates in the temperature range 10-45°C (50-113°F).[2] This is the stage favoured by earthworms and mesophilic bacteria and can take between a month and a year to complete. In hot composting, which is what we will be doing, an initial mesophilic stage is rapidly succeeded by the thermophilic stage above 45°C (113°F).[2] In this stage thermophilic bacteria take over for a brief period, rapidly consuming organic materials and oxygen. It is important for our needs because it burns out weed seeds and pathogens. As the compost cools down the mesophilic organisms gradually recolonise the heap and continue the composting process, breaking down what is left. Worms such as *Eisenia veneta*, *Eisenia fetida* and *Lumbricus terrestris* introduced in Chapter Four are particularly important as partly broken down material is inoculated by bacteria through their guts and released as highly fertile casts. The action of composting worms has a powerful and well-proven effect on composting material which cleanses the resulting compost[3] of whatever pathogens had survived the thermophilic process.

C:N Ratios of Composting Materials[6-10]

	Composting Material	C:N Ratio
Greens	Human urine (dry solids)[6]	0.8:1
	Blood meal[7]	4:1
	Bat guano (average)[8]	4.7:1
	Chicken manure (no sawdust)[7]	7:1
	Soil microorganisms[9]	8:1
	Comfrey leaves[7]	10:1
	Clover[7]	12:1
	Tomato leaves and stems[7]	12:1
	Cabbage heads[7]	12:1
	Pea and bean plants[7]	15:1
	Alfalfa (lucerne hay)[7]	13:1
	Cow manure[7]	15:1
	Mixed weeds[7]	19:1
	Horse manure[10]	25:1
	Seaweed (average)[7]	20-25:1
	Grass clippings[7]	20-25:1
	Vegetable peelings [7]	20-30:1
	Fruit wastes [7]	35:1
Browns	Leaves (mature)[7]	60:1
	Straw (general)[7]	100:1
	Paper [7]	170:1
	Sawdust (old)[7]	200:1
	Pine bark (fresh)[7]	500:1

At this stage it would be unwise to apply compost directly to crops because of phytotoxins and the fact some materials will be acting as nitrogen sinks until they are fully composted. Phytotoxins are unfavourable to crops and only exist in harmful levels in unmatured compost.[4] Finally, in the curing stage which lasts at least a month,[5] the phytotoxin issue is resolved and the compost matures to a crumbly, dark, earthy smelling material which is ready for use. A rule of thumb to indicate that the decomposition process has reached the right level for finished compost is that the compost will no longer be heating up shortly after turning.

Things to avoid include cat, dog and human faeces and material that has had contact with poisons including pesticides. Seaweed is fine as long as it is not polluted but it is important not to add too much. With regard to household leftovers such as non-organic vegetables or highly processed meals invariably a cocktail of chemical residues will remain. Presumably if you are prepared to eat it, you would be prepared to compost it.

Moisture

The correct moisture level in a pile[2] is 50-85% depending on the balance of materials. Understandably materials that are more absorbent such as sawdust and leaves can take more moisture than vegetable wastes. This translates to having the pile damp but not wet. If you soak and wring out a dish cloth, the material could be described as 'glistening' and this is exactly the kind of condition we want throughout our heap. During the critical thermophilic stage, moisture can evaporate very rapidly which must be replenished.

Aeration

This factor gives the composter quite considerable control over the time taken to produce compost. The more a heap is aerated the quicker the process becomes. The frequency of mixing and fluffing can range from daily to never for the home composter. For compost tea however, we need to thoroughly burn out as many pathogens as possible and thermophilic bacteria consume enormous quantities of oxygen during this process, so frequent turning in the initial stages is recommended.

Volume

In the early stage of the hot composting process it is important to keep to a heap over 3ft sq. ($91cm^2$) but under 6ft sq. ($183cm^2$). Any smaller and maintaining heat levels will be a problem, any higher and the lower parts may become compacted pushing out the air pockets and shutting down the essential aerobic processes.

Compost activators

These are materials that kick-start the composting process. Such a material either has a high nitrogen content in a readily available form or is something imbued with the right kind of microbial activity. When an activator is applied you can expect an immediate rise in temperature as the microbial processes are stimulated.

Examples of excellent activators include blood meal, comfrey and fresh manure. For microbial action, unfinished compost can be used. As discussed in Chapter Two, microorganisms have a need of many essential elements as do other organisms so adding a diversity of organic compounds even in small quantities can refine the process.

Another important, valuable activator, which is misunderstood and underrated in modern society, is human urine. It has been scientifically established that from a healthy person this is safe, almost totally sterile.[11] Applied fresh and warm to the compost heap it provides nitrogen to microbes in a highly available form. If you refer to the table of C:N ratios you may confirm that this activator is very much on the higher nitrogen end of this spectrum. More controversial is the notion that some activators can work in homeopathic proportions igniting life cycles and stimulating fertility. The most conservative and modern approach to this is in the use of aerated compost teas described in the next chapter. Stimulating the reproduction of aerobic microbes, then using them on an aerobic pile to stimulate the food web is a reasonable proposition, although there are few solid and truly scientific studies in support of this. It could be added that this could be the result of poor funding, flawed experimental design and oversights in practice.

Compost activators have a further avenue that is offered for interest and completeness. Rudolf Steiner, introduced in Chapter One, placed great emphasis on composting and developed a variety of preparations (BD 502-507) that were applied to composts to enhance the breakdown process and improve the compost. This included the use of certain herbs that were thought to have special properties. Yarrow blossoms, chamomile blossoms, stinging nettles, oak bark, dandelion flowers and valerian flowers were used in procedures related to the seasons in what was believed to harness flows of terrestrial and cosmic energy. A description of the preparation methods interpreted directly from Steiner's agriculture lectures, together with excerpts, can be found in Appendix Two at the end of the book and its value is left for the reader to decide.

Composting for aerated compost tea

What you will need:

Tools

Hammer, saw, measuring tape, pencil, set square, marker, compost thermometer, screwdriver, pliers.

Materials

5 pallets or the equivalent in boards
2 wooden panels: 75cm × 137cm × 2cm (29.5in × 53in × ¾in) plywood
2 planks: 400cm x 20cm × 2cm (157.5in × 8in × ¾in)
1 section chicken wire: 4m × 1.2m (157.5in × 47.2in)
2 sections chicken wire: 0.75m × 1.20m (29.5in × 47.2in)
38 galvanised fencing staples 20mm (¾in) (18 for back wire section, 10 × 2 for sides)
23 corner brackets 7.5cm (3in) with 2cm screws × 24 (secure panel and pallets to top plank)
56 bricks (32 for Mesophilic section and 24 for Curing section)
8 half bricks for Curing section
15 × 5cm round head wire nails
3m natural fibre string
20 assorted branches approx.: 2-5cm diameter × 160cm (1-2in diameter × 63in)
40 assorted branches approx.: 2-5cm diameter × 75cm (1-2in diameter × 29in)
3 wheelbarrows assorted twigs
350cm × 400cm (138in × 157.5in) muslin sack cloth/tarpaulin for Thermo section
200cm × 300cm (78¾in × 118in) for Meso section
150cm × 250cm (59in × 98.5in) for Curing section

Please bear in mind the infrastructure for this compost system is intended to be set up in a few hours with minimal equipment and preamble. There are more manicured projects that can be devised and there is considerable room for improvisation as long as the basic principles of composting are understood.

Heap construction

Set aside a 4m × 1m (13.1ft × 3.3ft) area and clear. The finished construction will be 1.2m (3.9ft) high so place it where it will minimise the loss of sunshine to your plants.

As pictured overleaf, this system is a refined version of the well-known three stage set-up.

The solid board in the middle partitions is to prevent the cross over of material at earlier stages which may cause contamination, however you can always add material from later stages to an earlier stage.

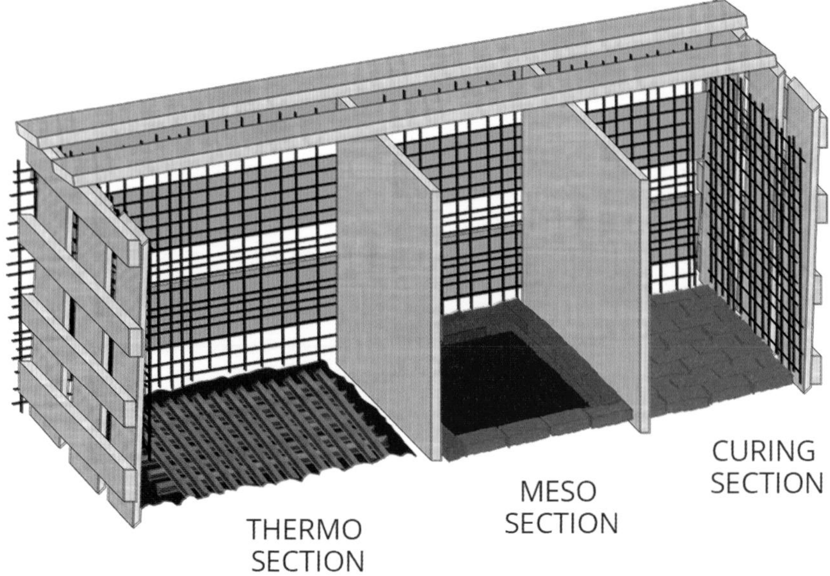

Diagram of the three stage compost heap.

After cutting the wood to size, the tape measure can be dispensed with. The key to constructing the heap is to get the back and outer panels constructed first then use the placement of bricks as a guide to the positioning of the central panels. (It is assumed that standard bricks are being used, 8.5in × 4in, otherwise refer to the measurements.)

1 Construct back section, secure chicken wire with galvanised staples.

2 Attach outer pallets to the back section tacked up with nails then four corner brackets (two on the inside of each side).

3 Attach back top plank with nails then three corner brackets to attach to back section and two brackets inside for each panel.

4 Attach upper planks tacked up with nails then with four corner brackets.

5 Secure the remaining two sections of chicken wire onto the outside pallets using the remaining galvanised staples as in diagram.

6 Curing Section. Lay the 3.5 bricks lengthways (74.73cm) and 8 deep (75cm) widthways as in the diagram. The point of using the bricks in this way is to give good drainage and a shield for encroaching weeds when the composting process subsides. If you have the time and spare bricks, you can dig out a section beneath for another layer to further resist encroaching grasses/weeds.

7 Position first central panel using six corner brackets. One each side to the top plank, and two each side to the back panel.

8 Mesophilic Section. Dig a shallow hole 12cm (4.7in) deep × 74.73cm (29.42in) long × 75cm (29.5in) wide, remove weeds, grass roots etc. Fork for extra drainage. It is slightly higher than the thermophilic section to prevent leachate from spreading over. Lay two layers of bricks around the edges, four lengthways and one brick at a 90° angle forming the first part of the next side as illustrated, this should form a square (96.52 × 96.52cm). These bricks act as a barrier to encroaching weeds and make a heat sink to assist in keeping the temperatures for mesophilic organisms within a favourable range. The inner area is open to the soil to facilitate increased access to soil organisms. You should aim to have the top layer of bricks just over the soil surface. This is constructed as you would construct a dry stone wall with no cement to encourage maximum colonisation.

9 Position the second central panel against the edge of the mesophilic section using six corner brackets. One each side to the top plank, and two each side to the back panel.

10 Thermophilic Section. This section is designed to take a larger amount of material and to accommodate thermophilic activity. You can expect the finished compost to be around half the volume of the original material which accounts for the reducing sizes in each stage. Original three stage set-ups had equidistant areas for each section which can only be justified for aesthetic reasons.

Drainage Area

The intention behind this is to provide a substantial amount of aeration when the heap goes thermophilic and provide high levels of drainage.

10a Dig a shallow hole 20cm (8in) deep × 193cm (76in) long × 75cm (29.5in) wide. Fork the base for extra drainage.

10b Fill up the first 7cm (3in) with a bed of assorted twigs.

10c Construct a matrix of branches with approximately 0.5cm (0.2in) intervals tying it in place with the natural fibre string to form a loose matrix. This distance is a compromise between good aeration and being close enough to prevent material slipping through.

10d Lay this matrix over your bed of twigs. The top of this should dip into the ground by around 5cm giving extra insulation.

Step-by-step guide to practical hot composting

As an ideal for hot composting, when someone wants to construct a compost heap, all the materials would be to hand in the right volumes and composting can be done in batches. At the other extreme, materials are added a little at a time as a continuous process. For the home composter/grower however, the reality is somewhere between these two extremes. There is gradual flow of materials from household waste throughout the year and a large influx of garden wastes early spring and autumn. Because of these considerations the system we propose is designed to accommodate these flows so that a maximum volume of compostable materials is being returned to your system. It is expected that you will want to use normal compost as well as compost tea and the chart at the end of the chapter illustrates one possible schedule to integrate your composting with the seasons and your growing activities.

1. SPRING. Fill up the thermo section of your heap with compostables in alternate equal volumes of greens and browns in layers 5-10cm thick up to a minimum of 1.2m high × 1.2m long and 0.91m wide (4ft × 4ft × 3ft). This technique gives you an approximation of the 25-30:1 ideal C:N ratio for composting.
   Pour on any material that has been rotting down in water over the winter; this can be used as a green layer. Animal based materials such as fresh manure are highly recommended here but only at this stage before the pile goes thermophilic. Add activators with each layer. Add some compost from the previous year if you have any and sprinkle on topsoil with each layer. This is a good time of year to make compost tea from some of the previous year's compost and use it as an activator. If you have extra compostables, expanding upwards and outwards will enhance the heating effect and will account for the inevitable shrinkage that may cause the heap to lose heat too early during this critical phase. Do not stack your heap higher than 2m (6.6ft) because you will risk compaction and an uneven destruction of weeds and pathogens. Remember to put your covering in place, which can be weighed down with whatever you have to hand. Hessian is best because it breathes but a tarpaulin is still acceptable.

2. After your heap is filled up try to turn it every two or three days until it goes thermophilic. This is the most critical phase; use your compost thermometer to make sure it does not get too hot (see troubleshooting p.117). All addition of animal products must stop now because of safety issues, but human urine is fine. If things are proceeding as intended you can expect a considerable loss of moisture and this must be topped up.

3. Aim to maintain the thermophilic stage between 55-70°C (131-158°F) for at least three days. It is important to turn the heap thoroughly at this stage so every part of the heap goes through the heating process within. If the heap

starts to get too hot take off your covering to cool it down, add cool water or reduce the volume. Expect to have to replenish a considerable amount of water from evaporation.

4. Now the most critical phase is completed you can relax a little. Transfer your heap to the mesophilic section and turn again in two weeks. Your compost should be really progressing now, colour darkening and solids mostly disintegrated. In ambient conditions (10-20°C / 50-68°F) this should be around 10 weeks. If you turn more frequently whilst maintaining the correct balance of moisture and nutrients this process will be sped up considerably but you will stand a greater chance of injuring and disturbing the worms: your most valuable asset. With worm friendly good practice in mind the use of a fairly blunt fork over a spade is preferable. If you really want to be indulgent and turn your compost in style, consider obtaining a composting auger as displayed below. Whilst the compost is still unmatured and there is still considerable microbial activity, this is a good point to use some for your aerobic teas. Phytoxicity issues will resolve from the enhanced aeration and burgeoning microbial reproductive action in the bubbler (see Chapter Nine).

5. Up to week four you may still add material that breaks down quickly such as vegetable waste, comfrey or brown paper screwed into balls and saturated in human urine. There are numerous small-lidded vegetable waste units widely available from local councils which can be employed very effectively in the household. When filling these up try to mix in equal

The composter's manual auger.[13]

quantities of browns as you are adding them as this keeps the C:N balance right. It cuts the smell right down and can be added directly to the top of your heap. Screwed up egg boxes and brown paper work very well in this way and if things are proceeding as intended you will be surprised at how quickly they disintegrate.

6. You may again start adding fresh materials in the empty thermo section for another crescendo in autumn or earlier if you have the materials. Because it is a three bin system you have the capacity to have your compost simultaneously in each of the three stages depending on your level of activity. You can add in equal proportions of brown and greens to the thermo or you could even hold the process back, adding a greater proportion of browns then compensating later in the year when you have accumulated enough materials for another thermophilic pile.

7. Transfer your fresh compost from the mesophilic section to the curing section and leave for at least one month. When the compost is nearing completion you will notice a considerable reduction in worm activity and the compost will cool down. Leaving compost longer further enhances its disease protection qualities which can be applied in your teas.[14]

8. AUTUMN. When composting later in the year the approach is slightly different; you may have more weed seeds and you have to take account of the colder conditions. If you have a considerable amount of recalcitrant perennial weed seeds or perennial roots thicker than a centimetre it will be better to throw them in a water butt after which you can add them to your compost heap the following spring. Be sure to use a cover because the aroma can take on an agricultural note after a few weeks.

   Because the breakdown process is slower in colder weather leave your compost in the Meso section until early spring. This will give the earthworms more opportunity to work on it and will be less exposed than in the curing section. If you are really concerned about the cold slowing down the process add straw and more covering to the top.

A year in composting

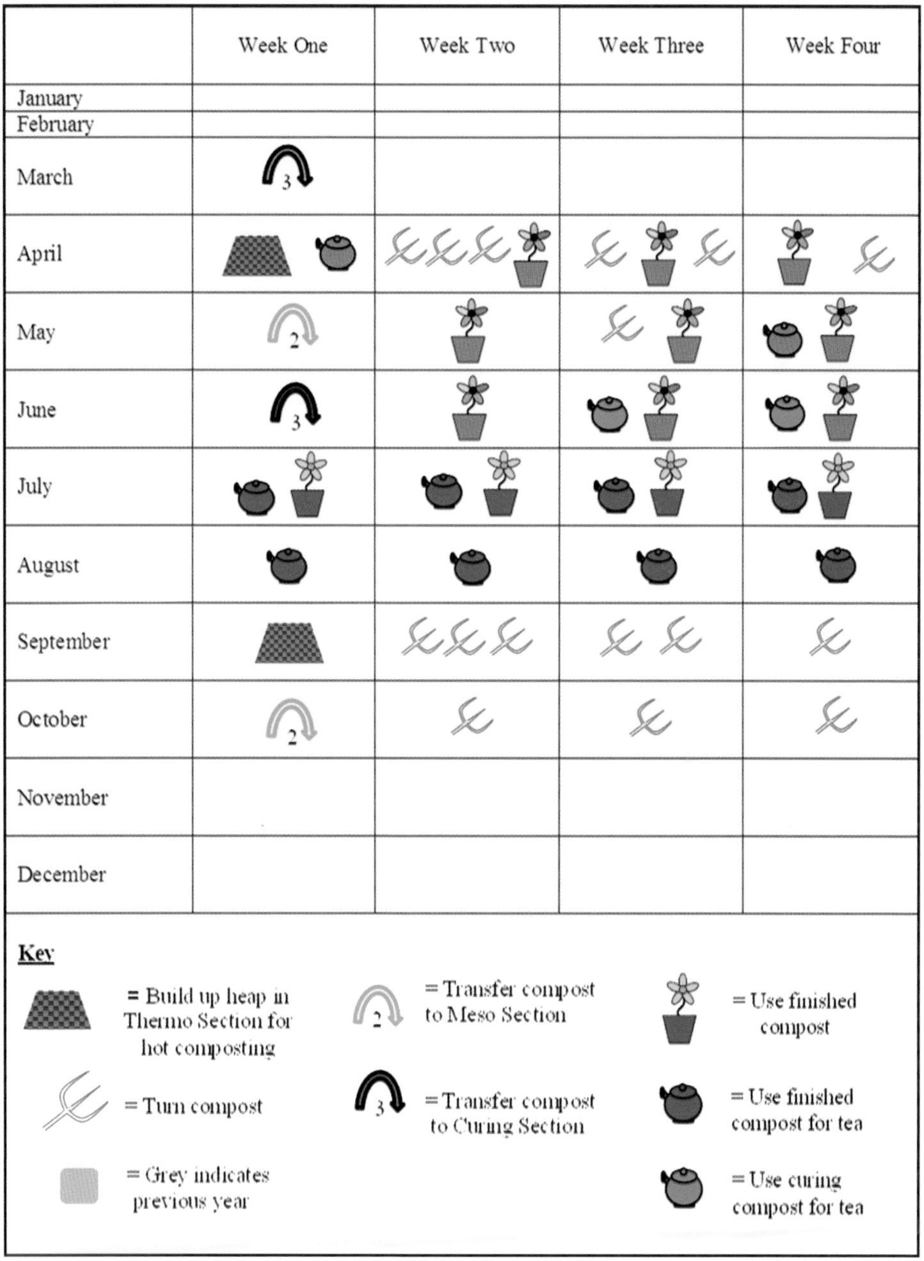

	Week One	Week Two	Week Three	Week Four
January				
February				
March	3			
April				
May	2			
June	3			
July				
August				
September				
October	2			
November				
December				

Key

= Build up heap in Thermo Section for hot composting

= Transfer compost to Meso Section

= Use finished compost

= Turn compost

= Transfer compost to Curing Section

= Use finished compost for tea

= Grey indicates previous year

= Use curing compost for tea

Twelve month chart for composting, compost tea application and compost use.

Lazy composting

It could be said that the hot composting set-up from p.109 is somewhat idealised and something of an overkill. This criticism cannot be levelled at the more basic system described here. It could also be suggested that because of the proven sterilising effect of worms,[3] the thermophilic stage is unnecessary for the destruction of pathogens, but it is the best method for breaking down weed seeds since the action of worms alone can be insufficient.[12] It is important to get the heap over 3ft sq. (91cm sq.) in volume as discussed previously. It is expected that this more relaxed method will still attain thermophilic temperatures but the process will be less controlled, so you may lose a few percentage points in weed destruction. What you will gain however is a more disease suppressive compost because of the extended time period of composting.[14] Alternately if you don't want it to go thermophilic and you don't want weed seeds, be careful not to incorporate any initially (e.g. in manures). Although quicker, the three stage hot composting method above loses more nutrients to the atmosphere than other methods. The lazy method is not intended to accommodate multiple batches in different stages of decomposition, however there is also no risk of contamination from pathogens in fresh batches of organic matter. The good practice described in the idealised first version can still be done but worms and fungus will take on a greater role here since their activities are disrupted less.

Left: A photo of the base of a windrow style compost heap for increased aeration.

Right: Completed heap with semi-permeable cover allows moisture in. In windy conditions rocks and bricks are placed around the top and edges. The use of windrows is well known in vermiculture as the darkness, increased warmth and moisture retention are excellent for worms. In this style compost is arranged in rows and new material is added to one end extending the heap. The above is very small compared to municipal operations but is still large enough to attain thermophilic temperatures. Its connection with the soil allows access to soil organisms, especially composting worms, and facilitates their retreat when things get too hot.

Find an area approximately 2m × 1m and clear to a depth of 15cm (approx. 6in). Break up wood and twigs then lay in a criss-cross fashion for drainage and aeration. After this is done build up the pile with whatever material is to hand, cutting up and breaking to increase the surface area and then cover. The use of carpets as a cover is not strictly correct because of the many unpleasant chemicals that have gone into making the artificial material. On the plus side they tend to be the cheapest, most available and most effective option. If they look like they are rotting it is best to find a fresher one that can last two or three years. The addition of composting materials can last a number of months but at some point there must be a cut-off point; around two months is fine. If you still have compostable material a new heap needs to be constructed out of the way of the original heap. Finally turn every two months for six months and leave till spring.

Troubleshooting for hot composting

Symptom	Probable Cause	Solution
Heap is smelling putrid.	Heap has gone anaerobic.	Turn heap thoroughly. Make sure heap is not too damp and is draining OK. If too damp mix in some more absorbent materials such as screwed up brown paper. Make sure you have not processed your raw materials too finely. If you have, mix in some larger materials.
Heap is full of ants.	Heap is too dry.	Add water working it in thoroughly. Make sure heap is properly covered.
Heap is not heating up sufficiently.	Not enough high nitrogen materials. Not enough aeration. Heap is less than 1m^2. Poor covering. Season too cold.	Refer to list (p.106) and work in some high nitrogen materials (greens). Make sure the heap has the right balance of moisture, is covered, is greater than 3ft^2 (91cm^2) and is thoroughly turned.
There is a smell of ammonia.	Too high a proportion of high nitrogen materials have been added.	Mix in more browns.
Compost has white powdery residues.	Too high a proportion of high nitrogen materials have been added.	Mix in more browns.
Compost heap too hot (>70°C).	Well done! The thermophilic process is in overdrive.	Take off the cover. Add cool water. Reduce volume if everything else fails, but try to avoid this.
Compost heap thermophilic for too long.	Well done! The thermophilic process is in overdrive.	Same as above.

Some of the results from my activities with seed balls described in Chapter Nine. I was pleasantly surprised at how in this instance practice exceeded theory. After only four days, germination processes in the balls were already well under way, then things continued swiftly as the balls established a multipurpose live mulch around my sweet corn before being transplanted to raised beds on Day 47.

CHAPTER NINE

Aerated Compost Tea

The compost teas in Chapter Six can be criticised because they encourage the build up of anaerobic bacteria and cause undesirable chemical reactions such as putrification.[1] In the case of uncomposted manure teas there are serious concerns over disease risks. The approach in aerated teas is fundamentally different. Whereas in Chapter Six you just supplied materials with plant foods already present, in this chapter you introduce something that is inherent in a greater food chain that has plant foods as byproducts. The aim is to nurture the soil. While it is accepted knowledge that most plant feeding is biologically mediated this process pushes the concept a step further. With aerated compost teas the intention is to multiply healthy aerobic microorganisms already present in good quality compost by supplying them with the right food and conditions to dramatically expand their population. This can then be used either to enhance disease protection or indirectly supply a source of food through an enhanced food chain. Detractors contend that the microbial food and other additives added to compost tea introduce an alien agent to a natural process and hold the potential to incubate diseases already present such as *E.coli*.[2] A counter to this argument is that *E.coli* grows best in anaerobic conditions and would be outcompeted by the vigorous aerobic organisms in an aerated brew. Furthermore a taskforce report concluded that compost using the 'hot composting' method as described in the previous chapter had no risk of *E.coli*.[3] If you engage in muck spreading for your crops you might as well discount most of the concerns over disease in your teas and put it down to the modern obsession with sterilisation. In America the governing body on compost tea application, the National Organic Standards Board (NOSB),[4] treats compost tea rather like manure and suggests that you need to wait 90-120 days after application to consume any produce.[5] This ruling is controversial and has been disputed by other official bodies and during their own internal wrangles. Part of the confusion lay in the irreconcilable views of the NOSB board itself.[6] This same organisation, guided by a trial into compost teas,[7] displayed a startling lack of coherence in its interpretation which was challenged recently.[8] Based on this performance they cannot be considered an authority regarding this research area.

Studies have shown that people living in overly sterilised environments are often more susceptible to certain diseases because their systems are less accustomed to dealing with them,[9] so in a sense a little exposure to disease pathogens can be a good thing. The existence of *E.coli* in human guts has been linked to a beneficial role in the immune system.[9] A parallel could be drawn between overly clinical farming practices and the obsessive pursuit of hygiene in technically developed societies. A recent review suggests that humans need contact with the 'right kind of dirt'[9] although we are yet to establish the finer details.

A further criticism has been put forward partly because of the variability of compost itself, that research findings have not been well replicated which makes it difficult to establish clear facts.[10] There are no universally agreed standards and this issue is still ongoing.

There are many large and powerful organisations who have vested interests in the failure of organic growing and related practices such as compost tea and these voices are often the most vociferous.[11] In grassroots style however, the growers themselves have taken the lead in this area and there are many good reports of positive gains in plant health and vigour. In truth the use of aerated teas is at an early stage of its development and any project is pioneering rather than an established practice. Seemingly minor errors in dealing with what should be a vibrant myriad of soil life may account for many of the failures and inconsequential results in crop studies.

An avenue that has not currently been explored as far as the author is aware is the potential of using canines to detect good teas or teas that are going wrong. To suggest that the sense of smell in combination with taste of some canines is awesome, is not an exaggeration when considering what they are capable of detecting – cancer cells, low blood sugar in diabetics, the onset of epilepsy, an approaching migraine, some pheromones and more well known, certain illegal drugs and explosives. With unlimited time, funding and enthusiasm the prospect of using canines to identify healthy aerobic smells or chemical signatures of anaerobic soil could lead the way to improving the quality of compost teas. Of course we could go considerably further down this avenue identifying chemicals that are considered indicators of certain organisms in the soil and so forth.

It is important to emphasise that using compost teas should be part of an integrated growing regime and is not intended to be done in isolation to other organic practices. We outline suitable complementary practices later on in this chapter with an emphasis on no-till and food web oriented soil conditioning.

Adding compost tea raises the carbon content of the soil where 3% is thought to be ideal for growing crops and increasing soil health.[12] A key point is that there is thought to be an ideal ratio between fungal and bacterial content in the soil, where a bacterially dominated soil is best for growing arable crops, and a fungally dominated soil is better for bushes and trees.[1] This notion may be rather a crude approximation for what is actually happening with plants actively involved in highly complex and sophisticated processes attracting guilds of beneficial organisms to develop or be destroyed by more powerful processes.

This chapter crosses over with Chapter Seven to some extent because the effect of aerated tea is a holistic process with benefits to plant health and vigour that are inseparable. The holistic approach as discussed previously is in sharp contrast to the reductive and prescriptive approaches in conventional growing. Science is still trying to come to terms with how to make consistently good teas. In part this may just be a problem within contemporary science itself as it attempts to wrestle with the holistic phenomena of a complex system that it cannot adequately reduce and simplify to produce positive results.

A significant problem that does not exist for the artisan grower is how to standardise compost tea for expansion to an industrial scale. This preoccupation within the literature obscures the artistry of the small-scale grower in making effective teas.

Without doubt there are unscrupulous companies who make exaggerated claims and also there are critics. The most publically vociferous compost tea cynic is Dr Linda Chalker-Scott. She is a strong adherent of single factor analysis that can be a rather inappropriate instrument in the life sciences with a tendency to produce simplistic, misleading and divisive results. Holistic phenomena and emergent properties may not be so amenable to this style of investigation. It is not without a sense of irony that the author had to abandon researching Chalker-Scott's material because of a lack of solid references![13] She unfortunately fell victim herself to the same criteria (below) she suggested for evaluating the vast amount of questionable material on compost teas. On a more positive note, her suggestions regarding this 'grey literature' are mostly reasonable and valid. To summarise Linda Chalker-Scott's criteria, which are her most useful and thought provoking contributions to the topic of compost teas, she asks *if the author (regarding compost tea accounts) is linked to a mainstream academic institution, whether the author refrains from selling a product, whether the information can be verified and whether the account appeals to reason rather than emotion.*[14] If the answer is yes to all of these questions then there is a good chance that the material is acceptable. Three qualifications to this acid test should be considered. Firstly, it would leave out a number of important writers, orators, 'doers' and independent scientists. Those not affiliated with any 'mainstream academic institution' may be less stifled in their expression and free to choose their own area of interest. This point is even more valid now crowd funding is becoming available. The possibility of an important breakthrough coming about through this channel cannot be discounted.

There are many examples of unaffiliated thinkers who have contributed greatly to humanity. This long list includes James Lovelock (Gaia Hypothesis), Buckminster Fuller (Geodesic Dome), Nikola Tesla (AC and Free Energy), Peter. D. Mitchell (Nobel Prize), Luis Leloir (Nobel Prize), Alfred Russel Wallace and Charles Darwin (Natural Selection).

Secondly, the direction of research in universities is governed by funding. If this was the only kind of research people did, many noble but unprofitable areas of enquiry could never develop. This links to the first qualification rather because the unaffiliated agents with passion and dedication sometimes break through in risky, potentially unprofitable areas despite a lack of funding.

Finally, the orthodoxy in mainstream institutions often have a bias against new approaches until they become established and thoroughly verified by the terms laid down by their own version of what science is. Such institutions are subject to the contemporary paradigm in science that can change over time. Where a phenomenon exists that is not so amenable to the favoured style of analysis certain factors can be overlooked. A science of emerging properties and multi-factor analysis is not so well advanced currently.

If you choose to abandon Chalker-Scott's criteria for any of the qualifications mentioned, you will have to rely more heavily on your own judgement. Another victim of these criteria is Elaine Ingham, a central supporter of the food web approach and aerated compost teas. Reading Ingham's accounts[1] her style is self-referential which yields a cultish impression. The focus encourages the reader to apply vast amounts of compost tea at every opportunity. Many of her assertions seem reasonable but do not appear to be backed by studies, so we only have her word for the validity of what she states and she does indeed sell her compost tea products and services. The author suspects however that she may have the last word. Her material is intuitive and insightful and her pioneering techniques to directly evaluate compost tea using microscopy should give a good indication of tea quality, although this may not translate well to solid proofs. In her works, Ingham presents ratios between fungal and bacterial dominance in relation to different kinds of crops. In the thought provoking interview between the author and Dr Gavin Lishman[15] (Martin Lishman Ltd.), he suggested that although it is difficult to ascertain how she came across these figures, practice suggests that they hold true broadly.

Activities begin for this chapter by constructing a small brewer and then continue to the preparation steps needed to make a basic aerated compost tea. A short section is included at this point to describe the equipment and methods you need to apply the compost tea bearing in mind that we are dealing with living organisms. Proceeding I look at preparing special brews for root drenches, foliar sprays, seed treatments, root dips and leaf digesters.

Continuing, I make a focus on soil improvement and how compost teas fit into the picture. As a background to the subsequent procedures we depart briefly from tea preparation and consider the topics of soil texture and structure then make a comparison of tillage methods with a view to reflecting on the best practices for our requirements. Whilst on this topic I make a comparison of important soil preparation methods which will dramatically increase your successes in developing your food web and offer your aerobic microherds the best possible environment to continue their expansion. I continue with a consideration of using compost teas as part of an integrated package of steps towards holistic soil improvement which are coordinated with other good organic practices as an illustration of how best to approach compost tea application.

Moving towards the end of the chapter I provide a troubleshooters section where some of the failures in compost tea research are considered, highlighting basic errors in procedure. Concluding I suggest some points with regards to fine tuning your brew with a view to producing a tea that truly lives up to its promising potential.

Constructing a five gallon brewer

Now that we have our most important ingredient (Chapter Eight), the next stage is to rig up the brewer. The five gallon set-up is suitable for an average sized plot and is recommended for first timers.

Brewer set up with bubbler for aerated compost tea.

Components

5 gallon bucket (19 litres)
2 metres (6.6ft) × 5mm (0.25in) tubing
Silicon sealant
Aquarium air pump (min. 240 litre/hour, dual outlet)
Stockings or aged burlap for suspending compost
Bamboo cane 32cm (12.5in) approx. (diameter of top of bucket + 0.5cm)
Cycle inner tube

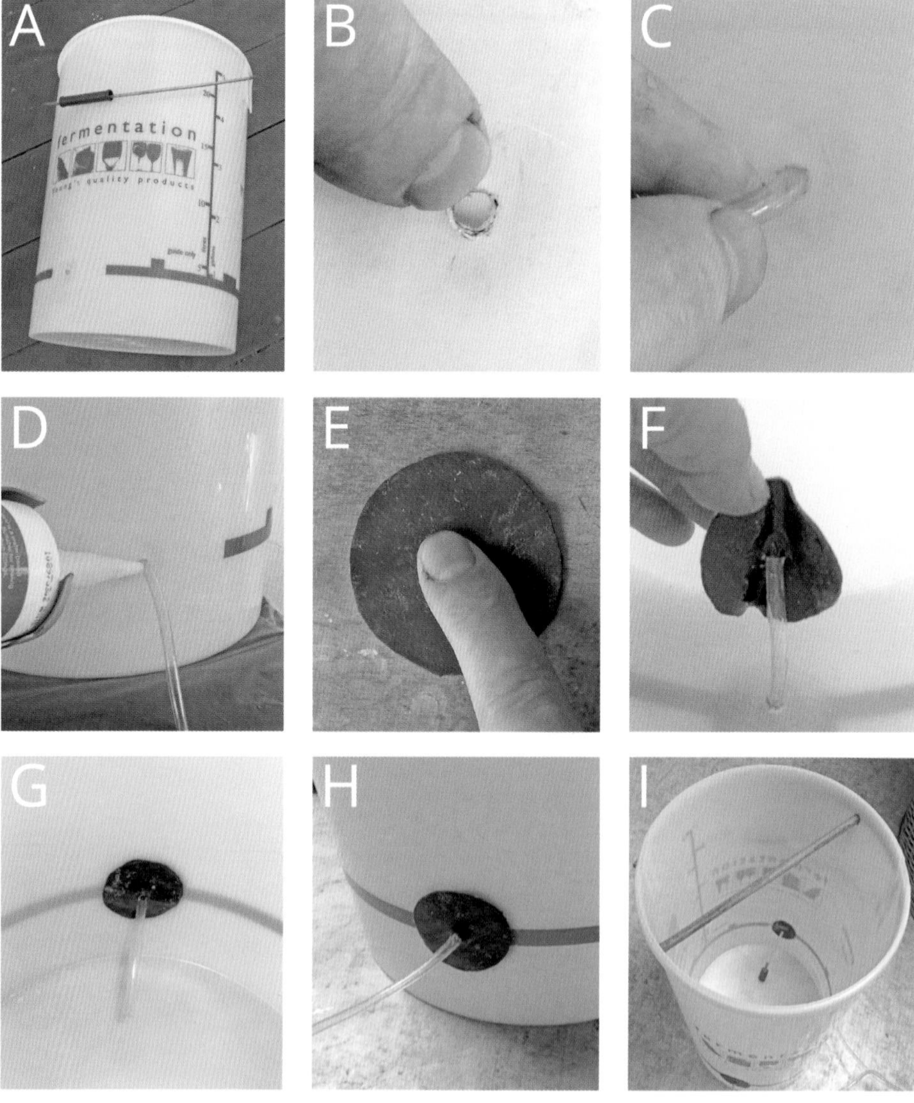

Brewer construction steps.

Tools needed for construction

Sealant gun
A sharp knife
A drill (not essential) with 5mm (0.2in) drill bit

Step-by-step construction

Each step will now be described with photographs to lead you through the process. You can consult the photos on the previous page at each stage:

1 Take a typical five gallon vessel [A] and drill holes at opposite sides approximately 8cm (3in) up from the base [A-B]. If you don't have a drill with a 5mm drill bit handy you can pierce your bucket with a skewer then carefully hollow out the hole with a kitchen knife to just smaller than the diameter of your tubing. Either way you will need to trim off the edges of loose plastic. The aim is to make as tight a fit as possible for your tubing.

2 Cut two 80cm (31in) lengths of 5mm tubing bearing in mind that you will want to have your air pump placed above the water level of your bucket. Thread the tubes through your newly drilled holes [C]. If the aperture is too tight a fit for your tubing you can cut your tubing at an angle which will get it in easier. The main thing is to not let your aperture be too loose or you might end up having to buy a new bucket and start over.

3 Apply silicon sealant to both sides of the holes smoothing around the edges and leave until it sets [D].

4 Get hold of a cycle inner tube and cut four rough circles [E] approximately 5cm (2in). Make a very narrow hole in the centre of the tubing by folding into a cone and making a very small cut to the central peak. Thread the rubber through your tubing, apply contact adhesive or epoxy resin to the inner tube and around the first side of your holes, and attach to bucket [E]. When applying the rubber tubing be sure to squeeze out all the air bubbles. While the glue is setting you can hold the rubber tubing in place with tape and give it plenty of time to set completely. Do this for the other three circles of tubing [F-H], so each hole is secured by the tubing circles inside and outside the bucket. In the literature on tea bubblers leaks were an issue and the set-up has resolved the problem in a similar way,[8] but care is needed at this stage.

5 Cut to size a length of bamboo 0.5cm longer than the diameter of your bucket. Drill two holes 2cm from the top of your bucket on opposite sides, using a drill bit the diameter of your bamboo (approx. 0.75cm), alternately you can just hollow out the hole with a sharp knife. Apply air stones to the tubes inside and connect the other ends to the air pump outside. Finally feed the bamboo through the holes in the top and you are ready to go [I].

Basic preparation steps for aerated compost tea

A well made aerated compost tea has good chances of success and many of the problems highlighted in the literature can be bypassed in practice by the artisan grower. Perhaps the skills needed for success in this area are more within the remit of the practitioner than the scientific researcher as suggested by its ever expanding grassroots popularity.

Certain scientific institutions, such as some within microbiology, are rather narrow in their scope, possessing a tendency to re-shape essentially interdisciplinary activities, such as rearing micro-consortiums, to fit within their own comfort zones leading ultimately to the project being compromised.

For the gardener, timeliness and a direct awareness, guided by experience, inform their actions and supersede book learning. The ideal compost tea maker would be a well-read and experienced composter-gardener, free from the artificial controls and protocols of the laboratory, but even so access to a good quality microscope would help such a grower develop a better experiential link between their composts and the microscopic world than what could be reached with normal eyesight and intuition alone.

Ingredients

1 tbsp (5.92ml) unsulphured molasses
1 litre (1¾ pints) fresh home compost
22.7 litres (5 gallons) rainwater or dechlorinated

What follows are the preparation steps for a basic aerated compost tea. Further on in this chapter I describe different preparations and make alterations to this basic procedure, applying the full spectrum of resources and processes described in the other chapters. The specific points where the processes are refined for specific goals will be noted by reference to the process steps below and overleaf.

1 Find a shady location at room temperature (15-20°C / 59-68°F). Switch on your connected air pump. Make sure it is above the expected water level of your bubbler or you may have problems with water forcing through your pipes the wrong way. Fill up your completed bubbler with water to 22.7 litres (5 gallons). Rainwater is best, but if you need to use tap water let the system run without any additions for one hour to remove chlorine [A]. Further discussion of chlorine and chloramine issues is referred to in Chapter Six.

2 Fill the stocking with your finest compost, minus worms [B].

3 Add 1 tbsp (5.92ml) of unsuphured molasses making sure it all gets dissolved. Tie your stocking so the ball is just above the air stones. Give your stocking a gentle massage to help liberate the microherds.

4 On day two add a further 1 tbsp of molasses and give it an additional stir

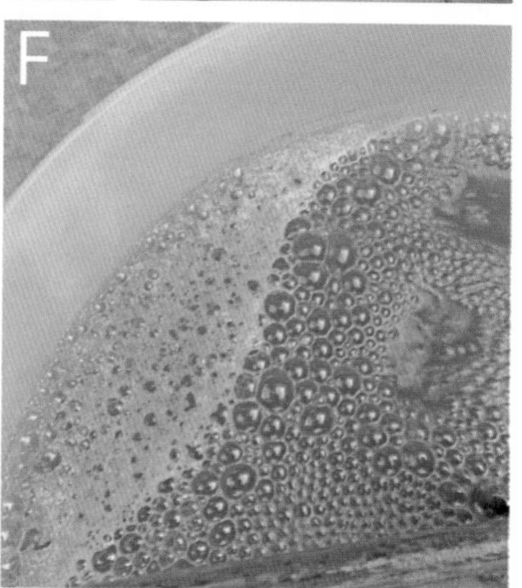

Steps for basic compost tea preparation.

A. De-chlorination phase.

B. Stocking filled with compost.

C. Tea brewing day one.

D. Tea brewing day two.

E. Tea brewing day three.

F. Close-up of froth.

making sure not to collide with the air stones. Give your stocking another massage. You may notice the water getting murkier [C] and a slight build up of slime around the water level of the stocking which is to be expected. The aroma from the bubbler should smell faintly earthy. If your bucket is smelling unpleasant there is something wrong. Either your compost is not of the correct quality or there is a problem with aeration.

5 On day three you can expect your mix to be getting more frothy [D+E] and this is an indication that it is ready to use. The aroma from the bubbler may be more intense but not unpleasant. Take into account that this process may take a day more or a day less given variations in compost, temperature and aeration.

Basic guidelines for compost tea: quality control, handling and application

This section is most relevant regarding any time you intend to use a spray to apply live compost tea. More specific guidelines have been added in each section but this section applies generally. Of course the considerations are mostly with regards to foliar spraying but circumstances may arise when you want to apply your amendment with a spray for other applications also. The most important considerations are gauze size, pressure and pounds per square inch (psi). It is common sense that forcing beneficial organisms at high velocity through an aperture that is too small for them is going to be counterproductive. Be careful of swift and unnecessary changes in temperature and minimise handling. Also with fresh tea you want to be keeping up the momentum to avoid creating anaerobic conditions and get it applied as soon as possible.

Using a watering can is the safest option but if you have a large area to spray this becomes impractical. It is also difficult to get to the underside of the leaves and if you have tricky areas in the centre of a patch and so on; you don't want to be trampling down your soil. A fine spray or a nozzle that has awkward angles will shred the larger organisms in the microherd you have just gone to the trouble of rearing. A spray size of 500µm and a maximum pressure of no more than 29 psi (pounds per square inch)[15] should keep disruption to a minimum. Try to avoid sprays with roller pumps, piston pumps and some of the more powerful centrifuge pumps.

The use of compost teas work well with other organic practices such as no-till approaches and mulching. There have been enough studies conducted to indicate that if the compost has been properly made and the resulting compost tea has been properly made, there is going to be an extremely high probability that you will be holding a very concentrated community of beneficial microorganisms. If you have access to a microscope you can confirm this to your own satisfaction. It may also offer indications if things are going wrong, for instance high numbers of ciliates may indicate anaerobic conditions.

Since you will want to sustain your beneficial organisms they will need some nurturing around your plants. It is common knowledge now that mulching improves soil productivity[16] and laying down an organic mulch before you apply your tea is an important proviso. If you have a 'brown earth' tilled field with impoverished resources for your tea organisms to consume, you will be wasting your time and effort. As previously discussed microorganisms tend to be very resilient and can manage extreme conditions and famines, but you won't be getting off to a flying start if you do not make the necessary amendments.

Protozoa tea (optimised for crop growth)

If you have good quality live compost, you will have protozoa in abundance so a longer but more accurate title would be 'Bacterially dominated compost tea optimised for protozoa'. Since protozoa are strongly linked to plant growth in research we have bypassed just doing any 'bacterially dominated tea' and proceeded straight to the brew with the best prospects of success.

To recap an enigma within the research literature introduced in Chapter Four, during studies it was found that even when there was an abundance of nutrients in the soil, protozoa still had a positive effect on plant mass although the mechanism is unclear. A possible analogy might be made to a computer system that is optimised to perform more efficiently bypassing certain unneeded features and streamlining the necessary ones. To return to the plant rhizosphere and bearing in mind that plants sacrifice a great deal of their nutrients to microbial agents, in this scenario the protozoa may be positioned as the optimising agent tinkering with the root system's metabolism and bacterial responses.

From the plant's side a related hypothesis could be that the protozoa act as indicator organisms, when the population gets past a certain level a chemical cue is expressed in the plant which may trigger an alteration of its metabolism to capitalise on this beneficial association and since there is no such thing as a free lunch, increased exudations for the protozoa herds continue their expansion.

In the aeration unit, protozoa increase in numbers rather later than bacteria and are higher up the food chain. Within the constraints of the bubbler they are probably the top predator. By day three, generally protozoa should be in their largest numbers. This makes sense given that protozoa have a lifespan of around two days so by day three they will be into their second generation, whilst the bacteria have begun their population boom almost instantly.

Soil protozoa are quite eclectic in their tastes so the plan for this mix is to fuel an expansion of a consortium of organisms lower down the chain. The abundance of live food will then fuel the protozoa's expansion. When the protozoa are at an optimum it is time to apply them with the rest of the tea to the rhizosphere of plants whereby a large part of the carbon nutrient supply will be taken up by plants in exchange for assorted benefits.

Please bear in mind that our focus in this chapter is to improve the fertility of the soil, not to attempt to feed the plant directly. Hence our intention in the bubbler in this preparation is to provide a balanced diet for healthy organisms as discussed in Chapter Two.

Ingredients

1 tbsp (5.92ml) unsulphured molasses

9.1 litres (2 gallons) multi-nutrient tonic (mix of comfrey, nettle and kelp meal, see Chapter Six)

1 litre (1¾ pints) fresh home compost

Preparation steps (with reference to steps 1-5 of the basic preparation)

1 As Step 1 in basic preparation (p.126) but add only 13.64 litres (3 gallons) of dechlorinated water and top up to 22.7 litres (5 gallons) with multi-nutrient tonic.

2-5 are the same as the basic preparation steps.

Application

We want to get as close to the rhizosphere as possible so a plant pot sunk into the ground is one simple option but care must be taken not to damage roots. Of course more sophisticated (and more expensive) options are possible with watering stakes, drip hoses and so forth but it must be acknowledged that permitting a lack of aerobic conditions, filters (<500μm) and overly high pressure (>29 psi) are extremely poor practices for this application especially since it has been optimised for the larger microorganisms.

Compost tea for organic orchards

To do compost tea for the orchard, the best activity you can possibly perform is to set up a vermi-heap by one of your favourite trees. This practice has an amazing effect on your tree, although it is important to not place the fresh decomposing materials in direct contact with the tree as this could cause problems. The organisms swarm, wriggle, crawl and walk to the heap leaving their frass, casts and guano for the tree to absorb in timely fashion. You can use the composting processes described in Chapter Eight and when the compost is ready for the bubbler you have also incorporated an indigenous element that you simply could not get commercially. We have another preparation for the indigenous practitioner in Chapter Ten where the topic of indigeneity is addressed further.

Orchard soil drench

It seems somewhat redundant to add aerated compost tea to a healthy, burgeoning, productive tree that has a compost heap next to it. If you are fortunate enough to have such a tree however, it is highly recommended that you take some soil from around this healthy specimen and include it into your compost heap for other less privileged trees that may not be doing so well.

There are many ways that trees can become stressed – drought stress, heat stress in high summer (or bonfires), a late spring frost, injudicious pruning, storm damage, poor soil, poor nutrition, root damage from rotovators and more. Often the effect of this stress is not immediately apparent. It may manifest as dying branches, early leaf senescence, a thinner canopy or an increased vulnerability to pests and disease. It is important not to give up on your tree too early because generally trees have formidable powers of recuperation, if the stresses are mitigated with good practice.

The drench here offers a tonic to the rhizosphere around your tree that can be used in conjunction with whatever other measures you are performing.

Ingredients

1 tbsp (5.92ml) unsulphured molasses
9.1 litres (2 gallons) kelp tea (see Chapter Six)
1 litre (1¾ pints) fresh home compost
Optional extra: mycorrhizal pellets[17] added to bubbler 12 hours before application

Preparation steps (with reference to steps 1-5 of the basic preparation)

1 As Step 1 in basic preparation (p.126) but add only 13.64 litres (3 gallons) of dechlorinated water and top up to 22.7 litres (5 gallons) with kelp tea.

2-3 are the same as the basic preparation steps.

4 is the same except add mycorrhizal pellets on day two, if required.

5 is the same as the basic preparation.

Application

Can be applied undiluted around the base of the tree and up to the tree's drip line (edge of the tree canopy) since this gives an indication of how far the tree roots extend. A good time for application is when soil temperatures are over 15°C (59°F) in early summer as it would increase the chances of the live microorganisms surviving. Otherwise, whenever the tree is looking lacklustre.

Please note if you have dug a small hole for the drench be careful not to damage your tree's roots and remember to fill the shallow hole afterwards to offer the mycorrhiza more protection.

Orchard foliar and vegetable leaf spray mix

The 'not so secret' weapon used in this mix is kaolin clay. Until recently this substance had been overlooked and only gained attention by default as more dangerous man-made chemicals fell out of use. The approach generally in this book has had a strong bias and emphasis on substances that can be made or found with minimal recourse to mass corporate technology, which for a small-scale tailored application would be disproportionately expensive without some communal enterprise such as crowd funding. Such guidelines are not unequivocal however and there is an exception in this instance, in the form of refined kaolin clay. It is better because standard clay contains impurities which are bad for human health and the finer particles of the processed clay are easier to use in spray form. Even though mild exposure via inhalation of pure kaolin dust is considered safe it is wise to avoid prolonged exposure[18] because of health uncertainties.

The primary purpose for using kaolin clay in this preparation is as a habitat and substrate for the natural fungal propagules in your compost, which will help them establish in the phyllosphere. The secondary use which is not to be underestimated is that kaolin clay is well-known as an organic pesticide[19] that affects some pest species directly by clogging them with fine particles. Other pest organisms may land on an area with kaolin clay then leave swiftly.[20] The reduction in pests also reduces fruit damage from bird activity as they hunt. Reports in the field suggest that beneficials are not affected by kaolin clay[21] however some caution is advised.

With regards to the yeast extract suggested below, a number of studies have explored the value of using it in foliar mixes and it is included here for this reason. It is well-known as a good fungal food and when delivered as a foliar spray for the phyllosphere has been shown to improve the growth of beans,[22] peas,[23] lettuce,[24] cucumber,[25] eggplant[26] and tomato.[27]

From these positive results, it would not be unreasonable to extrapolate similar responses in related members of the same families of Fabaceae (pea and bean), Cucurbitaceae (cucumber), Solanaceae (eggplant and tomato) and Asteraceae (lettuce), especially since there is already a strong background of success with this extract in a broad spread within these plant families as illustrated in the span of studies referenced.

Ingredients

1 tbsp (5.92ml) unsulphured molasses
25-100g/m^2 (source)[28] kaolin clay
1 tbsp (5.92ml) yeast extract
1 litre (1¾ pints) fresh home compost

Preparation

1 As Step 1 in basic preparation (p.126) except add only 13.64 litres (3 gallons) of dechlorinated water, add molasses and top up to 5 gallons with kelp tea.

2-3 are the same as the basic preparation steps.

4 is the same as basic preparation steps.

5 is the same as the basic preparation steps.

Application

Add soapwort 4.55 litres (1 gallon) solution (see Chapter Seven) at an equivalent temperature to the mix along with kaolin and remember to gently agitate to keep the mix in suspension during spraying.

Pralin root dips

This treatment tends to be used in conjunction with soil drenches. Dips may give you an advantage in disease suppression and root stimulation when potting on crop plants or transplanting seedlings. They can also be used to inoculate root fragments and tubers such as seed potatoes for subsequent planting. Traditionally an important function of root dips in tree nurseries was to protect roots from sunlight, drought and transit. This preparation works well for these purposes, with the additional benefit of maintaining a favourable rhizosphere prior to re-planting. It is highly recommended that this preparation is used in conjunction with the willow water rooting dip described in Chapter Seven.

This dip has been adapted from the French technique of pralinage.[29] Many may be aware that pralin is the combination of sugar and nuts enjoyed as confectionery, but pralinage traditionally refers to making root dips of a pralin consistency in the form of a thick adhesive slurry. This practice is not well-known outside of France although similar processes have undoubtedly gone on in the UK for generations. The author was shown pralinage from a very old and knowledgeable French farmer. In the time honoured process manure is used, however in this preparation matured vermicompost replaces it and compost tea replaces what was normally stream water. A fundamental but sometimes overlooked aspect of this process is the significance of filling up air spaces around the root with a pralin of the correct consistency. It is especially important that a well-matured compost is used as described in Chapter Eight since a compost that is not properly decomposed will burn the roots and be counterproductive.

Ingredients

1 tbsp (5.92ml) unsulphured molasses
6.82 litres (1.5 gallons) kaolin clay or locally sourced clay from the garden subsoil
6.82 litres (1.5 gallons) rainwater

For pralin container

6.82 litres/1.5 gallons home vermicompost (well-matured)
2.27 litres (0.5 gallon) local soil (fine)

Preparation steps

Steps **1-5** are the same as the basic preparation steps.

6 This step is executed in your pralin container of the dimensions necessary to fit your largest plant root. For this preparation we need to aim for a semi-liquid consistency that will facilitate a good adherence of the mix to plant roots. Break up the clay into crumbs and place it in your container. Add your local soil. Pour in 4.55 litres (1 gallon) of the freshly made compost tea (leaving 2.27 litres/0.5 gallon) and mix in. Add 4.55 litres (1 gallon) of compost (leaving 0.5 gallon) and mix it in. Leave for 30 minutes to allow the clay to absorb the live liquid.

7 Finishing the mix. There is something rather satisfying and anachronistic about getting your pralin just right. In practice it is somewhere between making dough and mud pies. What you want to aim for is an even film around your roots. Using your remaining compost and compost tea you must balance the consistency. If you place roots in the mix and the pralin just falls, add more compost. If the roots are difficult to insert and remove, and come out with great gouts of the mix attached, then add more compost tea.

Application

Dip roots in the pralin and leave for two hours. Plant shortly afterwards.

Seed treatments

Commercially very limited options exist for the organic grower to buy organically coated seeds[30] and what is available is very poor.

Although the author developed this method using aerated compost tea without initial reference to the scarce and rather sketchy contemporary literature, he subsequently found historical precedents that supported a related practice by none other than the late Masanobu Fukuoka, one of the pioneers of the no-till movement and alternative agriculture. Further on in this chapter we will take a closer look at no-till practices with regards to soil improvement and compost tea. The seedball method later combined with broadcasting ties in very well with this approach. Fukuoka describes his 'seed pellets' as preventing rotting and offering protection from animals.[31] Also difficult to germinate seeds such as carrots and spinach can improve with seed coating. The elegant idea suggested is to abandon tillage and broadcast coated seeds instead. This he asserts allows a greater amount of oxygen as compared with sowing within the soil, which improves germination[31] and of course greatly reduces the amount of effort required because of the removal of the need to insert seeds directly.[31] Interestingly he also suggests that the neat lines of furrows people like to prepare support the behavioural traits of certain seed eating arthropods that have a preference for moving in straight lines, that is, along the base of the furrow. Hence broadcasting seeds in a less linear, more chaotic way increases their chances further of not being consumed.

More recently clay is linked to being an important habitat for soil micro-organisms[32] and is linked to benefits in disease suppression.[33] The preparation on p.136 pays special emphasis on this feature whilst retaining the other benefits. The author would be interested in reader experience with seed coating and can be emailed via the address at the beginning of the book. To date, excellent results have been reported with carrots, parsnips, broad beans, runner beans, squash, chard, leeks and perpetual spinach.

Disease suppression

The theory regarding disease suppression is that a suitable live envelope of single or multiple beneficial organisms may outcompete disease pathogens in the rhizosphere for colonisation sites and resources. A number of mechanisms exist such as the production of chemicals toxic to the disease organism and stimulating the host seed to protect itself more effectively. Regarding the latter, two distinct pathways are known – Induced Systemic Resistance (ISR) and Systemic Acquired Resistance (SAR).[34] With ISR the beneficial stimulates the host to respond to an immediate threat which may induce the plant to produce defensive antibiotics[34] or alter cell wall composition[35] for instance. Interestingly SAR ties in with the discussion in the introduction regarding the overemphasis on clinical germ free procedures. With SAR the plant is exposed by the beneficial organism to a low level of infection which then helps it to develop a resistance when a real threat emerges. Currently the only way conventional science has succeeded in approaching disease suppression is through isolating specific beneficial organisms in highly sterilised environments. What is happening in Nature could be considered as almost completely alien to our current state of knowledge and research culture with around 99% of organisms being unculturable,[34] by standard laboratory techniques. The technology exists to observe consortiums of living organisms directly under the microscope but currently a chasm exists between what is actually happening in the soil and the divisive GM (Genetically Modified) dominated approaches of the lab.

Properly made compost has engaged thoroughly with the mystery we call Nature. We might not quite understand what is going on, but compost has a robust research history in suppressing disease.[36]

In studies with aerated compost tea, the most effective defence against damping-off developed when fungal nutrients were included.[37] This result was unambiguous with 13 out of 13 trials succeeding.[37] One concern not made explicit in the subsequent debates is that the 'damping-off' in this study was with *Pythium ultimum* only; in field conditions numerous other fungal pathogens can be responsible for what we generalise as damping-off.[38] The study listed is yet to be expanded outside the laboratory and as is often the case, the essential next step is left for pioneer practitioners to explore.

Seed germination

With larger more manageable seeds from 2mm upwards the author has found greatly speeded up germination from allowing seeds to become turgid by steeping them in water for at least one day. If the seeds begin fizzing you have left them too long. Placing your seeds in a small loose meshed pouch inside the compost in the bubbler a day before the compost tea process finishes, introduces the growth-stimulating and disease suppression potential at the earliest stage of the developing plant and has been found in practice to be an improvement on just steeping. If you are not doing the seedballs below, the next step might be to sow or place the seeds in a moist, dark, warm, well-drained, lightly sealed container sandwiched between untreated tissue paper. This technique has proved very successful in terms of both speed of germination and percentage of failures, but would need more tests for scientific validation. It has been tried with excellent results with chillies, squash, French beans, runner beans and broad beans.

Seedballs

To reiterate, in this chapter our focus is to create soil fertility that is passed onto the growing plant. In Chapter Two we illustrated what nutrients organisms need to be healthy; our intention in the bubbler is to give our beneficial organisms a balanced diet with all the nutrients that they may need in place.

With regards to the choice of clay as a cheat, cat litter can be used, but beware, the synthetic ingredients or other additions of certain litters are inappropriate for seedball repurposing. Organic, biodegradable and compostable litters are available, however some contain walnut shells (allelopathic) which would inhibit growth unless you were developing a walnut tree guild with plants that could handle this. Others use non-clay materials which would rather defeat the point. Another important feature to be aware of is that the best litter is of the clumping type for obvious reasons. So choose your cat litter wisely.

Many practitioners like to mix a number of companion seeds together when making their seedballs which can add considerable further interest to the process.

Ingredients

30g kelp meal (fungal nutrient and traces)
7 litres kaolin clay or 1 spit (as per photo approx. 20cm × 19cm × 10cm) locally sourced clay from the garden subsoil (habitat)
Previously steeped seeds as described (only if you intend to sow/germinate immediately)
1 litre (1¾ pints) live home compost
Optional: Use clay based, clumping, organic, biodegradable, cat litter
Optional: Mycorrhizal fungi, oyster mushroom spores

Preparation steps

Note: Use gloves at all times during these procedures because the oils and microorganisms from the skin may potentially compromise the process.

1 The larger seeds (>5mm) can be placed in a separate pouch in the bubbler a day before completion. Seeds 2-5mm can be placed in a fine mesh pouch. Any seeds smaller than this can just be folded into the clay when making the seedballs without any steeping.

2 As basic preparation Step 2 on p.126.

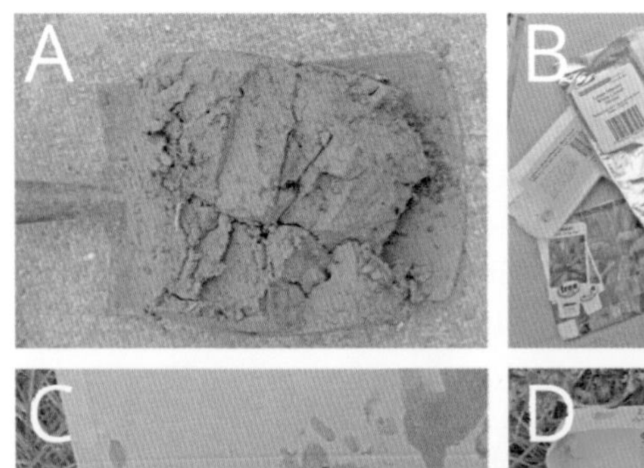

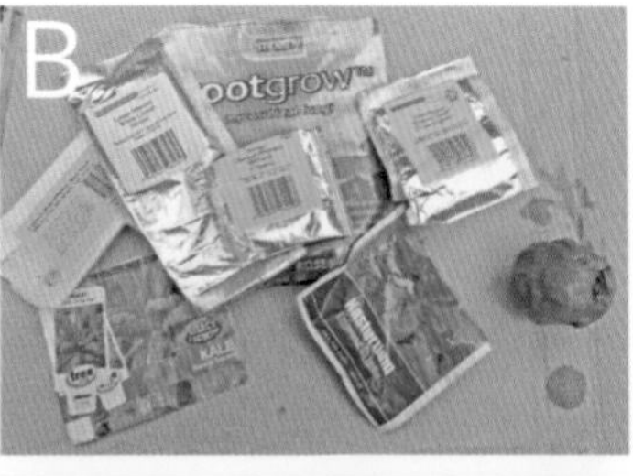

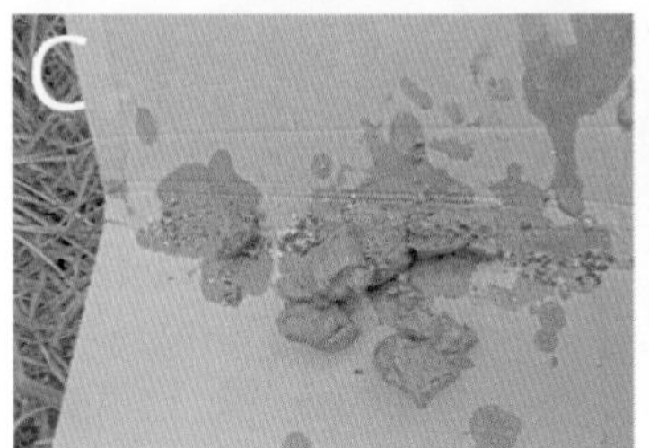

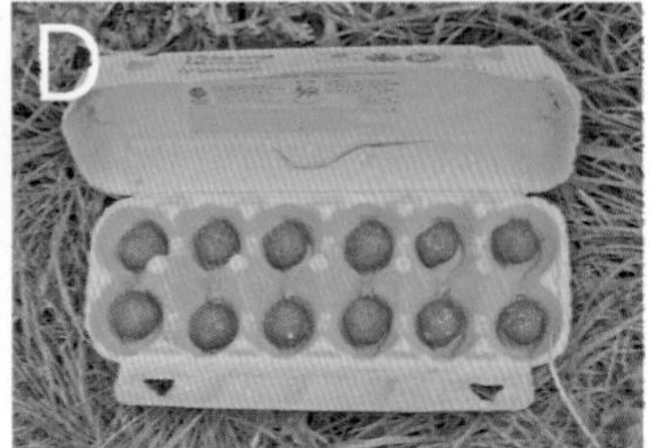

Seedball preparation.

A. A spit of clay.

B. Seeds and clay.

C. Balls making process.

D. Completed seedballs.

3 When the bubbler run has finished, remove the seed pouch and open into the small container. Pour some fresh compost tea onto these seeds, since there may by larger organisms in the tea that could not travel through the pouch to the seeds inside.

4 Break up a spit of clay (approximately 7kg) into 10 or more manageable portions and place in a five gallon bucket. Take a pointed object such as a skewer for testing cakes and make more holes in the clay to increase the surface area further. Pour on the compost tea until it is covering the clay. Leave for two hours somewhere with an ambient temperature; if the clay is too cold it will be difficult to work.

5 Lay out a workspace as shown above. Take a portion of steeped clay and lay out your seeds.

6 Work the clay until it's malleable then gently work in the seeds to form the finished ball as above; 1-3cm is recommended unless the seeds are particularly large. If the seeds are 0.5cm or more you can roll two to three seeds into small balls and more for smaller seeds. The spit of clay (7 litres/ approx. 7kg) photographed (20 × 19 × 12cm) above is enough for at least 350 seedballs at 3cm diameter. Making seedballs tends to send an adult back to childhood days with memories of playing with Plasticine. It is a very messy activity and kids love it. It is also an excellent opportunity to bring out your languishing and overlooked old stock. Seeds are known to hold a concentrated source of many nutrients to develop new life, so if they have are too old to be viable they will still be useful in the future when they break down.

7 Continue with steps 5-6 until you have run out of clay. Egg boxes may be used as a convenient storage medium.

Application

If you intend to broadcast your seedballs, rather than put them in plant pots just to see what they do, it is recommended that you do so as swiftly as possible after preparation because depending on your choice of seeds, germination processes can get underway rather quickly. Wherever possible, use seedballs in conjunction with no-till practice and after broadcasting, finish with a misting of compost tea.

It is a matter of taste whether the point of a seedball is to shatter on impact or remain a stable base for seeds to develop. If you happen to be a guerrilla gardener with limited access to your chosen site you may prefer to let the balls dry out a little more so they shatter more readily, leaving the seeds with remnants of clay still adhering. This latter practice however may offer a narrower window between broadcast and germination as understandably it would be poor practice to pitch fragile plants after they have emerged.

Leaf digestion and fungal disease prevention

This preparation is used to either initiate or hasten decomposition processes in the orchard or veg-tree polyculture. For many no-till growers having an excess of organic matter is very desirable, however situations may arise when a choice is made to bring down the level of organic material. In orchards during a very damp spring the risk of scab fungus (*Venturia inaequalis*) and many other fungal diseases is increased. Fallen leaves are the main habitat for scab infection initially before the spores rise to the tree canopy.[39]

Copper solutions are often promoted to combat fungal diseases including scab, however it must be asserted that these copper based solutions are not recommended for the food web grower. In many organic programmes copper has 'restricted' status[40] which means it is allowed with great care. As described in Chapter Two, copper is essential to plants as a trace and also to other organisms[41] but even in modest amounts it is toxic to earthworms and soil microorganisms.[42] Earthworms will avoid land with as little as 34mg/per kg.[43]

Copper in various forms is a major fungicide in agriculture and has been used for hundreds of years.[44] Initial success may be achieved with copper sprays but they indirectly contribute to scab fungus and other fungal diseases by blocking the natural flow of soil processes. Such practices interfere with the earthworms' natural activities of burying and assisting decomposition. Areas that have been sprayed tend to build up organic matter because the soil life has been too damaged to cycle it normally.[45] Copper can accumulate in the soil with yearly spraying[42] and may qualify large areas of land in conventional agriculture as contaminated for the purposes of food web development, being in serious need of remediation. If you happen to have the misfortune of having your land in an agricultural area, the food web may have been obliterated and you may need to consider bringing in earthworms to colonise your soil, which can be a difficult and time consuming process.[46] In temperate areas *Lumbricus terrestris* is a very good choice but it is worth researching your particular area to find out if there are any indigenous anecic worms (deep vertical burrowers) that perform a similar service if *L. terrestris* is not established.

Moving on from copper, it is important to bear in mind that a good mulch of leaves is important to the tree's ecology. It is linked to greater yields[47] but this must be weighed against the emergence of fungal diseases. Minor cases of scab are just cosmetic but they can be a prelude to more serious infection that may then also result in reduced productivity.

Although orchard soils are fungally dominated, a burst of bacterial activity is desirable in this instance. To increase decomposition, the surface area of the leaves are often increased by mowing or shredding although this may also disrupt surface dwellers. This process can be sped up further by covering the leaves, which helps to raise the temperature, retain moisture and facilitate darkness, thus encouraging earthworms. Furthermore a physical barrier will interfere with the rise of fungal spores to the tree canopy. Comparisons between organic and other practices suggest overall there is less disease in organic orchards but rot related diseases are worse. It was hypothesised that one important factor towards organic orchards scoring better was from having ground cover plants which interfere with the disease spores as they float to the leaf canopy.[48] An alternative to leaf digestion if you don't have a great amount is to set up a traditional leaf pile with a cover and make a problem into a useful resource. An area of 1.5ft (46cm) × 7ft (213cm) × 3ft (91cm) will easily hold 20 standard bin liners of leaves which only takes a couple of hours to collect.

Continuing with our plan for leaf decomposition, it is suggested that you integrate separate lime and manure applications with compost tea. The application of lime during leaf fall in autumn and a boost of rotted manure in spring is standard organic practice. Finely ground eggshell is also a possibility and is discussed further on in this chapter. Over time orchard soils gradually become more acidic and agricultural lime is used to raise the pH again. If you coat your leaves with aerobic tea and apply lime in a normally functioning food web, earthworms will incorporate the lime coated leaves into your soil together with their own calcium infused mucus. Most aerobic microorganisms[49] and earthworms[50] favour soil that is around the mid-range in pH. The 6-7pH range is best for most fruit or vegetable crops and it is no coincidence that in a purely chemical context, in moist, uncompacted soil, this range has the best availability of nutrients.[51]

It is suggested that you coincide aerobic tea application with this practice in four stages, two of which involve aerobic tea. In autumn when a substantial amount of leaves have fallen (<50% left on tree) you can coat your leaf litter with aerobic tea then apply lime straight after.

When almost all the leaves have fallen you can perform a further inoculation. In spring when you apply well-rotted manure or manure tea as described in

Chapter Six, leave it a couple of weeks then apply the second amendment of aerobic tea when the risk of frost has passed and soil temperatures rise to above 15°C (59°F) and ideally 20-25°C (68-77°F) in early summer. The gap between manure application and compost tea application serves two purposes. Firstly it will stop the microorganisms burning out in the first few days with the excess of free nitrogen and secondly, in ambient conditions with increased warmth they have a better chance of getting established.

Preparation stages

1 Autumn (during leaf fall) application. Protozoa tea as described previously before lime application.

2 After 90+% leaf fall. Kelp tea (see Chapter Six).

3 Spring. Manure applicaton/manure tea application (see Chapter Six).

4 Late spring. Application two weeks after well-rotted manure application. Basic compost tea as described previously.

Application

Watering can, or for larger areas a spray with an appropriate pressure and bore thickness as discussed previously.

Soil improvement and aerated compost tea

In this section we are not just trying to improve soil structure, that is not much of a challenge given the many resources at our disposal. We are aiming to create a nicely balanced and fertile soil even if your ground is very degraded (nutrient poor), compacted or too loose. A section on contaminated land is included later, but special technical, legal and safety concerns exist which need a full book in their own right to properly do these matters justice, so it will not be engaged with deeply.

Before proceeding with improvement plans some background to our activities is laid down in the next few pages with a brief review of soil structure, texture and tillage options, always with the smaller scale organic grower in mind. It is suggested that the reader also consult Chapter Two where directions for a variety of simple soil tests are outlined. From these simple tests and other indicators you will at least have a rough idea of primary nutrient levels and a basic knowledge of your soil proportions which will now be expanded upon further.

Soil texture and soil structure

Briefly for those new to this area of interest, soil texture is the balance of sand, silt and clay in a soil and it is classified by particle size. Sand is the largest (0.05-2mm), followed by silt (0.002-0.05mm) and then the smallest, clay (<0.002mm).

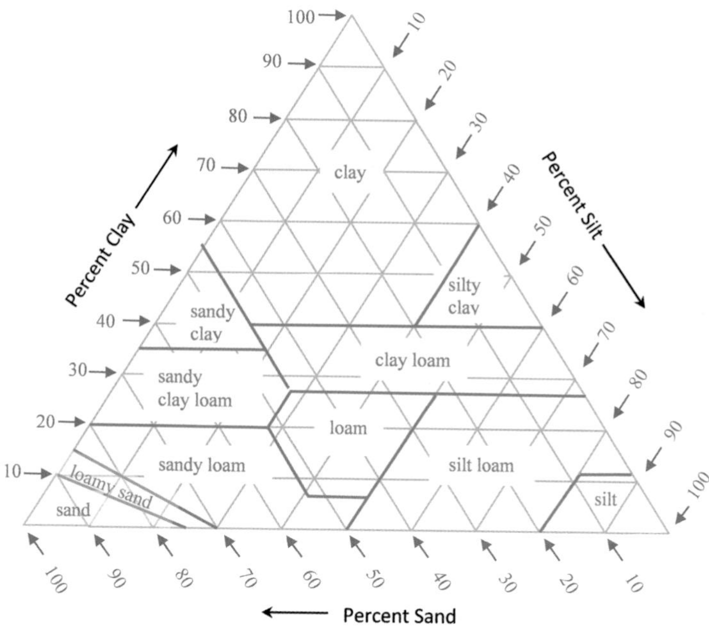

USDA Soil Texture Triangle.[52]

Observe the USDA soil triangle above. The ideal soil texture rather depends on what you want to do with your soil. For pond construction a high clay content, when making concrete, a high sand content and for foundations, if possible try to stay away from expanding clay soil otherwise you may need to add certain amendments to make the soil more rigid. If you happen to be growing vegetables, the best all round soil (loam) is around 30-40% sand, 30-40% silt and 20-30% clay.

Many naïve growers, and growers who should know better, who have clay soil have caused themselves problems by adding small amounts of sand to clay soil in a simplistic attempt to improve these proportions. Conversely a small amount of clay added to a high sand soil tends to improve soil properties because of greater moisture, nutrient retention and microbial presence, but consider the third point overleaf if this does not happen to be the case.

If a grower has a clay soil and attempts to add sand directly, so at least on paper they have improved the proportions of the soil, they will be faced with three problems. Firstly, adding sand to clay directly increases the soil density because the smaller clay particles tend to bond in between the larger sand particles to develop greater hardness, which we really do not want for crop growing. Such a grower may find an increased tendency for their soil to form caps which is poor for seedling emergence in spring. Following on from capping, increased silting and erosion may also be evident. Secondly, to overwhelm this issue, a proportion of over 50% sand would have to be added to clay for the 'ideal proportion' which can become financially and practically prohibitive. For instance consider a high clay soil bed 4 × 2m with a topsoil of 30cm; even taking into account say a 20% intrinsic sand component, you will still need around 48 large (60 litre) sacks of sand to bring the soil to its ideal proportion. So for this small space with sacks of sand at £5 ($7.65) the bill will come to a stunning £250 ($382.50). If this is expanded to larger areas the cost becomes astronomical. Thirdly, in real terms, the difference between the ideal loam and something someone has mixed up in a test tube to the same proportion is down to the way the clay, silt, sand, organic matter, air pores and other components are arranged, which is described as the soil structure. A natural sandy clay loam may have developed over geological time scales and have been involved in complex interlinked biological, chemical and physical processes to reach its current state. When large quantities of new amendments are added to the soil, the soil structure can be disrupted and the system may take a number of weeks or months to settle down.

Analysing sand, silt and clay proportions can be a useful exercise and allows soil scientists to predict certain soil properties. In real world situations however, organic matter and other factors can confound these predictions. Often biological factors and different types of organic matter make universal predictions based on soil texture alone very inaccurate.[53]

The presence of clay can have a profound effect on soil properties even in small quantities and the literature suggests that between the other fractions of soil texture, clay extends the most influence. As discussed in Chapter Three, clay improves the ability of a soil to retain cations(+) because of its net negative charge and a certain amount is a great benefit to plants.

Another aspect of the soil in real world situations is the presence of organic matter which also has a powerful influence on soil properties and can reasonably be considered the silver bullet for soil improvement. Most clay soils in temperate areas are of the 2:1 expanding type. Water can be drawn between these weakly bonded layers and organic matter can bond with the clay sheets to reduce its adhesion index or 'stickiness'. Conversely with sandy soil, organic matter assists in drawing soil particles together into aggregates. This is merely chemistry, but consider when there is a live biological component also involved, actively bonding soil into aggregates with their exudates and promoting fertility with active aeration.

Where compost teas enter the picture is that in liquid form live organic matter can penetrate or envelop a material such as soil colloid quicker and more completely than a less fluid amendment such as compost on its own. The brewing process has also supercharged its potential with the introduction of a diverse community of beneficial aerobic organisms.

Tillage options

From our review of soil properties on the previous few pages, tillage options naturally follow. We will now take a closer look at how best to create the most suitable soil environment to nurture your food web and provide the best environment for your crops. Below is a chart of the main styles of tillage for you to review, which is intimately connected with soil life and its survival.

Tillage comparison

	Intensive (e.g.plough/rotovator) 0% residue cover	**Conservation** (e.g. reduced tillage) >30% residue cover	**No-Till** 100% residue cover
Soil life	Least abundance[53] in most taxa compared[54] loss of biodiversty[55]	Raises biodiversity as compared to intensive [55,56]	Greatest abundance of micro and especially macroorganisms (e.g. earthworms)[53] in most taxa compared[54]
Soil condition	Worst. Poor fertility, erosion, disastrous long term sustainability.[57]	Improved soil porosity, soil water, organic matter as compared to intensive[56]	Best.[57] Great improvements over time in soil water, organic matter, soil stability, soil aggregation, soil fertility and sustainability.
Profit	Lowest[58,59]	Better than intensive in dry conditions[60]	Highest.[58,59] (Less inputs, less costs) after transition but can be more unpredictable.
Crop yield	Equal[58,61]	Equal[58]	Equal.[58] (Highest in low water/drought conditions e.g. Soane *et al.*)[61]
Weeds	Herbicides considered more effective than organic practices[62] however selective resistance[63] must be factored	Reduced as compared to intensive[56]	Organic practices possible but considered not as effective as herbicides for certain weeds (e.g. perennials) but restrict annuals better[62]
Diseases	Worst[64] but narrower spread of diseases[65]	Superior to intensive,[64] problems different with moist conditions[65]	Best after transition.[66] Problems mitigated with crop rotations[67] and enhanced habitat for beneficials. Transition is the most risky period.[66,62]
Pests	More pests overall than no-till[68]	Reduced as compared to intensive[56]	Less overall pests than intensive from better competition but more slugs and snails[68]

Intensive tillage on farms and allotments is the norm and is still called 'conventional' tillage, but in many parts of the world this attribution is a misnomer now since it has been superseded by conservation tillage. In this respect Europe could be considered a 'developing' country[70] as the arguments for no-till begin to outweigh the case for intensive tillage. To clarify, this is not the kind of tillage associated with the small-scale French intensive growing described in the book introduction, rather the style colloquially labelled 'brown earth gardening' associated with monoculture and industrialised farming. The amount of residue cover associated with this style of growing is around zero on both the large and small scale. It may be observed that whilst an enormous proportion of farms worldwide have moved on and converted to no-till, the norm in UK allotments is still a scaled down version of intensive farming replacing tractors with cultivators and continued use of agri-chemicals both legal and illegal.

With conservation tillage residues on the soil are maintained at 30%[71] or more and could be seen as a halfway house between no-till and intensive. No-till is where full coverage of the ground is aimed for.[72] In its various forms it facilitates the food web at a higher level and is recommended because it gives compost tea applications considerably more potential to succeed with increased shelter, more microbial food and habitat. Furthermore research suggests that with no-till your space becomes a beacon for earthworms; they have been shown to migrate to no-till areas[73] from intensive tillage zones. No-till has demonstrated its best performance in warmer climates where organic matter breaks down rather quicker and success or failure can be discerned more sharply. For a cooler temperate climate such as the UK, a sensible novelty is suggested here. A main drawback for these areas is that the ground is slow to warm in spring, which delays germination and plant growth. A minor innovation that can be applied on the small-scale to heat up no-till areas in spring is the simple provision of using warm water to help increase the soil temperature. One note of caution is to try this gradually using water only a few degrees above the soil temperature of nearby un-mulched soil since great fluctuations in soil temperature may be counterproductive. It is as straightforward as leaving a water butt in your greenhouse and then applying it to your soil. If you have the time and motivation to tinker you can lay down insulation at night and uncover it in the daylight. Because water has a higher heat capacity, it is slower to heat, but also slower to cool; this component is the reason that spring soils are slow to heat. Once warmer water is added the disadvantage is reversed and now heat is being retained by the mulches. It might be suggested that this is only appropriate on the small-scale but there is no reason that it cannot be applied on a larger scale through irrigation equipment that may already be installed. A solar powered pump circulating passively heated water through the soil and back again in underground pipes is not beyond the bounds of possibility.

For those who like technical fixes, black polyfibre is a possible option to refute the problems with the soil temperature in no-till. Its semi-permeable membrane allows water in and restricts weeds. If you don't want to pay the high price for this material, use black bin liners layered over your mulch. It is not glamorous but it works and the worms are clearly attracted to the increases in moisture and warmth. If you are a nature grower and dislike using poorly biodegradable products of the petro-chemical industry, a compromise between mulch layers and increasing soil temperature may be the best answer, which is known as strip tillage. In this style of tillage an area is left clear where your crops are to go in and the surrounding areas are left well mulched. In this instance you can just leave these areas clear in spring for a couple of months. Ridging the clear areas would intensify the heating effect as would the use of raised beds.

Recent studies suggest that the best soil structure is found with permanent living roots which is good news for growers who develop polycultures but bad news for both no-till and intensive tillage if they leave their land with no living plants for long periods of time.[75] The use of green manures as cover crops and occasional fallow periods with perennial grasses can help to balance this problem if you prefer to keep certain areas open for vegetable crop rotations.

We now proceed to a schema of the processes and steps we can take towards soil improvement shown below. Sandy and clay soils have been concentrated on, in contrast to silt dominant soils, because they constitute the majority of soils.[74]

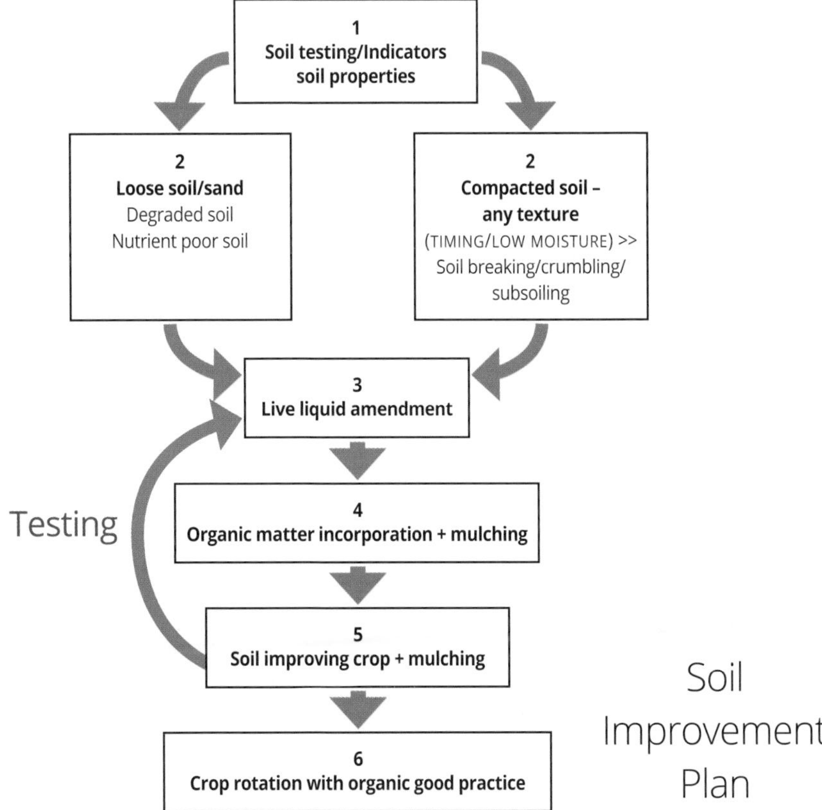

The schema can be seen as a loose guide to the soil renovation processes to follow with added nuances included in each section. What we are doing with this plan is stacking the cards in our favour. Much of it is well informed common sense based on an understanding of the food web and soil, but these are not the methods of industrialised growing, more its antithesis.

Compacted soil

Much of the literature on compaction comes from a highly mechanised farming perspective with heavy machinery and large areas as the norm. As discussed in the introduction our perspective is rather smaller scale, however a formidable body of knowledge has built up in farm scale operations that we can apply with a rather different attitude.

On this more pedestrian scale the most significant dangers of soil compaction come from repeated footfall, the action of raindrops, cars and possibly very small-scale machinery such as a garden cultivator. In communal allotments very often one end of a plot is given up to parked cars. If you have replaced your car with a bicycle and want to make this area of land a productive site for growing fruit and vegetables, then this section is for you.

An important process needs to be highlighted regarding compaction generally which is valid regardless of your choice of tillage style or soil type. Breaking up the subsoil has been indicated to be the most effective course of action to begin the decompaction process.[76,77] Subsoiling for the calorie driven gardener equates to the well-known technique of double digging. To clarify this involves temporarily removing a spit (approx. 30cm / 11.8in) of the topsoil in sections and then going down a level to break up the layer beneath with a fork, being careful not to mix the layers. A skilful gardener can do this without stepping on the areas they have just decompacted. With petrol driven equipment, various attachments can be applied to garden cultivators to do this, but it is rather worrying that these little machines with their fast rotating tines can compact soil and damage soil structure in more unfavourable ways than large tractors with disks.[76] This disparity is increased further when caterpillar tracks are used with tractors.[76]

The next stage up at the smallholding scale is the use of tractors pulling chisel ploughs. Unsurprisingly these loosely resemble chisels but with a more arced form. They can go approximately 30-60cm (11.8-23.6in) into the soil, 60-150cm (23.6-59in) apart[78] and compared to other mechanised tillage devices are less disruptive. They do not invert the soil but they are very energy intensive. A number of contraptions also exist to decompact the areas traversed by the compacting tractor tyres during decompaction runs!

In the same rare study[76] referenced above on small-scale compaction, a trial was made between areas that were hand tilled, tilled via garden cultivator and by tractor with plough. The unequivocal result was that double digging outstripped all other tillage methods including both tractor and cultivator in terms of crop production.[76] Such a study will never be referenced in the glossy brochures of tractor sellers displaying caricatures of farmers crassly brandishing bags of money, but it has pride of place here. Of course the historical advantage of petrol fuelled machinery is the speed, power and arguably cost effectiveness with which land can be prepared, rendering a high percentage of agricultural workers obsolete.

In studies with compacted soil it was shown that no-till crops underperformed in the first year when the subsoil was not broken up, but gradually caught up in subsequent years.[77] If you are a Jain gardener, a Buddhist or are an advocate of Deep Ecology, where the welfare of every living being is sacred, your priority will be to minimise damage to other beings. Therefore if you follow these beliefs or happen to be sympathetic to them, your most ethical choice is no-till, then you need to be very patient since without the disruption of subsoiling and initial soil loosening, renovation takes considerably longer. Regarding these ethical concerns it must also be considered that earthworms tend to avoid compacted soil[79] and subsoiling will actively assist in providing a better habitat for them.

If there was a degree of compaction just below the topsoil it was clear that subsoiling improved every crop whichever style of tillage was used.[76,77] If you are new to no-till and subsoiling has just been done in the correct way described, you can expect earthworm numbers to greatly exceed their previous levels.[80]

With the nature harnessing method of decompaction described on the previous page, the best time of year to begin your soil improvement programme is in late March, or as a second choice, August, when there is still time for some soil renovating crops to grow. These months are just preceding periods of high earthworm activity[81] in spring and autumn. Most importantly it is necessary to make sure the ground is not wet because working the soil in these conditions is likely to do more to damage soil structure than improve it. What we are aiming for is to reduce compaction to a degree that our friend *Lumbricus terrestris* (see Chapter Four) will engage with the soil. They function well in areas with a good level of calcium which they help to maintain. As discussed previously, 6-7pH is the most favourable for most crop plants and usually more acidic soils are found in temperate conditions rather than a tendency towards alkalinity. If you do happen to have alkalinity however, say between 7-7.5pH, there would be little need for concern, but past this level sulphur is often used as an effective but extreme option. Consult the discussion on woodchips overleaf for a softer option. A particular concern is that with more alkalinity phosphorus, iron and manganese become unavailable.[51] Organic amendments and the activities of earthworms bring the soil pH down to more neutral levels[82] in time. The nature grower may just consider going with the flow and concentrate on plants that thrive in alkaline conditions.

L. terrestris are known to require calcium in their diet which is important for metabolic processes in conjunction with their calciferous glands[83] and carbon fixing.[84] During the transit through the gut, soil particles are enveloped in a calcium rich mucus prior to excretion.[85] Earthworms can penetrate slightly compacted soils[86] and improve drainage but studies indicate that they will avoid very compacted soil.[79] Deep burrowers, they draw dead leaves into their vertical abodes which can go down up to 6m (19.6ft). After some degree of decomposition the worms consume the leaves[87] and excrete valuable, fertility enhancing casts. As detritivores they can consume both fresh and decomposing plant matter including animal manures, but they favour partially decomposed material.[87]

The burrowing activities from their powerful muscles combined with fresh casts make this an excellent worm for decompaction and soil improvement. They are widespread in temperate areas and unless you are in the middle of a conventional agricultural area as discussed previously, they will be drawn to your land if you create the right conditions.

Staying with our diagram, if after soil testing you found compaction then regardless of texture, the physical decompaction (Step 2 on diagram) would be conducted when the soil was not too wet to better enable fracturing. From this point we would then proceed to an amendment of live compost tea (Step 3 on diagram). Whilst the process steps remain the same there are slight differences in the type of liquid amendment and cover crop, as below, which are used that reflect the features of a clay soil or sandy soil.

Improving clay soil
(with reference to Step 3 of the Soil Improvement Plan, p.145)

If you happen to have clay soil then tree fruits, strawberries, raspberries[88] and many vegetables such as Brussels sprouts, leeks, cabbages, daikon, beans, peas, and certain potatoes or kales do rather well. Certain cultivars are well-known for doing better than others such as Sharpes Express and Kerr's Pink in potatoes and Pentland Brigg or Pentland Dwarf in kales.[89] In truth a determined and creative grower can grow anything they want, but some choose not to swim against the tide. Before taking the time, effort and expense of altering your soil properties, and exposure to the seasons, consider if you would just prefer to concentrate certain areas on plants that do well in the prevailing environment.

To recap, high clay soil can have poor drainage, poor aeration and is slow to heat up in spring. You may have come across the garden centre fix with the superficial approach of just mixing up various soil improvers bought at the same time to solve the issue. You may also have found that the year after, you need to buy more because after winter the soil appears just the same as the year before. The fundamental failing in this 'shaker maker' approach is to treat the soil as a short term chemistry exercise, rather than a long-term biochemical one. The inconvenient, messy, unpredictable and volatile state of life has been overlooked.

Before adding your compost tea as on p.150, work 3-4in (up to 100mm) of composted bark into the soil. Composted hardwood bark is best in comparison with softwood bark because of concerns over acidity. It is important that the hardwood bark is composted because the high carbon to nitrogen ratio and the speed at which it breaks down[90] can encourage microorganisms to metabolise a large amount of nitrogen from the soil. The USDA recommends the ratios to avoid this should be below 30:1 (C/N).[91]

With particular reference to softwood bark[90] be aware of the pH in the chips; if you are not sure grind up a small amount with a minimal amount of water then leave overnight. In the morning perform a pH test with litmus paper or a meter. Avoid any chips under pH6 or over pH7.5 unless you are attempting to unlock certain toxins during remediation. Certain trees such as conifers (*Pinophyta*) are well-known for their acidity and make fantastic mulches for blueberries. For people who want a soil closer to neutral, well-ground eggshells are an established alternative to horticultural lime and can be used with good effect to balance pH levels.[92] Finely ground eggshells which are composed mainly from calcium carbonate albeit in a more organic state can be used in a comparable way to agricultural lime. If you are considering using eggshell to raise pH levels it was highlighted recently that assessments used to approximate how much eggshell you must use to get the same effect as lime (Effective Calcium Carbonate Equivalent) are around 2½ times too low.[92] Since eggshell is virtually pure calcium carbonate it is reasonable to assert that the volumes should be more or less equal to horticultural lime and only dependent on its fineness and purity. In personal communication with an industry expert[93] this view was further confirmed in direct contradiction to the official 'E.C.C.E' position. Interestingly in a large field study between eggshells and industrial lime there were differences in performance with eggshell working more quickly to raise pH then levelling out. In contrast, the agricultural lime was slower but worked over an extended period to raise the pH slightly more.[92] After the incorporation of 2,000lbs per acre (907kg per 4,047m/sq.) of eggshell, it was found that adding more had little effect. The pH was raised with eggshell approximately 0.8 of a pH point after six months (5.2-6.4pH) and 1pH point after 18 months. As the pH approached neutral it became increasingly more difficult to raise the pH further with this method. To raise your pH to the same level as in the study, other factors being equal you will need 6.27g of eggshells per square metre. Based on a standard chicken egg at around 57g (11% eggshell by weight) you would need to make roughly 18 (two egg) omelettes to get the necessary amount of eggshell byproduct. To apply eggshell to a standard 75 × 15m plot you would need to make 20,250 omelettes. Therefore, considering the volumes of eggshells[94] needed to raise pH effectively it might be worth considering contacting a custard factory or some other related industry as day-to-day household usage is clearly not sufficient.

The compost tea chosen for the live liquid amendment (Step 3 on diagram) is the basic tea preparation (p.126). We want to concentrate on the smaller organisms in this mix because as discussed with regards to kaolin, bacteria can gain better access to small spaces within clay and it would be expected that their key predators, the protozoa in the brew, can occupy the newly made spaces in the clay on the perimeter of the clay platelets.

Compost tea application/preparation (Steps 3-4)

A. Application of composted hardwood bark (or neutral pH softwood bark), dug in.

B. Basic tea preparation. Diluted 1:3 (enough to drench) 4.55 litres (1 gallon) per m^2 making sure the soil is moist beforehand.

C. Application of more bark or straw (2-3in) immediately after compost tea application. Laid on top.

If the conditions are particularly cold it would certainly be prudent to wait two to three weeks longer before beginning activities. If you have trouble getting crops going and you are indeed only dealing with a smaller area, starting your seedlings off indoors initially could be considered. With clay soil especially it is important to make sure the soil has warmed enough to allow a soil improving crop to germinate (Step 5 on diagram). A great multi-functional combination could be a mix of fenugreek, field beans and sweet clover which can handle heavy clay and improve the soil properties, in combination accumulating organic matter, pushing through clods and fixing nitrogen. After this point some more soil testing would be in order. If you are still not happy with how your soil is improving you may then consider repeating steps 3-5 again, but your soil would have to be in an exceptionally poor state for this to be considered.

Later in the year this could be dug in with further additions of manure and compost followed by winter tares (Step 6 on diagram) to prevent nutrient leaching. From this point you can begin rotational activities with vegetables should you wish.

Improving sandy soil

(with reference to Step 3 of the Soil Improvement Plan, p.145)

With regards to sandy soil the same principle applies. If you prefer growing garlic, parsnips, beetroot, onions, turnips or certain potatoes and aromatics then proceed no further with sandy soil worries unless we are talking about a desert environment. Potatoes Desiree, Picasso or Red Rooster are known to favour high sand conditions as do marjoram, rosemary and sage in the aromatic category. You may even know a friend who has clay soil and you could barter your produce to get the best of both soil textures with minimum effort.

In sandy soil there is always a concern regarding nutrient leaching, particularly nitrogen and phosphorus. If your carrots are looking rather anaemic in your sandy loam then this could be the reason, but do not despair because we can take remedial action to encourage better nutrient retention which will make a massive difference in quality. Consider the left side of our diagram where sandy soil is just not holding together properly. It may have been over tilled with the soil structure completely devastated or just have an extremely high sand content with little soil life. The magic of organic matter and more specifically live compost is that whilst ameliorating the disadvantages of a clay soil it also helps sandy soils improve moisture retention, nutrient retention and aggregation.

In the compost tea preparation here for a sandy soil, a greater fungal aspect is emphasised, to help aggregate sandy particles around the network of fungal hyphae and stable, aggregating products of soil life such as the glomalin 'soil glue'[95] as produced by such fungus. Add to this increasingly favourable earthworm conditions, a soil conditioning crop, and with time success is almost guaranteed. As discussed in Chapter Four it is not without a certain irony that the very organisms you may wish to discourage on your plot, the gastropods, make great contributions to the supply of such useful related substances.

Compost tea application/preparation (Steps 3-4)

A. Same as Orchard Mix but diluted 1:3 (enough to drench). 4.55 litres (1 gallon) per m^2 making sure the soil is moist beforehand.

B. Optional extra: Add proprietary mycorrhiza to bubbler 12 hours before application. Optional extra: Add oyster mushroom spores to bubbler 12 hours before application.

C. (Step 4 on diagram) Mulch of hardwood chips to 8cm (3in) applied immediately after.

It is important to add the amendment of solid organic matter using the mulch above as quickly as possible after the live liquid amendment bearing in mind the short lifespans of many of the organisms involved and to minimise the chances of emergency survival responses, which would slow down the process. The same points with regards to the pH of wood chips and clay soil amendments are also valid here.

A spring legume mix that includes lupin, red clover and crimson clover would be suitable, fixing nitrogen in the soil and adding organic matter (Step 5 on diagram).

A second soil improving crop later in the year from August may include a general winter mix of Italian ryegrass, mustard, red clover and crimson clover for excellent results the next year.

Regarding both clay and sandy soil a degree of patience is required now combined with further soil testing to gauge the soil improvement process. Because our techniques are nature friendly it is very difficult to fail too severely.

Soil remediation and compost tea

Many of the practices described regarding soil improvement are generally applicable in this section also, however in soil remediation, conditions can be rather more extreme. With suitable soil testing and identification, a number of specific and proven actions are possible which fall out of our remit for compost teas and the context of this book. A useful, nature oriented and non-academic text for further reading in this area is Leila Dawlish's *Earth Repair*.[96] Also, the residents of Gentilly[97] did some remarkable work after the hurricane in New Orleans. Their account is a great case study and example of how compost teas were used in conjunction with other processes such as mycoremediation (fungal remediation) using oyster mushrooms.

The chemical fixes often prescribed in conventional literature to solve specific problems can be rather simplistic often resulting in other knock on effects. A key technique in this area is to raise or lower the pH level and in that way unlock what is sometimes a problem metal for removal. In bioremediation, the use of living organisms to remediate soil is an area of activity favoured by survivalists as well as community pioneers and self sufficient growers. A great deal of research needs to be conducted into the history of the site to help establish the type and degree of contamination including the various clues onsite. Sometimes such as with asbestos contamination even breathing masks need to be worn and authorities must be informed. Tests for contamination need to be conducted which generally fall into tests for fuel compounds, heavy metals and toxicity testing. After the right information has been acquired and the necessary steps taken, further tests need to be performed to establish that the site is safe to use for its intended purpose.

Understandably it is important to establish that the ground has indeed been remediated to a safe level before crops are grown. In a recent informal discussion with an English council official specialising in such matters, the author was surprised to find that the officer was rather casual regarding these legal issues with a go ahead style of approach. He was also seemingly very willing to co-operate in the process which was somewhat contrary to what had been imagined. The council worker requested his identity and county were not stated. Subsequently it was found that a number of councils, universities and community groups today are very interested in using compost teas for remediation. Quite a body of activities have built up around disused areas close to motorways, inner city developments and brownfield sites in conjunction with community projects where compost teas are often suggested as a safe and organic first step towards regeneration, aiding in soil aeration and stimulation of toxin processing bacteria.

Keeping within the organic remit is entirely possible using phytoremediation with certain green plants[98] that are known to absorb specific toxins. For instance copper toxicity was discussed on pp.138-9 and a number of known accumulator plants exist as indicated in Chapter Two including yarrow, nettle and sunflower. In agricultural land an increased amount of sodium from fertilisers can contaminate the soil and again a number of common plants can be utilised to remediate such as clover and dandelion. The residents of Gentilly[97] added compost teas to their plots around their sunflowers and Indian mustard a week before they pulled them up and discarded them. The intention was to make the heavy metals contaminating the soil more available for their accumulator plants to absorb.

Other forms of bioremediation also exist with many types of contaminant having fungal or bacterial organisms capable of breaking down or otherwise rendering the contaminant harmless. The Gentilly residents also used compost tea quite frequently when they were aerating the soil to help break down petrochemical contaminants. Generally speaking the more complex and varied the contaminants the more difficult they are to deal with. The Soil Improvement Plan (p.145) could still be applied but with the added stipulation of more research and testing plus a greater emphasis on suitable plants.

In researching this book a theme kept emerging that combining activities can be considerably more effective than doing just one activity and again a similar theme presented itself with regards to remediation. Often because of the nature of today's research specific fixes are tested and the potential of synergistic events often fall through the selective gaze. The particular study referenced here[99] however offers a clear observation that combining mycoremediation (fungal remediation) using liquid mushroom compost leachate with phytoremediation was much more effective than either on their own. Mentioned in Chapter Four regarding crop protection, the oyster mushroom again emerges as an ally to humans[100] when applied to contaminated land.

If a little patience is applied, research also suggests how underestimated our earthworms truly are. There is a considerable body of research exploring how earthworms are capable of remediating contaminated land. A significant recent study from the University of Reading[101] described how over a number of generations, miraculous evolutionary processes selected for traits that support remediation. The broader teleological debate regarding how such a life affirming process can come about without some guiding order is the bane of evolutionary biologists and this is just one more example of this unresolved dilemma.

Within the study referenced the earthworms still endured and enveloped the toxic metal in a special protein which facilitated its absorption by plants which could then be removed from the site. The final revolution projected regarding valuable metals in this cycle would then be to refine these scarce resources for further use.

Troubleshooting

Moving on from remediation, we complete this chapter with a return to a key discourse around compost tea regarding its effectiveness in practice.

Reviewing a number of the research papers from well-respected sources that produced negative or inconclusive results, it is difficult to question the researchers' dedication and genuineness to produce unbiased trials. A key concern however is that it is possible that their own standard procedures are a touch institutionalised and certain oversights undermined their good intentions. Overleaf is a list of suggested flaws in procedure that may guard my readers against making the same mistakes. If you wish to follow up the results of individual papers you can refer to the references at the end of the book:

- Unsuitable spraying equipment. For instance high psi,[102] spray aperture too fine, overly vigorous mixing, CO_2 pressurisation.[103]
- Using poor quality compost. Compost that has been made in a way that has few living organisms in it.[1] Compost that contains chemicals that inhibit growth of beneficial organisms.[1]
- Sterilising live tea or filtering too finely.[104] The practice of sterilisation is in direct conflict with diversity and that of beneficials outcompeting disease organisms. When we reviewed 'tillage options' the above research suggested that more beneficials and pests were better for crops than devastating microorganisms with artificial chemicals.
- Using poorly stored compost. Compost that has been stored anaerobically in sealed or in poorly ventilated plastic bags[105] often observed in standard commercial compost. A comparatively dead or dormant mass as compared to the seething bundle of life which is the bespoke compost of the artisan, fresh from the heap.

Fine tuning

When you have made good quality compost for your tea and you get the brewing right there is every chance that you will have an effective amendment for your crops. Research suggests however that things can go wrong despite the best of intentions. One way to stack the odds in your favour is to invest in a microscope. Good quality equipment can be rather expensive, however if you happen to have a computer you can get scopes at less than £60 that plug straight into your ports. Whilst these products are not up to a professional standard they work remarkably well for our purposes.

In Chapter Ten we will investigate certain areas that are less accessible to conventional scientific enquiry but still may hold validity and further out on the fringes we engage with the folklore that challenges and often confounds scientific rationality, so be warned.

CHAPTER TEN

Alternative Practice and Compost Tea

Fresh worm casts collected locally.

Before embarking further into this chapter, the author would ask the reader to suspend judgement for a moment to consider the following paragraphs.

Over 80 years ago the famous experiment was conducted where protons were fired through a slit and the pattern recorded on a screen to indicate either a wave or a particle imprint.[1] The odd thing about it was that the result seemed to be affected regardless of whether it was observed or not. When it was not observed a wave pattern was indicated and when it was observed the pattern seemed to collapse into a particle signature. There were subsequently many refinements made to this belief challenging study and as the theories altered the results remained very consistent.

Why does this matter?

In conventional science there is a hard line between the observer and the observed. If this line did not exist it would open the door to the existential soul searching that science left behind in the Middle Ages. It is a very convenient line, and has worked very well in countless studies and has been applied in countless applications. For the next 80 years conventional science tied itself in knots trying to maintain the concept of the separation between mind and matter in various forms, but the weight of evidence[2] is going against this orthodox view now. The real questions are the scope of the degree and its implications. Whilst it is extremely inconvenient to accept, we are now entering a post-enlightenment phase where our place in the universe needs to be re-assessed. The ideas that fuelled the industrial revolution and led onto scientific materialism served a purpose in their time but also led to a dogmatic and prosaic view of nature and the world. Arguing that mind does not affect matter is now a non-rational position.

The dominant ideology in science today is that of 'scientific materialism', a superficial type of monism. To such a materialist this is not an ideology at all but just how it is. The materialist asserts there is just matter and that is all, but what really distinguishes this type of monism from the other types is a willingness to stay as close as possible to what seems reasonable in Western culture. For instance, a materialist may state that all we are, are our physical bodies and there is nothing beyond that. Any theory of mind, if it cannot be totally denied, is the result of a physical process that could some day be located in the physical mechanisms of the body. There is no proof for this view other than a common sense belief and other theories tend to be discounted without investigation because they seem too outlandish. The superficial quality of this materialism as suggested by the author is because the nature of reality is barely even entertained, such is the confidence of the materialist.

If the theory stays as close as possible to day-to-day reality, that is sufficient to continue regardless. The problem that exists today is that if the solid peer reviewed studies are considered rationally without reference to our cultural milieu, the results suggest our day-to-day experience of the world is a long way from what our scientific findings in the slit studies suggest.

An important assertion of the materialist is that we know almost everything there is to know about science and it is just a simple matter of tying up a few loose ends. Such a view is very dependent on the current stage that science is at. If the materialists are correct, then this observation is less significant. Despite this confidence, in a recent poll of top experts regarding in part the 'measurement problem' briefly described on the previous page, there was little consensus.[3] There are numerous examples of science's profound limitations in what is known and there is no point listing every one, but it is difficult not to highlight from a related context the fact that Dark Matter has only recently been officially acknowledged to almost certainly exist which is a monumental oversight given that it is reputed to fill around 80% of the universe. The reason for the uncertainty is that there is no direct proof, only very strong indirect data.[4] Also it is still rather unclear what Dark Matter actually is. Without wishing to digress into the various nuances of this labyrinth, thus far is sufficient for the current discussion.

The cycles of the Moon had a strong influence on the growing practices of the past and many growers today are rediscovering its subtleties. The waxing or waning Moon can be discerned in a glance by extending an imaginary line, to cut across the dark side of the Moon, either making a 'p' (Waning) or a 'd' (Waxing) as here.

In contrast the 'Post-Materialist' view that the author confesses to endorse, suggests there is still much to be known and that if matter exists, in the way we normally perceive it, it is indeed affected by mind. Such a view does not necessarily contradict the previous solid science that has been established, rather it is viewed with a different hue. A further contention that may find an appropriate context here is that our human intelligence, culture and perception may limit us in some way from ultimately understanding the mysteries of nature. A similar suggestion was put rather more eloquently by Werner Heisenberg, one of the founding pioneers of quantum mechanics theory: 'Not only is the Universe stranger than we think, it is stranger than we can think.'[5]

Over the last few decades many high profile scientists have broken ranks and endorsed a post-materialist view but they are still very much in a minority today. Any break from the institutional view tends to be professional suicide. To add further fuel to this anti-materialist stance none other than Nikola Tesla, the father of modern electricity, suggested:

> The day science begins to study non-physical phenomena, it will make more progress in one decade than in all of the previous centuries of its existence.[6]

(It was a tragedy that this visionary died in poverty and did not succeed in applying the full potential of zero point free energy to liberate humanity.)

The quote above is not intended to convey that all post-materialist thought is necessarily dualistic, hence separating mind from matter. For instance, certain variations of post-materialism suggest that consciousness is a type of physical substance that would help account for the seeming anomalies in the slit studies. The author's own position would stand not far from some Buddhist thought where a kind of panpsychism exists with consciousness existing in varying degrees throughout the universe. Again for the sake of brevity and the discussion in hand the author need only to step thus far into this ontological wilderness to develop his point. No unequivocal answers are offered regarding the metaphysical and practical implications of this view.

A rationale for alternative practice

Moving past the rhetoric it is clear that Western science at present has not unravelled every mystery in nature and is a long way from establishing that consciousness fits in with its worldview. No bold claims are made in this rationale, only rather in the manner that a lawyer in a legal case may argue that a reasonable doubt exists to overturn a case, so a suggestion here is made that because the fundamental nature of science, reality and consciousness are so unclear it is rather naively hypocritical to cut out unequivocally many non-rational practices on the grounds of them being unscientific. If they don't work, that is another matter!

The butterfly effect and compost tea

The studies on intent, consciousness, mind and so forth affecting matter seem counter-intuitive in the extreme to a community brought up on classical mechanics where everything seems to be understood and things fit together neatly. Because such findings fly so severely against many people's worldview, it is one of the many reasons there has been so much resistance towards them, not least in terms of research funding where conservatism rules. A long term study at Princeton University was conducted that endeavoured to persevere in spite of this institutionalised prejudice, established that human intent could seriously affect complex machinery[7] and conclude that in the future such systems should be made to take this into account.

Now consider the concept of the butterfly effect. It is well established in conventional science[8] and could be described as the phenomena of very small forces having profound effects on comparatively massive systems. The classic example is the notion of a butterfly wing setting the forces of a storm in motion. Astute readers may already have an inkling of what the author is alluding to now within the context of the compost tea bubbler. To reiterate, there is nothing controversial with regards to the notion of complex systems being affected by very small-scale forces even at the quantum level[9] and science has also established that consciousness, at the very least, can affect matter on very small levels. When these two notions are combined, however, the possibilities are endless and highly controversial which is why, despite appearing logical, it has been placed in this chapter that is dedicated to fringe related practices.

Following this line of thinking in the complex and chaotic system of the compost tea bubbler, when brewing, the previous peer reviewed research findings can infer that cultivating a confident, positive approach and really paying attention to the process might be better than the standard scientific approach of scepticism and *attempting* to cultivate detachment. Whilst these suggestions certainly cannot do any harm to the brew, there may exist a metaphysically challenging opportunity that it could give the process an extra something.

This chapter is split into two parts. In the first we describe a method of making Indigenous Microorganism Tea, IMO for short. In the second part we look at working with natural cycles with regard to fertigation (portmanteau of fertilisation and irrigation) and composting. Attention to the rhythms of nature and how different organisms in the food web respond is an intrinsic part of Natural Farming approaches.[10] The holistic water test overleaf moves away from the procedures in conventional science.

Of course we are very familiar with the influence of the Sun on the seasons but the effect of the Moon is less obvious at least for modern man. Since the Moon has a profound influence on the movement of water on Earth it is reasonable to postulate that it also has an influence on plants that are approximately 80% water and have adapted over many millennia to exploit to their best advantage whatever conditions are available to them.

Anyone opting for a regime of synthetic chemicals for crops could dispense with the end of this chapter because it appears that the nuances that exist in plant metabolisms from the influence of the Moon are obliterated by non-organic practices.

> I came to the conclusion that mineralized soils hardly reacted to these cosmic rhythms and their fine influences, whereas a humus-filled soil of whatever soil type was a good mediator of these forces.[11] Maria Thun (1964)

In Chapter One photographs of biocrystallisations of organically grown produce were compared with non-organic produce. A further test might provide interesting results by comparison of organic produce and organic produce grown according to the lunar practices discussed later. Would the patterns in the produce from lunar sensitive gardening become even finer and more elaborate?

The Natural Farming approach has strong biodynamic influences, the basic idea in IMO preparation being to multiply beneficial fungal moulds from the local area which are then made into tea. Since it was developed in the Philippines the choice of ingredients and processes have been altered so that they are more applicable to a temperate climate. Improvisation and flexibility are encouraged within this system and so as long as the principles of this approach are still applied there is considerable room for experimentation.

According to advocates of the Natural Farming approach such a tea:

- Increases the decomposition rates of soil organic matter leading to better nutrient availability and a more nutritious crop,
- Improves the food plant's yield,
- Decreases plant pathogens,
- Increases natural plant defences.[10]

To get the full benefit of such an approach it must be integrated with other organic practices.[10] Natural Farming advocates growing a multiplicity of cultivars and variations in crops in order to foster biological niches to host a diverse range of organisms. They suggest that such a system is more resilient than the monocultures favoured by conventional agriculture.

A holistic water test

In cultures that understood water in a dynamic way, the way it flowed gave an indication of its uses. For instance, when the Maoris of New Zealand looked at water in their streams, rivers and lakes, they allocated different roles by the health and status of the 'Mauri'.[12] This conception is similar to the Steinerian emphasis on vitality and translates to high productivity and life in the areas with healthy Mauri. The importance of water as a vehicle of Mauri is of key significance:

> Maori conceive that each waterway carries its own mauri, guarded by separate spiritual guardians and tribal caretakers and having its own status or mana. A water body with a healthy mauri will sustain healthy ecosystems, support cultural uses and mahinga kai (food sources), and be a source of pride and identity to the people.[13]
>
> Immanent within all creation is mauri – the life force which generates, regenerates, and upholds creation… elemental energy derived from the realm of Te Korekore (Potential Being), out of which the stuff of the universe was created.[14]

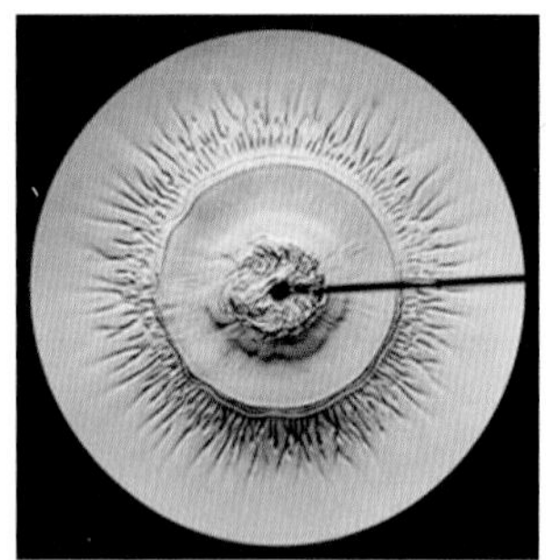
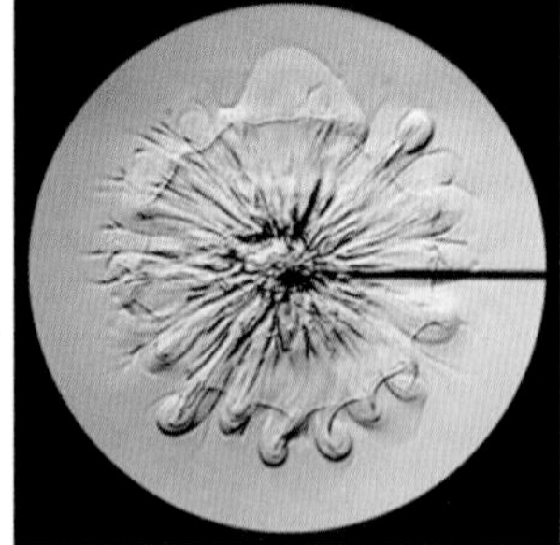
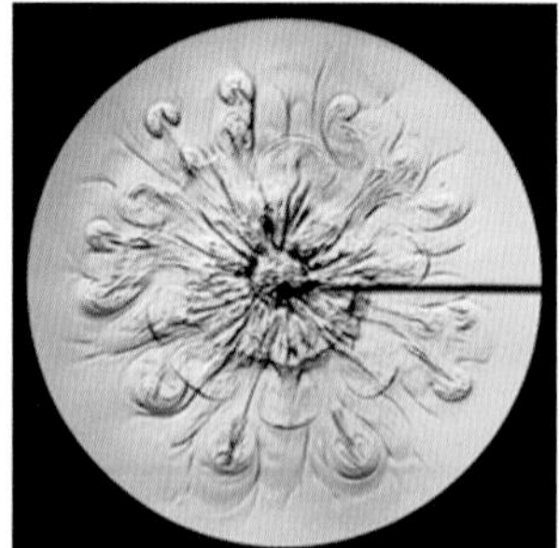

From left to right, drop pictures of water with detergent, processed tap water and spring water.[15]

Masaru Emoto, the originator of the drop pictures above, found that if water is dripped onto a solution of water and glycerine, certain patterns emerge that vary greatly with its history and level of contamination. Whilst Emoto destroyed his own credibility with unsound conclusions[16] the value of his results invites further enquiry. Viewed intuitively, the way water flowed in the spring water sample showed a greater integration. In the processed water the patterns of flow were more discontinuous and in the picture with the detergent indications of flow appeared absent. What the implications are for the way finely involuted flows or restricted flow behaviour impact soil and plants is an open question. In the soil, a restricted flow pattern might indicate a less than uniform distribution of moisture and in plants a growth that is more restricted. Returning to Chapter One, only a small mental leap is needed to link the flow forms on the previous page with the biocrystallisation patterns of organically grown produce. What these flow forms could mean for agriculture is rather incongruent to the blinkered mental processes of conventional planners and industrial processes related to the agricultural infrastructure of today, but they can be useful to those taking another direction.

It is easy to do this test yourself to get a subjective indication of water quality. When using water for compost teas we undoubtedly need water that makes the finest, swirling and most interconnected patterns. A dropper, petri dish and glycerine are easily obtainable from most pharmacies and the test is simple to perform (but awkward to photograph). Try the tests with water directly from streams, springs or water butts and see how chlorinated drinking water performs.

Preparation instructions for IMO tea

Ingredients

500g (17.6oz) of local soil with worm castings (see overleaf)
250g (8.8oz) powdered rice bran
1kg (35.3oz) unsulphured molasses
7 litres (12.3 pints) local rainwater

Equipment

1 bucket
1 stick, 2cm longer than the diameter of the bucket (topbar for suspending the mixture)
20 × 20cm of thin cloth (e.g. material used for lace curtains)
50 × 50cm cloth (e.g. burlap)
Clean gloves
50cm cleaned stick
Large clean wooden spoon

Keep all utensils away from cleaning products with harsh chemicals. Baby soap or the soapwort solution as described in Chapter Six can be used, but make sure your equipment is thoroughly rinsed, ideally in rainwater.

Finding the right local soil

Firstly soil needs to be collected close to the intended place where you will be growing crops and this soil needs to be high in microorganisms. Also the area should not currently be under cultivation. Under cultivation in this instance was taken to mean 'just been dug'.

In line with other thinking, an excellent indication of high beneficial microorganism activity is the presence of earthworms. In the original script it was suggested the key ingredient, worm casts, should be taken from around a bamboo tree. Further research revealed that the bamboo has its own worm that can only be found around bamboo trees. Whether the casts of the bamboo worm have special properties not found in other casts is yet to be investigated.

Since bamboo is not the most prevalent plant around in temperate areas the Jerusalem artichoke (*Helianthus tuberosus* below) in its dormant phase was chosen as an alternative. Originally native to North America, it has naturalised across temperate zones in both northern and southern hemispheres. Invariably given good organic practice worms congregate under the vigorous root systems of these tubers as indicated below in the photo. The reasons why are not totally clear but it could be speculated that because worms have a preference for darkness and partially decayed organic matter, artichokes are a good choice. They may be attracted to the high inulin content of the artichoke that sweetens the soil around the root and fuels microbial activity. During winter the artichoke's bulky stems and leaves break down over a few months providing worm food and an insulating mulch. When you consider the worms find refuge from the light under the artichoke tubers and the proximity of the roots to the surface it is understandable why they favour artichokes.

A good time for preparing this tea would be very late winter just before the artichokes come out of dormancy when it can be put to good use shortly after. If you are making tea in the summer, when artichokes are very much out of their dormant phase, you can instead use compost at the end of its mesophilic stage (see Chapter Eight) as most of this material will have passed through the gut of a

Freshly upturned artichoke tubers with worms.

worm many times by then. Assuming the compost is connected with the earth as recommended, the edges of the compost heap are a good place to look as it will be less disturbed and have more fungal activity. It is also possible that these areas show activity from *Lumbricus terrestris* (common earthworm), a worm that adorns its holes with small pieces of organic matter.

All the processes can be performed outside and no power is needed. This has some advantages to it in areas where access to electricity is poor such as allotment plots, but it may be impractical to prepare outside in freezing weather. An unheated greenhouse or shed is usually a few degrees warmer and this should be acceptable except in the coldest periods where in any event a compost tea would not be of much use outside.

Preparation method

1. Use gloves because the material should not be handled with bare hands since the microscopic organisms that naturally live on human hands are thought to transfer to the substrate and interfere with the IMO processes. Break up the soil you have collected so the material is smooth and uniform. If there are stones use a garden sieve to screen them out. Mesh size 3mm should be fine.

2. Mix your soil with the rice bran so it is of a uniform consistency [A, overleaf]. Using rice bran as a growing medium is well-known outside of Natural Farming also.[17] Add enough water to the degree that you should be able to form a ball with your mixture, but when you apply pressure it should crumble easily. The mixture should still feel somewhat dry.

3. Wrap the mixture in a thin cloth and place it somewhere cool and shaded for three days [B-D]. It can be tied underneath to prevent the culture from falling out. The original text recommends placing it under a mango tree, however since mangoes are tropical plants and cannot grow in temperate areas, one of our indigenous apple trees or hazels would be an excellent choice. Alternately if you have a fig tree you could place it in the shade of the large leaves. The important thing is to not let your package freeze or get too dry as this will interfere with the process.

4. On day four inspect the mixture for moulds. Noticeable from the photo is that the areas of the culture more exposed to air have the most vigorous moulds; compare the top of the bundle [E] with the base [F]. According to Jenson *et al* white moulds are best, orange and blue moulds are still good, but black moulds are undesirable.[10] A few are acceptable as long as visually they are not the most dominant strain. If all the mould strands are not very noticeable leave up to another couple of days, but if you leave it too long the culture will become overly anaerobic which will be noticeable by the foul smell.

5. Now use the clean stick, place the mixture on a clean hard surface [F]. Using the stick break up the mixture to pieces of 2cm diameter [G-H]. This

Steps A-H for IMO Tea preparation.

is thought to stimulate the IMO process, aerates the culture by increasing its surface area and keeps the material distant from the organisms on human skin. You will notice that the culture is considerably more flaky in texture especially around where there are greater concentrations of mould.

6. Make sure the mixture is not too dry; sprinkle water over the surface if necessary.

7. Return the pieces back into the thin cloth being careful not to compact it, re-tie and return it to a cool, shady place.

8. Prepare your bucket as in Chapter Nine with a top bar for suspending your culture (p.127). We are not planning to use a bubbler on this preparation so the other holes described are not required.

9. Take your bucket and mix up 7 litres of rainwater with the 1kg of molasses. If you don't have any rainwater fill up your bucket with tap water, stir vigorously and leave overnight.

10. Take the cloth with your culture and suspend it under the topbar so that it is ¾ submerged in the solution of molasses. The use of covering also serves to trap carbon dioxide, reduces contamination, minimises drying and stimulates vegetative growth.[18]

11. Cover the top of your bucket with a large cloth. This will help to protect your mixture from the attention of flies et cetera that will be attracted to the sugary solution.

12. Stir for 10 minutes, two times a day for 10 days. Stir in a clockwise direction gradually increasing speed until a vortex forms then stir anticlockwise to form another vortex [I]. The same stick for breaking up the soil in step five can be used. In practice it is easier to remove the top bar temporarily and hold the cloth to the side during stirring. This is better than simply removing it because on a purely practical level more fungal strands will be liberated from the cloth. If you feel that the cloth is interfering with the symmetry of your vortex or the chaotic flow forms emerging momentarily when the stirring direction is in flux, you can alternate days of putting it to the side and removing it temporarily to a sealed container. If you like to operate in what science has proved only, then the stirring is still good practice because this provides excellent aeration of the mix. It is important during this stage to be on the alert for the mix turning too anaerobic which is undesirable for this process. If it is starting to

I – Clockwise vortex. J – Intermediate stage.
K – Anticlockwise vortex. L – storage for live IMO.

smell unpleasant it means you are not stirring the mixture enough.

13. After this period strain the liquid through a fine cloth into a glass bottle. Because the liquid is live it is important to not fully seal the bottle, as gas needs to escape. Using greaseproof paper with an elastic band, rather in the manner of jam preparation, works very well [J].

14. You can keep the solution live for up to six months with the proper treatment. Maintain aeration by opening it up and shaking it every week and feed it with molasses to 20% of the volume of the liquid every month. If the solution starts to give off an unpleasant smell dispose of it onto the compost heap.

Variations on this theme could involve using soil from around favoured crops for the same type of crop, collecting from forests which may be useful for introducing mycorrhiza, collecting from beside streams, under rocks, or taking selections of soil from various areas in your plot and combining them.

Fertigation and watering with the cycles of nature

> On the days of the Full Moon, something colossal is taking place on Earth ... these forces spring up and shoot into all the growth of plants, but they are unable to do so unless rainy days have gone before.[19]
> Rudolf Steiner (1924)

In this section we move from the fringes of science into folklore. On the further reaches of this divide is an extract below from *The Lost Secrets of the Garden* which suggests the Moon alters the qualities of light reflected back to Earth from the Sun and the cosmic sources. The author, Greg Willis, wrote this account as if he were interviewing the Moon; whether this was a narrative device, or he believed himself to be channelling the voice of the Moon, is unclear.

> This mixture of light provides a special life force for plants and animals. This special, processed and mixed light works as a growth regulator and enhancer for plants and animals. Light from all the planets in our solar system is reflected off the Lunar Body onto the Earth. Light from other solar systems is also reflected. This mixture is then processed by the Lunar Body, the Moon, and reflected back to Earth. Certainly the light from Mars, Venus, Jupiter, Saturn and the others is reflected directly on to the Earth too. But the Moon is able to provide a special influence with this mixture. That is one role of the Moon.[20]

Many of the traditional lunar practices came about when people were less insulated from the cycles of nature and more directly dependent on them. There is a consensus around the world with similar approaches to using the cycles of the Moon existing in places that are unlikely to have had any direct communication.[21]

In conventional Western science there is also a growing body of knowledge that lunar cycles can indeed be used to enhance crops. We will outline the most clear cut areas that can have a powerful impact on your growing practices and some of the areas that are more controversial.

There are a variety of cycles associated with the Moon and agriculture which we will now describe to disperse any misunderstandings which have often made this topic more confusing than it need be. From this point we will proceed to how we can apply these cycles.

Synodic cycle

The synodic cycle is usually 29.53 days[22] and is the most noticeable cycle as it is related to the amount of reflected sunlight coming to Earth from the lunar surface. It is illustrated below and is a very significant one for traditional growers.

The synodic cycle with compost tea suggestions for lunar synchronised applications

	Days 1-7	Days 8-14	Days 15-21	Days 22-29
Moon phases	●●●●●●◐◐	◐◐◐◐◐◐○	○○◑◑◑◑◑	◑◑◑◑●●●●
Moonlight	Waxing: increasing moonlight		Waning: decreasing moonlight	
Lunar gravity	Decreasing	Increasing	Decreasing	Increasing
Application	½ foliar spray ½ root drench	Foliar spray	Root drench	Root drench
Compost tea example	High potassium Low phosphorus Increased liquids Kelp tea (Chapter 6)	High phosphorus Low potassium Increased liquids Root and shoot stimulator (Chapter 6)	Reduced liquids Feed the soil (Chapters 9+10)	Reduced liquids Feed the soil (Chapters 9+10)

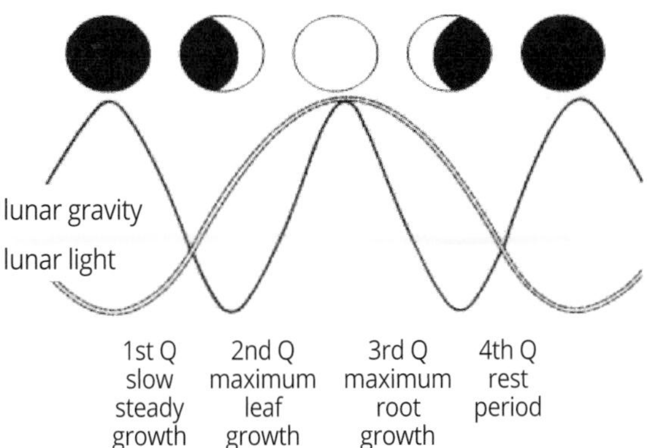

Right: Comparison of lunar light and lunar gravity with the synodic cycle adapted from 'Lunar Growth Cycle' by the Oregon Biodynamics Group.[23] 'Q' represents lunar quarters of the cycle.

Plants absorb most of their nutrients in liquid form, so variations in water intake and metabolism will also reflect their nutrient usage. In studies of carrots and potatoes[24] there are statistically significant peaks in their metabolism in the days preceding full Moon and especially around two days before. Extrapolating these findings to fertigation regimes, an emphasis on feeding and watering crops rather more during the waxing Moon (up to full Moon) than the waning Moon, should have positive results on your root crops.

When considering major nutrients (NPK) some useful results were found for potassium and phosphorus. Interestingly Lai[25] discovered plant uptake of phosphorus was highest during full Moon and lowest during new Moon, but potassium uptake was highest during new Moon and lowest at full Moon, so positive results can be expected factoring in these results into fertigation regimes as suggested in the table on the previous page.

Looking at research by Chow and Brown on beans and water uptake indicates peaks in water uptake at each quarter of the Moon's phases[26] shown on the previous page, so this would be the best period to apply liquid feeds.

In earlier chapters we discussed different approaches. One key departure since most organic plant nutrition is mediated by microorganisms was a shift in emphasis from feeding the plant to nurturing the soil and food web. Considering the peasant wisdom, the consensus is to make soil amendments over the waning Moon. Over the waxing Moon plants and organisms are more active so it would follow that reserves could be built up in the slightly quieter waning period, in preparation for the next burst of activity.

Syzygy inequality cycle

Closely connected to our introduction of cycles is the related topic of conjunctions. This is where in one planet's orbit a straight line is passed in between Earth typically and another planet. A classic and visual example of a conjunction is an eclipse. One of the most significant for tides is the Sun-Moon conjunction, which causes the highest tides[27] and is known to be exploited by certain marine organisms during their life cycles.[28] An interesting enigma is that this lunar sensitivity is also exhibited by many terrestrial organisms that have less obvious tidal associations.[29] One to three days after each full Moon and new Moon there is a Sun-Moon conjunction. The highest tide alternates between full Moon and new Moon every seven months which is syzygy. The highest tides of the year can be found during the spring time.

Although no crop studies have been found that relate to the effect of syzygy, since plants are influenced by the tides it would follow that this may be significant and may be an overlooked factor in some crop studies that have produced equivocal results.[30]

The sidereal and tropical cycles

The ancients mapped what they saw as the passage of the Sun and cosmos as it revolved around the Earth. We came to revise this more recently. As observed from Earth the stars were divided into the 12 constellations which returned to roughly the same place every year. This formed the basis of what is called the Tropical Zodiac today. The Aries constellation marked the beginning of Spring as the Sun's inclination increased in relation to the Earth. This was followed by Taurus, Gemini, Cancer, Leo, Virgo, Libra, Scorpio, Sagittarius, Capricorn, Aquarius and finally Pisces. This in itself was not necessarily a superstitious concept, and closely relates to what we now know as the 12 months of the year. These symbolic associations became hardwired in the collective psyche of the West and remained, immutable, for thousands of years. The only problem with it was that the relation of the Earth to the cosmos changed by precession. Every year the Earth wobbles very slightly and the alignment of our equinoxes to the constellations alters by a fraction of a degree. It was Hipparchus who first noticed this in second century BCE, but nobody did anything about it, so today the constellations are out of alignment with the Tropical Zodiac by around 25°. What was Aries when the zodiac was first conceived is now Pisces. A swashbuckling, albeit heretical alternative would be to retain these ancient archetypes, but join up the dots in the sky in a different way to realign them to the sky of today, then every few hundred years redraw them again. In this way the ancient symbols of the zodiac stay in the same place every year but each symbol transforms in its outward appearance.

In the Sidereal Zodiac below, the link with the actual constellations as the Sun and Moon pass in front of them is retained. This is the zodiac of the Eastern traditions in various forms and the choice of many modern astrologers. The point of departure from strict factual observation towards folklore was the attribution of metaphysical qualities to the different signs – Earth, Water, Air and Fire as shown overleaf. A powerful case has been made for seeing nature in terms of the medieval elements or more spiritually the 'formative forces'[31] however it would be expected that the association becomes increasingly specious when plants sown or planted on a certain day acquire astrological qualities associated with that day. What is surprising is that a number of studies[30] have verified that crops do show greater yields when planted according to this non-rational practice. It is of little surprise that this bears the trademark of Steiner's influence of being effective but not very explainable and points to a greater science that is only partially formed currently.

It takes the Moon 27.3 days[32] to make the above passage and return to the same place in the sky and this forms the basis of the Sidereal Cycle as used in many lunar gardening calendars.[33] In this system the constellations are divided equally and each lasts between two and three days. At first sight this might appear contradictory but the day difference is accounted for by the fact that the Moon orbits elliptically around the Earth.

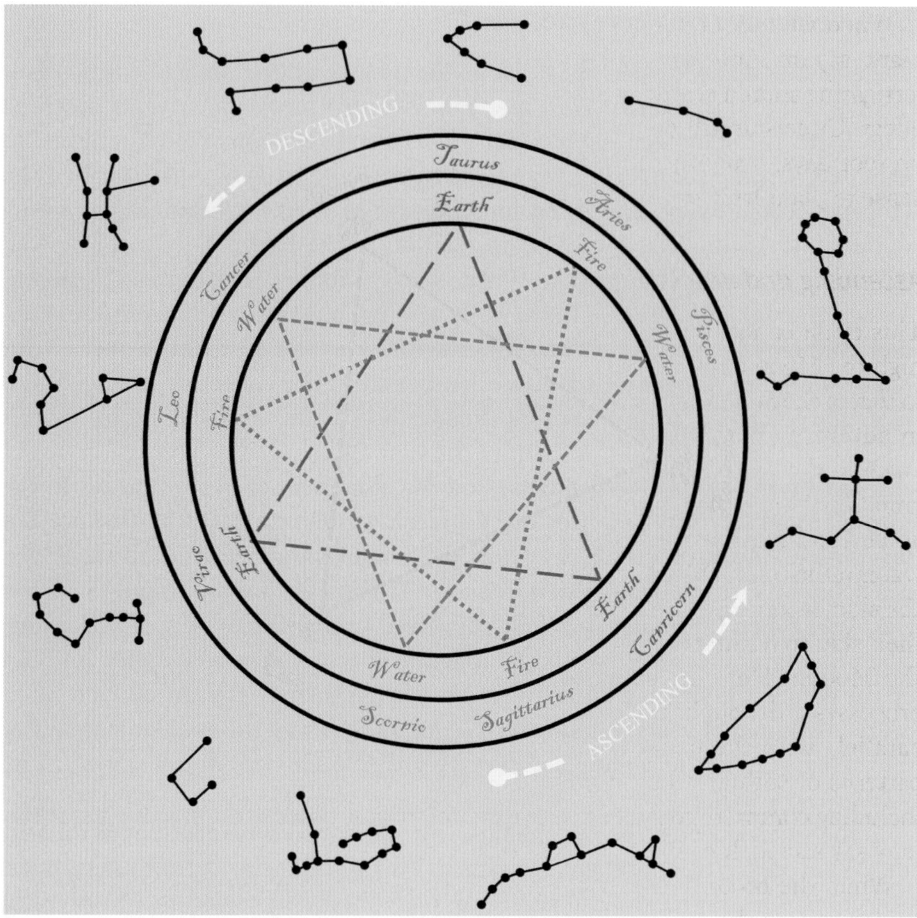

The Sidereal Zodiac with associated constellations. Ascending and Descending positions are marked in white.

In the biodynamic calendars[34] most popularly altered by Maria Thun, the well-known lunar researcher, the divisions of the zodiac are unequal which attempts to astronomically reflect more closely the varying sizes of the constellations.[32]

The 12 constellations of the Western Zodiac (both Tropical and Sidereal) with elemental attributions and planting days

Metaphysical Quality	Constellation	Planting Days
Earth	Taurus, Capricorn, Virgo	Root days
Water	Cancer, Scorpio, Pisces	Leaf days
Air	Gemini, Libra, Aquarius	Flower days
Fire	Aries, Leo, Sagittarius	Seed days

When we talk of planting days the suggestion in the literature is that if you plant, say on a root day, activities associated with that plant such as watering or fertigating should be done on another root day which is supposed to reinforce the effect. Understandably from this perspective we would want to plant root crops on root days, leaf crops on leaf days and so on. For example it would not make sense to plant fruit trees on a root day if you want good harvests of fruit.

Ascending and descending Moon

This cycle is not the same as the waxing and waning Moon, and refers to how high the Moon climbs in the sky. The ascent and descent of the Moon is similar to that of the Sun except whilst the Sun takes a full year (rising in spring, descending in autumn), the Moon takes only 27.3 days.[32] Alluding to normal spring and autumn this is sometimes called Lunar Spring and Lunar Autumn. In a rather grandiose way it has also been described as the Earth 'breathing in' during the ascendant and then out when descending.[35] A further analogy that emphasises water rather than air might be to compare the lunar influence on soil processes to the skin of some amphibians. These creatures have a strong affinity to water, use their skin in respiration and some have been proven to align their life cycles with the Moon.[29] Here moisture is key to staying healthy and facilitates aerobic processes. Although beyond the scope of this account there exist intriguing parallels with the hexagonal structures evident in dehydrated soils, the flow patterns of spring water and the skin structures evolved by many amphibians including common frogs, which are very much associated with this unique, some may say sacred, life giving substance.

When the Moon is ascending, water and dissolved nutrients are thought to be drawn into the aerial parts of the plant.[36] Any foliar sprays applied over this period could have an enhanced effect because the nutrients inside the plant are more easily available to combine with the sprayed nutrients for use in metabolic processes and so on. This would be especially good for feeding plants that you grow for their aerial parts such as beans or seeds.

The descending Moon is a suitable time for soil nurturing amendments, such as described in Chapters Nine and Ten. This process is also acknowledged in the sidereal calendar as the Moon passes in front of constellations that are seen to have different inclinations in the night sky, highlighted in the diagram overleaf. According to Michel Gros, of lunar calendar fame, the best period for watering is when the Moon is descendant in Virgo, Gemini or Libra.[36]

It is important to note these effects are only relevant in the Northern and Southern hemispheres and diminish in the tropics where the synodic cycle is more pronounced.[36]

Now that the main cycles of note have been outlined, there are some special points relating to the Moon's passage which we will look at.

Moonrise and moonset

Aside from the numerous anecdotal reports, only one serious study was found on moonrise which gave a strong indication that it has an important impact on plants. In Colin Bishop's rigorous study[37] it was found that radish sown in the first hour of moonrise had the highest yield after six weeks with lesser peaks when the Moon was at its lowest point and highest point. Furthermore he also found that sowing on root days (Sidereal Calendar) gave better yields than seeds sown on flower, leaf or seed days which gives further credence to other related findings.

The lunar day, from moonrise to moonset, is approximately 24.8 hours[27] and is an important period for other lunar events simply because the influence of the Moon is less obstructed. During new Moon, the moonrise comes up near sunrise, and the Moon sets around sunset. Each day during the Moon's waxing it rises on average 50 minutes later throughout the day up to full Moon. When the Moon starts to wane, moonrise occurs after sunset and gets progressively later until we are back at the new Moon. Towards September and the autumn equinox the Moon only rises around 30 minutes later each day until we have the Harvest Moon where the Moon is brightest and rises just after sunset which traditionally aided the farmers.

North and south nodal points

Otherwise known more colourfully as the Dragon's Head and Dragon's Tail, these latter labels have a dynamic connotation since the Dragon's head indicates the point when the Moon begins its ascent and the tail its descent. This is sometimes depicted as a dragon continually devouring itself from tail to head in a never ending cycle.

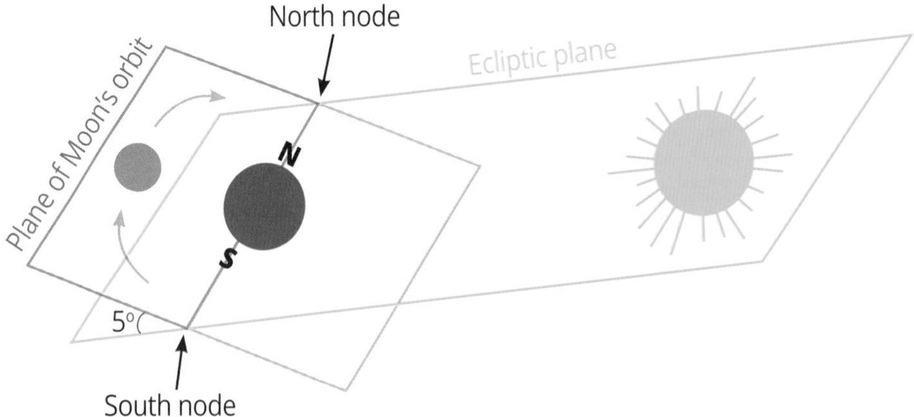

The two nodal points in the Moon's orbit. Also known as the ascending (S >N) and descending (N>S) node.

These nodes are possibly one of the least relevant and most difficult aspects of the Moon cycle to visualise, without a diagram, but it is useful as a marker to indicate a shift in the Moon's inclination. There is little evidence for the impact on growing plants as the Moon passes through these points, but it often appears in lunar calendars. According to Maria Thun and other agricultural astrologers, either node can be unfavourable to plant growth[32] so seed germination or any other plant related activities should be avoided during those periods.

Apogee and Perigee

The orbit of the Moon around the Earth is elliptical and takes approximately 27.3 days.[32] The perigee is the point where the Moon is closest to the Earth and the apogee the furthest. When the Moon is closer to the Earth it understandably appears larger but not to any great degree.

With the Ascending and Descending Moon we have the North and South Nodal Points. Perigee and Apogee are similar to the latter points except we have no corresponding terms for the continuum of proximity which we will now name Approaching Moon and Retreating Moon. At Perigee there is a moment of flux where the Moon alters its relation to the Earth from approaching to retreating and the opposite applies to Apogee. Manuals on lunar gardening suggest that this period is unfavourable to gardening practice[38] by around three hours either way of these points. This would make sense in the same way as the Nodal points are unfavourable but solid evidence is very limited.

When considering the effect of approaching and retreating Moon this could be thought of as a volume control to whatever other effects of the Moon are occurring; when the Moon is closer the effects increase and further the effects lessen. There is very good evidence on the approaching Moon leading up to perigee which is indicated to be the most consistent[39] lunar aspect of all, with greater yields in all crops tested. This included positive results for beans, potatoes and radish.[39] With the retreating Moon leading up to Apogee the least growth was found.[39]

Cycles combined

Considering each cycle in isolation can give a distorted view of the Moon and how plants and other organisms interact with its energies. In the past certain special periods in the Moon's traverse were holistically labelled in a way that revealed its relation to solar cycles. The Harvest Moon occurs only once in a solar year and other Moon names are often very informative and cogent within their respective cultures, such as the Samoan Palolo Day. A strong basis in fact was found for this festive day, which gave a statistically significant[28] indication of the possibility of Palolo worm swarms in the third quarter of the Moon during autumn, which the local fishermen sustainably harvested as a special delicacy.

We finish this section with a consideration of combined cycles when using sprays or drenches that could have the synergetic effect of assisting the draw of water and nutrients into the roots, the aerial parts of plants or when soil processes gain prominence. These are extrapolations based on available data of lunar influence and unproven by rigorous scientific method.

Combined cycles

		Enhanced
Dual	Waxing and Ascending	Foliar sprays
	Waxing and Moon High	Foliar sprays
	Waxing and Approaching	Root drenches
	Waxing and Moon Low	Root drenches
	Waning and Moon Low	Soil amendments
	Waning and Retreating	Soil amendments
	Waning and Descending	Soil amendments
	Sun-Moon Conjunction (Highest Tides)	Foliar sprays
Triple	Waxing, Ascending and Moon High	Foliar sprays
	Waxing, Descending and Moon Low	Root drenches
	Waxing, Approaching and Moon Low	Root drenches
	Waxing, Descending and Approaching	Root drenches
	Waning, Retreating, Descending	Soil amendments
Quadruple	Waxing, Ascending, Approaching and Moon High	Foliar sprays
	Waxing, Descending, Approaching and Moon Low	Root drenches
	Waning, Descending, Retreating and Moon Low	Soil amendments

The composter's moon

> The Moon showeth her power most evidently in those bodies which have neither sense nor lively breath; for carpenters reject the timber of trees fallen in the Full Moon as being soft and tender, subject also to the worm...[32] Plutarch (Roman Historian A.D 46 - 120)
>
> There is more water in trees and grass and all plants towards the full moon than towards the New Moon. They would not cut towards Full Moon, the timber was not as good. If we cut hay we also cut as much as possible towards the New Moon and not towards the Full Moon. We get much better hay that way.[40] Alex Podolinsky (Biodynamic Farmer, 1990)

Research on felling periods,[22] the traditional wisdom of the wood cutters and folklore on harvesting, suggests that material harvested in the waning Moon stores best, so by default the waxing Moon is worse.

The implications for composters were not considered. For the would-be lunar composter, matters are considered from the opposite direction: he or she wants composting material to be a beacon for organisms in the food web and the composting material to break down effectively. Following this line of thought if there is more water, metabolic and microbial activity approaching the full Moon, when living material is collected during this period, there will be more activity from decomposers and hence the composting process will get off to a flying start. At least 600 organisms have been found to have intimate associations with the lunar cycle in feeding or reproduction. A topical example is that of the bark beetle, *Pitogenes* spp., which has been proved to preferentially target trees felled just before the full Moon.[39]

Certain research suggests that on a full Moon, because of the increased visibility, many small mammals run for cover to minimise predation;[38] what better place for them to hide than a covered compost heap? In tropical areas research material contradicts these findings, asserting that many organisms do indeed venture out.[41] We suggest here that a different kind of dynamic may be in operation. In these areas there are an exponentially greater number of organisms on the same trophic level all around at the same time and they like to come out in the full Moon.[41] We suggest that a process of inundation is working as in the well-known theory involving numerous organisms of the same species arriving/emerging/hatching simultaneously, predators are overwhelmed and significant amounts of their prey survive.

With regard to composting activities we recommend doing your weeding a few days before the full Moon and finishing constructing your heap just before full Moon while it is still waxing. In this way there will be elevated moisture levels in the material, elevated potential for decomposition and greater numbers of small organisms stimulated but undercover ready to begin the process.

Returning to the Sidereal calendar and planting days we introduced earlier, extending this same conception, the metaphysical element most associated with the compost heap is Earth, so we want to set up a compost heap on an earth day and turn it on an earth day. If we use compost activators we collect, prepare and apply on a Fire day. For increased rotting also collect fresh composting material on Water days. This was originally suggested by Gros as a Moon sign to avoid in the context of harvesting.[36]

When we come to our heap and find for whatever reason it is too wet, then we turn on an Air day, too cold – use a Fire day, too dry – use an Air day, but always default to Earth days.

If we want to take a further step down this astrological avenue, we build our compost heap on a day with a high level of chaos and destruction, but no disease. Possibly the most inauspicious period to set up a compost heap is when Saturn, a planet traditionally thought to have agricultural influence, is in conjunction with the star Algol because it is associated with evil and pestilence, which we would not want for good compost and especially if you are composting manure!

At such a period the Foot-and-Mouth epidemic hit which resulted in the medieval, insane and unnecessary mass extermination of our much loved cows. Furthermore we should avoid eclipses because it disrupts the natural behaviour of organisms.

Traditionally the time of destruction and death, hence composting, is towards the end of our astrological year, autumn, or if you will, when the Sun is moving into the house of Libra. In terms of maximum destruction, autumn is unrivalled, however we may not want to go to that level. If we pull things up in autumn, we take habitat away from beneficials and disrupt their normal hibernation patterns, so if you only do one pile a year, for the organic grower the best is early spring where the compost heap may fuel the food web's growth and expansion. On the full Moon in March, over this time of year, the old farmers' almanacs of America call this special period the Full Worm Moon[42] as this is when the Earth begins to warm up, worm casts start appearing, and the robins become hard to avoid.

APPENDIX ONE

Glossary

This glossary was included to flesh out some of the ideas introduced in the main text and to clarify the usage of certain terms. From work on this part of the book, certain other terms needed to be introduced to better describe those terms and the glossary took on an extended life around the topics of this book.

Abiotic In contemporary Western science and culture, what is normally considered to be the non-living aspect of the environment as distinguished from the biotic part.

Acaricide A substance that kills mites and ticks. Because mites and ticks are in the same class as spiders (Arachnida) these might also be affected.

Acid An acid has a pH of less than 7. In solution an acid has more free hydrogen ions (H+) than hydroxyl ion (OH-). See also Base, pH, Ion and Ionic.

Aerobic Requiring oxygen. Used with reference to decomposition processes and aerobic organisms. See Anaerobic.

Aggregates In the soil groups of particles that clump together and bind with a stronger force than the surrounding particles. Pore space forms in between aggregates which can be an important aspect of soil fertility.

Agroecosystem An ecosystem that is harvested. Such ecosystems are adapted intentionally or unintentionally in varying degrees by humans for the production of fruit, vegetables, meat, medicine or other resources.

Alkali A base in solution. Alkalis have a pH over 7. See also Acid, Ionic.

Allelopathy This is a process employed by certain plants that produce and release chemicals which are toxic to most other plants growing around them. A classic example is the production of juglone by black walnut trees. Sometimes the effects of allelopathy can be abrupt such as sudden death of a neighbouring plant, or subtle such as impaired growth and suppressed germination. See Phytotoxin.

Allium A genus of flowering plants that includes onions, garlic, leeks and chives.

Anaerobic Not requiring oxygen. Used with reference to decomposition processes and in contrast to aerobic organisms. See Obligate Aerobes.

Anecdotal A second hand account of events or any information that has not undergone rigorous scientific investigation. Anecdotal information does not have the reliability of scientific evidence but can give useful indications as a pointer for more rigorous investigation. See Scientific Method, Folklaw and Grey Literature.

Anion An ion that has a negative charge. Anions that have gained electrons causing them to have a greater proportion of electrons to protons.

Annual A plant that naturally lives up to a year. Some perennials become annuals in colder weather than they are accustomed to.

Anoxic Conditions that have an extreme depletion of oxygen such as could occur in waterlogged fields. See Denitrification.

Anthracnose Or Canker. A general term for a large number of plant diseases that are characterised by similar symptoms such as areas of dead tissue which expand over years sometimes. Cankers can be caused by bacteria, fungi, viruses and mycoplasmas. They generally hit a single species of host. Fruit trees and vegetable crops can be affected.

Anti-feedant Chemicals that certain plants exude to discourage organisms from feeding on them. For example when horseradish is crushed the enzyme myrosinase is exuded which causes a reaction leading to the production of allyl isothiocyanate, which is harmful to the organism attacking the plant and the plant itself. See Phyto-toxin.

Apex Predator Or Top Predator. The predator at the top of a food chain that is not predated but predates other consumers at lower trophic levels.

Apogee The closest point of one planet's orbit around another planet such as the apogee in the Moon's elliptical orbit around the Earth, which recurs over around 27.3 days. See Perigee, Approaching Moon and Retreating Moon.

Approaching Moon After the apogee has been passed and leading up to the perigee.

Archaea A domain of single-celled microorganisms distinct from bacteria and eukarya. They are found in many habitats and are known to perform numerous functions in elemental cycles notably carbon, sulphur and nitrogen. For instance, *Methanogen archaea* function in anaerobic environments removing hydrogen and aiding in the decomposition of organic matter. *Nitrosopumilus maritimus,* an Ammonia Oxidising Archaea (AOA), along with Ammonia Oxidising Bacteria (AOB), convert ammonia into nitrite.

Ascending Moon The Moon ascends rather in the same way as the yearly ascent of the Sun, but much more quickly. To reach its highest point in the sky takes approximately 27.3 days as it passes in front of the constellations of Sagittarius, Capricorn, Aquarius, Pisces, Aries and Taurus. More fluid in aerial plant parts is indicated. See Descending Moon and Constellation.

Aspect (Astrology) The angle between one planet and another, usually with the Earth as a reference point. In astrology certain angles between the planets are considered important. In the well-known system originally introduced by Ptolemy (1 CE) there are Conjunctions (0°), Trines (120°), Squares (90°), Sextiles (60°) and Oppositions (180°). Mainly aspects involving the Earth and Moon are referred to in this book. See Syzygy.

Asexual Without sex/fertilisation. See Vegetative Reproduction and Parthenogenesis.

Asteraceae Or Compositae family. The main feature of plants in this family is a type of capitula (calathid) with surrounding involucral bracts. Includes sunflowers, artichokes, lettuce, chrysanthemums (such as Pyrethrum), marigold, pot marigold, dandelion, yarrow and *Echinacea*.

Atom A basic unit of matter consisting of a nucleus inside a cloud of negatively charged electrons. The nucleus has a mix of neutral (with some exceptions) neutrons and positive protons. See Ion.

Autotroph Can use energy from sunlight or inorganic compounds to produce complex organic compounds in contrast to heterotrophs that are unable to do this. There are two main types: photoautotrophs (use sunlight as energy) and chemotrophs (use inorganic compounds). In food webs autotrophs are labelled as producers. Plants, algae and many bacteria fall under this category.

Bacteria A domain of Prokaryotic microorganisms as distinct from the domains of Archaea and Eukaryotes. Amongst the first organisms to exist on Earth and present in almost every habitat. Function as mutualists inside the digestive tract of many organisms including humans and play important roles in decomposing organic matter. See Mutualism.

Bactericide A substance that kills bacteria. See Pesticide.

Base Any substance that has a pH greater than 7. In solution a base has fewer hydrogen ions (H+) to hydroxyl ions (OH-). See also pH, Acid, Ionic and Ion.

Beneficials A widespread colloquialism used within the text to mean any organism that has a beneficial influence on crops. Used often to mean beneficial plants or predators.

Beneficial Elements In plant nutrition elements that can have a positive effect on plants but are not essential to their survival. For example silicon [Si] and sodium [Na]. See Essential Elements.

Biocide A chemical that has the potential to kill living organisms, e.g. pesticide.

Biocontrol Is a method that attempts to control one group of organisms, usually ones that are a threat to crops, with another group of organisms. This can include parasitism, predation and herbivory. It is an important facet of ecological approaches to growing crops. Also known as Biological Control and Biological Pest Control.

Biocrystallisation Otherwise known as Copper Chloride Crystallisation. A picture forming technique introduced by Ehrenfried Pfeiffer which is not fully understood. Dendritic structures are formed when plant extracts are added to an aqueous solution of dehydrated copper. Proponents make the still controversial claim that it can be used to show how fresh a plant is and that it can be used to distinguish between holistically grown and conventionally grown produce. Dr Pfeiffer was a collaborator with Steiner and a co-developer of Biodynamics. Underlying many of his activities, as this technique above, was a search for scientific justification of vital energy. See Vitalism.

Biodynamics An influential and spiritual approach to farming introduced by Rudolf Steiner in 1924 at the request of a group of farmers whose crops were failing from early modern approaches. Intrinsic in this approach is a respect and reverence for all nature where balance and soil fertility are strived for. It is controversial because amongst numerous other criticisms, many of the techniques are based on a belief in vital energy and subtle energies of the Sun, Moon and cosmos, which are unproven by today's scientific methods. See Vitalism.

Biological Pesticides Naturally occurring substances that control pests, as distinct from something that is factory or laboratory prepared. For example: pyrethrum (natural sources) as compared with pyrethrin (lab). See Pesticide and Botanical Control.

Biosphere All the ecosystems on Earth combined. See Homeostasis.

Bioturbation The turning and mixing of soils and sediments by organisms using processes such as ingestion and excretion of soil particles, creation of galleries and burrowing. Common terrestrial bioturbators are earthworms and rodents. Bioturbation can be an important biological process involving the mixing of different soil horizons. See Krotovina and Gallery.

Biotic Considered within its own lexicon, the part of nature that is born, reproduces and dies. See Abiotic. More broadly the biotic-abiotic dichotomy was introduced by Western biological science to delineate a clear boundary between what is and is not 'life'. In other more spiritual cultures the distinction is less clearly drawn and life may have a broader meaning to encompass rocks, water, mountains, the whole of the Earth and the cosmos. See Homeostasis, Organism and Superorganism.

Botanical Control Many plants possess special qualities that can be used to deter, weaken or kill plants, bacteria, fungus and animals. Some of these substances can be processed and used as insecticides, fungicides and herbicides et cetera. For example Sun Hemp and Neem can be used in this way. See Phytotoxin.

Botanicals A colloquial term used to describe a substance that is derived from plant sources.

Bract A modified leaf often associated with reproductive formations in plants. For instance the leaves found under some flowers or inflorescences.

Brassica Is a genus of plants in the mustard family (Brassicaceae). Includes Brussels sprouts, cabbages, kales, broccoli, turnip, kohlrabi, charlock and mustard. Also known as Crucifers and Coles.

Broad Spectrum Referring to the application of pesticides. A wide range of organisms will be affected and this may include 'pests' and beneficials. Affected organisms (often insects) may have physical features in common, e.g. soft bodied. Examples of broad spectrum pesticides include pyrethroid, carbamate and most organophosphates. See Target.

Bulk Soil The part of the soil that is not in direct contact with plant roots, as distinguished from the rhizosphere.

Calathid Or Calathidium. Loosely referred to as a capitulum. A flower head that looks like a single flower and is a feature of plants in the asteraceae family.

Calcium Carbonate Equivalent (C.C.E) Expressed as a percentage, is a benchmark standard where the neutralising power of a calcium carbonate product is gauged against pure calcium carbonate. A product with magnesium carbonate included such as Dolomite can have a C.C.E of greater than 100% because magnesium is lighter than carbon. See E.C.C.E.

Canker Or Anthracnose. A general term for a large number of plant diseases that are characterised by similar symptoms such as areas of dead tissue which expand over sometimes years. Cankers can be caused by bacteria, fungi, viruses and mycoplasmas. They are generally tied to a single species of host. Fruit trees and vegetable crops can be affected.

Carbon Fixing The reduction of carbon dioxide mainly by autotrophs to produce organic compounds. The most important type of carbon fixing is photosynthesis. Some autotrophs and heterotrophs fix carbon chemically (Chemoautotrophs and

Chemoheterotrophs) without sunlight, which account for minuscule amounts of carbon fixing by comparison. Also certain organisms fix carbon from sunlight using a pathway that is different to that of autotrophs which arguably does not count as photosynthesis such as Halobacteria (Obligatory Heterotrophic Phototrophs).

Carnivore An organism that eats animal flesh either partly or totally. See Omnivore.

Cation Positively charged ions. Cations are atoms that have lost electrons and hence have a greater proportion of protons. See Anion and Ion.

Cation Exchange Capacity (CEC) The maximum amount of cations that a soil can hold. Gives an important indication of fertility and how well a soil can retain nutrients. Expressed in Meqs (milliequivalent of hydrogen per 100 g of dry soil) or Cmols (Si unit centi-mol per kg). See Soil Fertility.

Celestial Sphere An imaginary sphere encompassing the Earth, which is used to mark the movement of the planets rather in the way Earth is navigated using longitude and latitude. See Constellation.

Cellulases A group of enzymes that break down cellulose. Produced mainly by bacteria, fungi, protozoa and some insects.

Cellulose Is the most abundant organic compound on Earth. Is used as a structural component of many living organisms such as part of the cell wall of green plants. It is a polysaccharide chain of between hundreds to thousands of glucose units. Ruminants and termites can digest this organic material and use it with the aid of microbial symbionts.

Chitin The most abundant organic compound on Earth after cellulose. This is a structural polymer composed of glucose derivatives similar to cellulose and is found in exoskeletons, the cell walls of fungi and the radulae of molluscs.

Chlorophyll An important pigment in plants and other organisms and a primary vehicle of photosynthesis. Chlorophyll absorbs the red and blue part of the electromagnetic spectrum and reflects the green. There are different types of Chlorophyll – Chlorophyll a, Chlorophyll b and Chlorophyll c which varies between different organisms such as Terrestrial Plants (a and b), Green Algae (a and b), Diatoms (a and c), Kelps (a and c), Red Algae (only a).

Chlorosis From the Greek khloros meaning pallid, pale or pale green which reflects the symptoms. Chlorosis comes about from plant leaves not producing enough chlorophyll. Many factors may contribute to this condition such as deficiencies in iron, magnesium or nitrogen, unsuitable soil pH, poor drainage, damaged roots, ozone injury, exposure to sulphur dioxide or pesticides.

Class An intermediate level taxonomic ranking. Along with Order and Family one of the three most vague and arbitrary groupings in standard taxonomic rankings. A taxon that is very much at the mercy of the whims of individual taxonomists.

Cole Any plant in the brassica genus. Includes Brussels sprouts, cabbages, kales, broccoli, turnip, kohlrabi, charlock and mustard. Also known as Brassicas and Crucifers.

Colloid A particle roughly between 0.1 and 0.001 µm. Used in the context of soil and clay particles because of their comparable dimensions. Smaller colloids can be mixed with liquids that remain in solution. See Suspension and Emulsion.

Conjunction (0°. Astrology/Astronomy) With at least three planets lined up, a straight line can be drawn through the centre of each planet with the reference planet at one end. It differs from Opposition because the reference planet is not in between the other two planets. A conjunction is a kind of syzygy. During spring the conjunction between the Sun and the Moon causes the highest tides. This is a favourable period for uprooting weeds to use in hot composting. See Solar Eclipse.

Constellation A group of stars in the celestial sphere that is usually symbolised in a way that is culturally relevant. The divisions between the constellations are unequal which are accounted for in biodynamic gardening calendars but not tropical and other sidereal ones. See Zodiac.

Consumer These are members of a food chain that consume other organisms to get their food and energy. They differ from producers because they can't generate their own energy from the sun or other processes. Typically many consumers are predators but there are also parasites and detritivores in this category.

Contact Poison Poison that is absorbed through the skin on contact.

Cosmopolitan In the context of biogeography, when a taxon is cosmopolitan its range extends over most of the Earth. For instance humans, lichens of the species *Parmelia sulcata*, the mollusc of the genus *Mytilus* and many bacteria important to the composting process. Obviously some common sense is required when interpreting this definition, for instance a killer whale has a cosmopolitan distribution but you won't find many inland.

Crowdfunding An alternative stream of funding for projects that accesses funding directly from typically large groups of people who become stakeholders. It is separate from the traditional funding streams and their limitations. Its recent expansion into the mainstream is closely linked to the internet and related technologies.

Crucifer Any plant in the mustard family (Brassicaceae). Includes Brussels sprouts, cabbages, kales, broccoli, turnip, kohlrabi, charlock and mustard. Also known as Brassicas and Coles.

Cucurbit A plant in the Cucurbitaceae family which includes squash, pumpkin, cucumber, gourds and luffa.

Dark Matter A mysterious substance or substances that are reputed to make up around 80% of the universe. Its main feature is that it does not interact with electromagnetism which extrapolates to no light being emitted, reflected or absorbed making it rather difficult to detect. There is very strong evidence for its existence but to date it has not been observed directly. From the 1970s it gradually became accepted by the mainstream, although there is still a great deal of uncertainty around it. See Electromagnetism.

Decomposer Organisms that consume dead organic matter. Includes certain bacteria, fungi, actinomycetes, algae, flies, snails, woodlice and of course worms. Decomposers play an important role at the base of the food web and speed up nutrient cycling. The commonest and quickest decomposers are aerobic.

Dendritic A branched form resembling a tree.

Denitrification This is a process that is mediated by microorganisms and is the return of nitrogen from soil or water to the atmosphere. It is a natural part of the nitrogen cycle that occurs in conditions where there is a severe depletion of oxygen. This can occur in poorly drained or waterlogged fields and is unfavourable to growers since it depletes nitrogen. Heterotrophic bacteria (e.g. *Paracoccus*) and some autotrophic bacteria (e.g. *Thiobacillus denitrificans*) are known denitrifiers.

Descending Moon The Moon descends rather in the same way as the Sun's yearly descent, but much more quickly. To reach its lowest point in the sky takes approximately 27.3 days as it passes in front of the constellations of Gemini, Cancer, Leo, Virgo, Libra, Scorpio. Less fluid in roots is indicated. See Ascending Moon and Constellation.

Detritivore See Decomposer.

Domain The highest taxonomic rank. There are currently three domains: Archaea, Bacteria and Eukaryotes. The possibility of non-cellular life is currently not embraced in this system.

Dualism In the context of contemporary discussions in science, dualism refers to a philosophy where mind and matter are ontologically distinct. Dualism was favoured during the Enlightenment by figures such as Rene Descartes in 'Cartesian Dualism' before giving way to Monism which stands in contrast to it. See Ontology.

Eclipse A celestial body is obscured by another celestial body which casts a shadow onto the reference planet. Eclipses involving the Sun and Moon are considered unfavourable in lunar gardening calendars. See Lunar Eclipse, Solar Eclipse and Occultation.

Ecliptic Plane The planets of our solar system revolve around the Sun in roughly the same plane called the Ecliptic. As would be expected this is the plane where eclipses occur.

Ecosystem A community of living (biotic) organisms together with the environment (abiotic) that they interact in. An 'Ecocommunity' was suggested (Murray Bookchin) as a replacement term to ecosystem to highlight that such an entity is an integrated whole with emergent properties that are not reducible to individual components. See Biosphere, Holon, Food Web, Watershed, Reductionism and Holism.

Effective Calcium Carbonate Equivalent (E.C.C.E) Expressed as a percentage and combines the C.C.E (Calcium Carbonate Equivalent) with a very fine particle size (< 250 micrometres. Equivalents: < 60 mesh size or < .001 millimetre) to use as a benchmark standard for comparing other calcium carbonate products that may not be as fine or pure. Finer particles offer a greater surface area which facilitates quicker reactions in the soil. See C.C.E.

Electromagnetic Radiation Are waves of electromagnetic force with an energy inversely proportional to the wavelength and proportional to the frequency.

Electromagnetic Spectrum Electromagnetic radiation that spans from approximately 5 nm (Cosmic Rays) to 1012 nm (Radio Waves). In between these extremes is visible light (380-750 nm). Green plants use chlorophyll to absorb light radiation mainly in the blue (460-480 nm) and red (640-680) parts of the spectrum. See UV radiation.

Emergence Emergence is the manifestation of a complex system from numerous simple events or interactions happening simultaneously. See Holism.

Emergent Phenomena (or **Emergent Properties**) This can be either a property or a behaviour and can manifest when two or more simple or complex agents interact generating more complex phenomena collectively. See Synergy, Holism and Flow Form.

Emulsion A mixture of at least two liquids that are usually immiscible. e.g. milk (emulsified milk fat and water). In time emulsions normally separate, returning to the state before they were mixed. See Colloid and Suspension.

Enlightenment Today the meaning of this term tends to be overshadowed by what it means when applied to the 18th century European movement bearing its name. In this context it marked the ascent of science and logic which was believed to give humanity greater understanding and knowledge than other's ways of understanding. Considered broadly it is a state of having insight revealed with often spiritual connotations. For instance in Buddhism it suggests an awakening and spiritual transformation. See Scientific Method, Post-Enlightenment, Dualism and Reductionism.

Entomopathogenic Fungi Or Predatory Fungus. Such fungus can kill or seriously disable their insect hosts that are then used for sustenance. Can be used as a form of biocontrol. For example certain taxa in the order Hypocreales and order Entomophthorales.

Entomopathogenic Predatory Nematodes (EPNs) Such nematodes can kill or seriously disable their insect hosts, which are then used for sustenance. Can be used as a form of biocontrol. For example certain taxa in the *Steinernema*, *Heterorhabditis*, and *Neosteinernema genera*.

Entopathogenic An organism that preys on insects. Often used to refer to predatory microorganisms such as predatory nematodes or fungi.

Essential Elements (According to Arnon and Stout's three criteria.)
1. The plant will not be able to complete its life cycle without it.
2. An essential element's function cannot be replaced by a different element and;
3. The element must be directly involved in the plant's reproduction and growth (primary metabolism).

There are 17 essential elements currently: oxygen [O], hydrogen [H], carbon [C] nitrogen [N], phosphorus [P], potassium [K], calcium [Ca], sulphur [S], magnesium [Mg], iron [Fe], manganese [Mn], copper [Cu], boron [B], zinc [Zn], molybdenum [Mo], chlorine [Cl], nickel [Ni]. See Beneficial Elements.

Exoskeleton This is an external skeleton and has a number of functions including protection, sensing, support and resistance to desiccation. Organisms that have exoskeletons include spiders, insects, snails, crabs, lobsters and certain bacteria and fungi. Many exoskeletons contain chitin and when calcium carbonate is included in its makeup the shell is much stronger.

Exudates The oozing of liquids in various states of viscosity from a variety of glands, wounds, tears, pores, or other orifices into the outside world. Regarding plants, exudates include saps, gums, latex and nectar.

Eukaryotes More formally referred to as Eukarya or Eukaryota. One of the three domains in taxonomy. Includes the kingdoms of Animals, Plants, Fungi, Amoeba, Chromalveolata and Excavata.

Facultative Aerobes Or Facultative Anaerobes. These organisms can alternate between aerobic (using molecular oxygen) or anaerobic (not using oxygen) respiration. See Aerobic, Anaerobic and Obligate Aerobes.

Family A lower level taxonomic ranking. Along with Class and Order one of the three most vague and arbitrary groupings in standard taxonomic rankings. A taxon that is very much at the mercy of the whims of individual taxonomists.

Fertigation Adding fertilisers or other amendments dissolved in water using the same infrastructure as would be used for irrigation. Can be a complex set-up of pipes and timers or as simple as using a watering can with plant pots sunk into the ground.

Flagella A whiplike tail or organelle. Flagellates are found in bacteria, protists, spermatozoa, fungi such as the primitive Chytrids, plus some cells of non-flowering plants – mosses, ferns and green algae.

Flagellates Organisms that have one or more flagella.

Flow Form The shapes that occur in materials that come about from flowing. Flow forms were championed by Vicktor Schauberger and other visionaries to investigate the emergent properties of water. Flow forms are normally associated with the natural movement of water in streams and rivers, however they are also associated with gases and solids under the right conditions. Ice rivers provide a stunning example of solid flow forms. See Vortex.

Fluffing In the context of composting, fluffing involves increasing the surface area and the amount of air in between the composting material in a compost heap, by the process of turning, ideally with a worm friendly implement such as a blunted gardening fork. The process akin to puffing up a cushion.

Foliar The leaves of any plant. Mainly used in the text within the context of foliar sprays.

Folklaw (Alt. spelling **Folklore/Folklor**) Also called Peasant Wisdom and is closely related to anecdotal evidence. A kind of knowledge usually passed down that can be part of an organised belief system or fragmented pieces of confused information. Some folklaw can be valid and reliable but a problem can exist distinguishing good information from that which is plausible but false. For example in Lunar Lore, many similar views cross cultures distant in both time and space, might suggest a basis in fact but this could still be misleading. Rather than reveal something about the Moon itself it might only reveal a tendency inherent in human nature. Notwithstanding folklaw can provide a rich basis for further investigation. See Anecdotal.

Food Chain This shows a single pathway in a food web where nutrients and energy flows between one organism and another. There is one organism per trophic level with a producer at the base and a top predator at the apex, e.g. Plant (producer) > Mouse (herbivore) > Owl (top predator). See Food Web.

Food Pyramid An early model of feeding dynamics between different organisms. It illustrated statistically that there is less energy and food available as you ascend the pyramid. On the bottom there are autotrophs (producers), that are eaten by heterotrophs (consumers) who in turn are eaten by second level consumers and so on up to the apex predator at the top. By this model the terrestrial food pyramid can only go up to around four levels before the resources run out. As a model it is somewhat simplistic and does not deal well with omnivory. See Food Chain, Food Web and Trophic Level.

Food Web A food web comes about when a number of food chains operate together which is usual in natural systems. See Food Pyramid, Trophic Level, Producer, Consumer, Autotroph, Heterotroph, Herbivore, Carnivore and Ecosystem.

Formative Forces An alternative view of natural growth processes which was revisited by Steiner and has deep links with biodynamics, lunar gardening and the natural philosophy of a distant past. Developed further by Bockemühl, plant growth was linked to the four metaphysical elements (Earth, Water, Air and Fire). See Trigons.

Fumigation A type of pest control that inundates an area with a gaseous pesticide. Used in closed spaces and may also penetrate to some degree wood, soil, grain et cetera. Might be used in a greenhouse or a certain area may be enclosed or 'tented' for the gas to work.

Fungicide A substance that kills fungus. See Pesticide and Botanical Control.

Gallery Certain worms such as the common earthworm *Lumbricus terrestris* create small caverns in the soil up to 8mm across which they occupy. The worms take plant litter into these areas which helps to work organic material into the lower level of the soil. Such soil is known to be considerably higher in microbial activity than the surrounding areas. See Bioturbation and Krotovina.

Galls Abnormal, usually rounded, outgrowths from plants caused by a variety of fungi, parasites, bacteria, insect and mites. These growths are harmful to plants and are not the same as the alterations in structure caused by mycorrhizal interactions, e.g. fungal *Plasmodiophora brassicae* and root knot nematodes *Meloidogyne* spp.

Generalist Such species can use a variety of different resources and are more able to deal with different environmental conditions. This is contrasted with specialists who excel in exploiting less varied resources and conditions, but are put under pressure when conditions and potential food sources rapidly change.

Genius Loci From ancient Greek 'Spirit of Place'. The notion that each place has a unique essence or spirit.

Genus One of the lower taxonomic ranks used to classify organisms. A useful taxon when species are difficult to distinguish.

Green Revolution This took place between the 1940s and 1970s and was pioneered by Norman Borlaug. It was a global event where modern techno-agricultural knowledge, agricultural technology, management skills and new super productive cultivars were exported worldwide. It marked an important shift from farming for sustenance to farming for business. Over a span of 30 years, famine worldwide was reduced but population levels exploded even more. Local farmers were disempowered, employment reduced because of mechanisation and local agroecosystem polycultures that supported indigenous population were destroyed. In agricultural time spans, 30 years of success is a very small period of time for an inherently unsustainable enterprise. The reliance on fossil fuels, ruthless global economics and reduced food quality does little to improve food security. Today in the original area of the Punjab where the Green Revolution was pioneered there has been an unprecedented increase in cancer reports linked to pesticide and agricultural chemical usage.

Grey Literature Literature that is not peer reviewed and is of dubious credibility. Many of the topics broached in this book are on the fringes of science and sometimes the positions in contemporary science are open to question, so something that is genuinely impartial is almost impossible to find. However some organisations, especially within the field of organic growing, blatantly engineer a fantasy world which bears little relation to genuine truths and can be far more detrimental to a pursuit of knowledge. View with suspicion any material, reports, research that is not put up for peer review (unlike well-respected scientific journals) such as material that just circulates within the organisation's own periodicals and community. See Anecdotal.

Glomalin A type of protein found in the soil named after fungi in the order Glomerales by Sara F. Wright at the United States Department of Agriculture. It is produced by endomycorrhizal fungi, is important in soil aggregation and improves soil fertility.

Guild Any group of differing species that exploit the same resources in the same location. Can include groupings of fungi, plants and animals. Guild members may not be resident at the same location permanently and any co-operation is not necessarily fixed, e.g. nurse plants may provide shelter for smaller seedlings but then become overwhelming. Members of a guild may use the same resource but according to different rhythms and uses. There may be competition for resources such as light or space, however there are advantages such as increased shelter, enhanced pollination or increased access to nutrients. See Ecosystem, Mutualism and Symbiosis.

Half-Life A term used in agriculture to describe how persistent a pesticide is after application. If the pesticide has a half-life of three days its effectiveness will halve every three days.

Hardener With regard to horticultural practice a substance that is used in a spray to harden the aerial parts of a plant so it can better resist pest attack. Plants used in hardening sprays include seaweed and marestail. A high level of silica is usually the most important basis for these sprays. Although not widely known bamboo shoots contain the highest level of silica and could be a useful alternate source for plant hardeners.

Hardiness The degree to which a particular plant can withstand freezing temperatures. Plants can be tender (can't survive any frost), half-hardy (survive a few frosts) and hardy (survive a full winter). Hardiness zones are used to gauge how well a particular plant will do in an area. Understandably most tropical plants will not survive temperate conditions.

Hardpan This is a layer beneath the soil's surface that becomes solid, which restricts plant roots and may become waterlogged. It can be caused by heavy machinery, repeated footfalls or even the impact of raindrops on open soil. It can also result from applications of superphosphate, which create a cementing effect on calcium rich soils.

Hard Water This is water that has a high mineral content. Hard water does not form suds as easily as soft water. Water is softened sometimes to stop it clogging equipment with mineral particles such as in the formation of lime scale (calcium carbonate). Water hardness is gauged by the concentration level of multivalent cations that have charges greater than 1 such as typically Mg^{2+} (magnesium) and Ca^{2+} (calcium).

Haulm The stems of various crop plants such as potatoes, beans, peas and cereals.

Herbicide A substance that kills plants. See Pesticide and Phytotoxin.

Herbivore An animal that is adapted to eating plant matter (autotrophs) in contrast to carnivores. Such animals are usually part of the lower end of a food chain as first level consumers.

Heterogenous Can be used with reference to agroecosystems, ecosystems, habitats, soil and populations of different organisms. A heterogeneous ecosystem is one that has many varied habitats such as rivers and solid land. It is also used to describe combined substances that are dissimilar such as those in a suspension. Heterogenous is used in contrast to Homogenous.

Heterotroph These are organisms that are unable to fix carbon which makes them dependent in various ways on autotrophs or other heterotrophs (that at the lowest producer level have eaten autotrophs) to live. In food webs they are labelled as consumers. All fungi and animals are heterotrophs. Also some parasitic plants are partially heterotrophic.

Higher Plants (phylum Tracheophyta) Also known as Vascular Plants and Tracheophytes. This phylum of plants is distinguished from non-vascular plants by having vascular transport tissue of two kinds, the xylem and phloem. The xylem is dead lignified tissue that moves water and certain nutrients around the plant. The phloem is living tissue that carries soluble nutrients mainly associated with photosynthesis around the plant. In trees the phloem is the innermost bark. Higher plants include flowering plants, clubmosses, ferns, horsetails and conifers. See Lignin.

Holism The concept of holism emerges frequently within the text of this book and is closely linked to organic practice. Aristotle's old adage, 'the whole is greater than the sum of its parts' is a central feature of holism, although today it may be more accurate to replace the use of 'parts' with another term because it tends to infer a reductionistic frame of reference. In many ways holism contradicts the theory of reductionism. The term holism was originally introduced by J.C. Smuts in *Holism and Evolution* (1926). In his second edition he suggested that holism can be viewed as a theory of the universe as can Spiritualism and Materialism. See Holon and Reductionism.

Holon Originating from Arthur Koestler's book *Ghost in the Machine*, something that functions as a discrete entity and yet is intrinsically part of a greater whole which too could be described as a holon. This term is not in general usage and could be used in a new way to describe guilds, ecosystems and superorganisms. For example a superorganism could be described as a holon of organisms. See Holism.

Homeopathy A form of medicine that can be applied to any living organism. It uses small amounts of extremely diluted preparations that are thought to help an organism develop an immunity much in the same way as vaccination. In terms of this principle homeopathy is synonymous with vaccination which is a well established process. Unfortunately homeopathic practitioners fell down in their method of achieving this, with no solid evidence in support.

Homeostasis Regulation of an internal environment involving complex biochemical reactions to maintain certain properties such as chemical balances, temperature or pH. This is one of the criteria which distinguishes a living organism. Interestingly there is strong evidence in support of homeostatic phenomena at the biosphere level which could lead to 'life' being redefined. See Ecosystem, Superorganism and Organism.

Homogenous Can be used with reference to agroecosystems, ecosystems, soil and populations of different organisms. A homogenous ecosystem is one that has very few significant variations such as river networks, higher altitudes, changes in geology, et cetera. Homogenous is used in contrast to Heterogenous.

Honeydew A sugary viscous fluid secreted by some scale insects and aphids to aid in feeding which can be a vector of fungal diseases. The production of honeydew can also be a feature of symbiotic interactions with other organisms which can be useful as a source of food. See Symbiosis.

Hyphae (Singular **Hypha**) The branching, filamentous structures of fungus and actinobacteria. A large mass of hyphae together is a mycelium.

Identification Keys Are used in scientific fields to identify entities such as plants, animals, fungi, soil types, minerals or diseases. A path is followed where features of what you want to identify are used to narrow down the possibilities, until you arrive at an answer.

Inflorescence A cluster of flowers on a single stem that is part of the shoots of seeding plants. Each of these terms in different plants can be described as inflorescence types: Corymb, panicle, raceme, cyme, composite head, umbel, spike, spadix. See Pedicel, Peduncle, Bract and Involucral Bract.

IFOAM An acronym for International Federation of Organic Agriculture Movements. An influential international umbrella organisation that has its roots in the organic movement and biodynamics. Founding Members: Kjell Arman representing the Swedish Biodynamic Association, Pauline Raphaely representing the Soil Association of South Africa (Organic Growing), Jerome Goldstein representing Rodale Press (Organic Growing Publisher) of the United States and Lady Eve Balfour (a Theosophist/Anthroposophist). IFOAM Head Office: Charles-de-Gaulle-Str. 53113 Bonn, Germany. See Organic Agriculture.

Inoculation This is the act of placing a microorganism in a substrate where it has the potential to grow and reproduce. This can be in vivo (in life) or in a test tube (in vitro). Vaccination is a special type of inoculation.

Insecticide A substance that kills or interferes with insects. This could be something that is target or non-target.

Instar This is an immature development stage in arthropods. Arthropods must shed the exoskeleton of a particular instar when they are developing to the next instar or maturity. The number of instars before maturity varies with environmental conditions and species.

Invertebrate Animals that do not have a backbone. It is a term used for convenience and not a taxonomic division. Approximately 97% of all animals with the exception of the subphylum vertebrata. Includes worms, snails and insects. See Vertebrate and Taxonomic Rankings.

Involucral Bract A term that is used collectively to describe a number of bracts that occur in a whorl that subtends an inflorescence. Is common beneath inflorescences of various plants in Asteraceae, Apiacaeae, Polygonaceae and Dipsacaceae families.

Ion A molecule or atom that has a net electric charge from the gain or loss of electrons. Ions have an unequal number of protons (+) to electrons (-) giving either a positive or negative charge.

Ionic An electrostatic bond between two ions from the transfer of one or more electrons.

Kingdom Second only to domain in the taxonomic rankings. By English classification, popularly there are five kingdoms; further distinctions and divisions are being heavily debated. 1. kingdom Monera: Single-celled Prokaryotes. 2. kingdom Protista: Single and Multi-celled Eukaryotes. 3. kingdom Fungi. 4. kingdom Plantae. 5. kingdom Animalia. However...the American system splits the kingdom Monera into Eubacteria and Archaebacteria stressing substantial genetic differences, giving six kingdoms: Animalia, Plantae, Fungi, Protista, Archaea and Bacteria.

The variations in these different systems and others illustrate the confusion both within and outside the discipline of taxonomy which is in a state of considerable flux.

Knock-Down Effect Used in the context of pesticides, will paralyse or otherwise incapacitate an organism. Knock down does not infer death and after the knock down the organism may recover if its temporary vulnerability is not exploited by another organism. Pyrethroids are well-known for this effect.

Krotovina (Alt. spelling **Crotovina**) An obscure term describing a burrow made by an animal using material from a different soil horizon. See Bioturbation and Gallery.

Larvae (Singular **Larva**) Is a juvenile form undergone by some animals before they transform into adults which includes insects, cnidarians and amphibians. The larvae can look very different from the adult and operate in different environments. For instance the larval form of frogs are tadpoles which cannot survive out of water, unlike frogs.

Leaf Senescence In the context of trees this type of senescence results in the ordered degeneration of the leaf organ giving rise in temperate climates to the colours of autumn where nutrients are drawn away to sustain the tree.

Legumes Plants in the family Leguminosae otherwise known as Fabaceae. A very important family of plants because of their ability to form symbiotic relationships with nitrogen fixing bacteria. Includes familiar crops such as peas, beans, alfalfa, clover and lupins. Also there are many trees such as the Siberian pea tree, Guomi, *Cymopsis*, *Leucaena*, *Sesbania* and *Albizia*. See Symbiosis.

Lignin Is the most abundant organic compound on Earth after cellulose and chitin. It is found in certain algae, the cell wall of plants and especially wood. Lignin is one of the last substances to break down on a compost heap. It is unaffected by animal enzymes but can be broken down by certain fungus and bacteria that secrete ligninases e.g. wood rotting white rot fungi or litter decomposing basidiomycete fungi such as *Pleurotus ostreatus* (Oyster Mushroom).

Lunar Day As observed from Earth the period between two successive lunar days when the Moon is at its highest point in the sky. Approximately 24.8412 hours, being on average 50 minutes longer than a normal solar day. This inequality between lunar and solar days has the appearance of the Moon moving progressively more eastwards on successive days. See Synodic Month.

Lunar Eclipse This occurs when the Moon is in opposition to the Sun, happening when the Moon passes through the Earth's Shadow. See Syzygy and Solar Eclipse.

Macrofauna Soil animals that are greater than 10mm long e.g. earthworms, snails, slugs, millipedes, centipedes, beetles and certain larvae.

Macroorganism An organism that is large enough to be see using normal eyesight (= >.01mm). See Microorganism

Materialism Is a type of Monism which is dominant in Western society and science. It asserts that matter is the fundamental substance of nature which supersedes anything else. In this view phenomena such as mind and consciousness have a purely physical basis that exists only in the body's physical processes. Wherever possible materialists tend to adhere to viewpoints that seem most congruent with day-to-day reality. See Scientific Realism and Post-Materialism.

Maturity With regards to composting, the level of decomposition at which compost is free of phytotoxins and biological activity has subsided. When compost is mature it is ready for standard usage. See Stabilisation.

Megafauna The largest soil dwelling animals e.g. moles, voles, mice, shrews (vertebrates).

Mesofauna Soil animals between 0.2 and 10mm long e.g. earthworms, nematodes, some insects.

Metaphysics Is a branch of philosophy that explores the most fundamental principles of things such as knowing, being, identity and abstract theories that may, or may not have any direct link to reality. See Ontology.

Microfauna Soil organisms with animal-like qualities. They are approximately less than 0.2mm including protozoa, rotifers and nematodes.

Micrometre One millionth of a metre (10^{-6}metre=1µm).

Microorganism An organism that is too small to be seen singly with the naked eye (<0.1mm).

Midden Historically a midden is a kind of dump used by humans. In the text middens refer to the casts and plant litter deposited by earthworms at the entrance to their burrows. Such areas are highly fertile and hotspots of microbial activity and diversity. See Bioturbation and Gallery.

Mineral The definition of a mineral is contested and under debate. According to the International Mineralogical Association: 'A mineral is an element or chemical compound that is normally crystalline and that has been formed as a result of geological processes.' Normal in this instance would be 25°C. See Non-Mineral.

Miticide A pesticide used to kill mites and ticks. Spiders are part of the same class so they too may be affected.

Molecule A group of two or more atoms that are held together by shared electrons.

Monism Considered broadly monism is the view that ascribes a oneness to something. Materialism is a type of monism and the dominant viewpoint in Western society. Pantheism is also a kind of monism in a slightly different context that suggests that the universe is an all encompassing divinity. See Dualism.

Monosaccharide A simple carbohydrate such as glucose and fructose. Can be used as an energy source or as a building block of larger molecules such as cellulose.

Mutualism A type of symbiotic relationship. The relationship between two different species that both benefit from their mutual interactions. Ruminants are mutualists with the microorganisms in their gut which break down cellulose. See Symbiosis.

Mutualist A member in a mutually beneficial relationship. See Symbiosis.

Mycelium A large mass of hyphae together is collectively known as mycelium.

Mycorrhiza Types of fungi that form symbiotic, usually mutualistic associations with vascular plants. Includes fungi in the order Ericales, order Glomerales, phylum Basidiomycota, phylum Ascomycota, and phylum Zygomycota. See Symbiosis.

The fungi gain carbohydrates from the plants and the plants gain better access to water and nutrients, especially phosphorus. There are two main categories: Ectomycorrhiza that grow around the plant roots and Endomycorrhiza that penetrate within. See Symbiosis.

Necrosis This occurs when parts of a plant wilt and darken. The term is used with reference to plant diseases and deficiencies. This can come about from fungal, bacterial infections and nematodes. With fungal diseases necrosis may start as spots on leaves and gradually develop.

Nectar Nectar is a rich sugary liquid produced by plants to attract pollinators and feed beneficial animals. Nectaries within the flower are for pollination and extrafloral nectaries feed mutualists which can assist the plant in a variety of ways such as defence or essential elements supply, that would not be available otherwise.

Nematicide A substance that kills or interferes with the functioning or behaviour of nematodes.

Nephridia An organ found in invertebrates that has a similar function to kidneys which remove wastes from the body. The two main types are: 1. Protonephridium found in nemertea, rotifera and platyhelminthes. 2. Metanephria found in annelids, arthropods and molluscs. Nephridia are found in each segment of an earthworm's gut and can harbour symbionts such as Verminephrobacter.

Nitrogen Fixing Nitrogen gas from the atmosphere is converted into ammonia. There are three main processes.
1. Atmospheric fixation such as lightning strikes.
2. Synthetic fixing in the Haber Process.
3. Fixation by living organisms – Bacteria, Fungi and certain Archaea.
See Mutualism and Symbiosis.

Nitrogen Sink A place where nitrogen is held for a period of time. Globally the humus layer of boreal forests are significant sinks since the organic matter in conifers takes a long time to decompose. In a horticultural context nitrogen sinks in soils are strongly linked to biologically available carbon which helps retain nitrogen in the soil through microbial activity.

nM A nanometre. A billionth of a metre (1x10-9m).

Node Either of two points on a planet where the plane of an orbit is intersected by the reference plane e.g. the plane of the Moon's orbit intersects Earth's North and South Nodes on the ecliptic plane.

Non-Mineral The definition of a mineral is contested and hence a non-mineral is under debate also. A broad definition based on what falls out of the International Mineralogical Association's definition of minerals, is something that does not have an ordered crystalline structure at 25°C and is not formed from geological processes. Such a definition would include the shells of molluscs and the oxalate crystals in some plant tissues. It would also include the 'mineralised' products of bacteria in the nitrogen cycle. The most significant, and uncontested, non-minerals for horticulture are carbon, oxygen, hydrogen from air and water which are involved in major plant processes. See Mineral and Essential Elements.

Non-Systemic This term applies to the use of pesticides and means that the substance will just work on the surface. See Systemic.

Non-Target The application of a pesticide will hit a broad spectrum of organisms, which may include beneficials or other organisms. See Target.

Nymph An immature form in certain invertebrates that differs from larva because the nymph resembles a smaller version of the adult and the nymph undergoes moults. Nymphs go through a number of developmental stages called instars. Includes shield bugs, grasshoppers, dragonflies and certain mites. See Pupae and Larvae.

Obligate Aerobes Requiring molecular oxygen for respiration. See Aerobic, Anaerobic and Facultative Aerobes.

Occultation Where one celestial body is obscured by another apparently greater one. See Eclipse, Celestial Sphere and Transit.

Officina Medieval Latin for storeroom for medicines and useful herbs. See Officinalis.

Officinalis From the Latin 'belonging to an officina'. An officina was an area in a monastery where medicines and special herbs were kept. It is noted in this glossary because many plants that were useful to people in the past has this epithet in the species name. Spellings may vary e.g. *Taraxacum officinale* (dandelion), *Saponaria officinalis* (soapwort), *Valeriana officinalis* (modern form)/*Valeriana officinallis* (valerian).

Omnivores Eat a variety of animal and plant food. See Carnivore.

Ontology A branch of metaphysics that investigates the nature of reality and being. In comparison to metaphysics it is rather narrower in scope but still addresses fundamental questions regarding the nature of existence. See Metaphysics.

Opposition (180°. Astrology/Astronomy) An aspect of 180° between two planets and the reference planet. An opposition between the Moon and Sun is considered very unfavourable in lunar gardening e.g. Lunar Eclipse.

Order A taxonomic ranking towards the lower end of the hierarchy. Along with Class and Family one of the three most vague and arbitrary groupings in standard taxonomic rankings. A taxon that is very much at the mercy of the whims of individual taxonomists.

Organic Agriculture Or Organic Farming, is a type of agriculture that uses organic techniques such as composting, crop rotation, companion planting, biocontrol and green manures. Pesticides and synthetic fertilisers are excluded or strictly limited. According to the currently most dominant global organic umbrella group IFOAM: 'Organic agriculture is a production system that sustains the health of soils, ecosystems and people. It relies on ecological processes, biodiversity and cycles adapted to local conditions, rather than the use of inputs with adverse effects. Organic agriculture combines tradition, innovation and science to benefit the shared environment and promote fair relationships and a good quality of life for all involved...'

Organic Compound Molecules that contain carbon. A number of organic compounds such as diamond, graphite and carbonates are labelled as inorganic for historical reasons.

Organism According to Random House Dictionary, unabridged: 'a form of life composed of mutually dependent parts that maintain various vital processes'. A stricter definition may assert that for a living system to be an organism it must be contiguous, that is, connected at a genetic level. Organisms include all the domains of life and generally must be able to reproduce, perform homeostatic functions as a whole and respond to stimuli. See Superorganism and Biotic.

Oviposition The process used by certain insects to lay eggs with an ovipositor. Important beneficial parasitoid wasps use this process to lay eggs in crop damaging hosts.

Ovipositor A specialised tubular organ that extends from the abdomen of certain parasitoid wasps and is used for ovipositing eggs. See Oviposition.

Oxidation A loss in electrons by an ion, atom or molecule. Equivalent to an increase in the oxidation state. See Reduction, Oxidation and Redox.

Oxidation State The degree of oxidation of an ion, atom or molecule. Usually represented in whole numbers and can be negative, positive or zero. The strict definition is the hypothetical charge of an atom if all the bonds are 100% ionic (in contrast to covalent bonds), except between identical atoms. See Reduction and Redox.

Panpsychism Is the view held in Buddhist thought, certain branches of neuroscience and philosophy respectively that consciousness exists in varying degrees throughout the universe. Today this concept is gaining considerable mainstream attention because of unresolved ontological issues in the interpretation of quantum theory. Neuroscientists (Giulio Tononi with Christof Koch) have used magnetic pulses to assess the level of consciousness with subjects who are awake having the clearest reverberations. The author suggests that this may also infer that consciousness is not as self-contained as we normally tend to perceive it to be within ourselves. See Monism.

Parthenogenic A type of asexual reproduction where embryos occur and develop within an organism. Such a process gives aphids a jump start when expanding their numbers because it is quicker and less resource intensive than normal reproduction. Its main drawback for the organism is that large groups of parthenogenic offspring are equally susceptible to diseases because their genes are identical. Organisms that do parthenogenesis include Bdelloid rotifers, nematodes, aphids, water fleas, certain bees, various fish, amphibians and reptiles.

Pathogen Is a microorganism that is an infectious agent causing disease in the host.

Pedicel (Alt. spelling **Pedicle**) A type of stalk that connects single flowers to an inflorescence.

Peduncle The stem which holds a whole inflorescence.

Peer Review When material is put up for peer review it is reviewed independently and impartially by experts in the publication's field. It can be controversial when established views are challenged since the author and reviewer may not work from the same beliefs and knowledge base. The reviewers tend to represent the mainstream view but they are still subject to beliefs held within their own cultural milieu which they may consider to be unassailable truths. See Grey Literature.

Perennial A plant that naturally lives more than a year as distinguished from an annual.

Perigee The furthest point of one planet's orbit around another planet such as the perigee in the Moon's elliptical orbit around the Earth which is approximately 27.3 days. See Apogee, Approaching Moon, Retreating Moon.

Permaculture An influential holistic system of ecological design developed by Bill Mollison that mimics natural systems and is used to create sustainable and self maintaining agriculture. See Polyculture.

Persistence The period of time a certain chemical remains active in the environment. Some pesticides are very persistent and are designed to stay active longer to have more of an effect. Others have low persistence and break down immediately on contact with soil or sunlight. A useful measure for describing a chemical's persistence is the observance of the half-life of the pesticide. A refinement of this concept is the use of 'compartments' to consider persistence in different features of the environment which will have different chemical and ecological implications in each case: water, air, soil and within organisms.

Pest An organism that is considered a threat to growing crops. Used within the text as it is in common language, a more appropriate replacement for this term could be 'problem organism' which does not hold such negative connotations. From a holistic perspective the extreme proliferation of such an organism is considered as an indication of a broader flaw in growing practice. Organisms thought of as pests may indirectly benefit crops in subtle ways and their eradication may damage fertility in subtle ways. In polycultures pests have less significance because there is less potential for them to develop to plague proportions.

Pesticide A broad term describing a substance that kills or in some way interferes with the usual functioning of pests. For example: Nematacide (kills nematodes), Rodenticide (kills rodents), Acaricide (kills mites and ticks). This term is also used to refer to Herbicides (kill plants/weeds) and Fungicides (kill fungi). See Botanical Control.

Pesticide Resistance After an application of a pesticide to a population of pests, a few survive which are more resistant. They go away and reproduce, then the next time the pesticide is applied it is less effective. Historically this phenomenon is upwardly mobile both in terms of the species of pests gaining resistance and in the escalating resistance of the target. The contemporary view is that this illustrates natural selection in action.

pH This is a scale that measures the strength of an acid or base in solution. It is exponential, measuring the hydrogen ion concentration and goes from 0 to 14. Less than 7 is an acid and greater than 7 is a base. See also Acid, Base and Ionic.

Phases (of Plant Growth) In the life cycle of plants certain stages occur during their development. In flowering annual plants this generally includes germination (if not, vegetative reproduction), leaf budburst, first flower, first ripened fruit, seed set, drying and death. In plants where the crops are fruits/seeds the development stages are often generalised to three phases:

1. Vegetative Growth: Germination to the beginnings of inflorescence initiation.
2. Inflorescence initiation to Flowering.
3. Flowering to Ripening.

Phenology The scientific discipline involved in measuring the timing of phases of the life cycles of the varied domains of life and how the environment (seasonal changes, climate, soil et cetera) influences those events.

Phoresy The process of smaller organisms catching lifts on larger organisms e.g. the carrion feeding Silphid beetles of genus *Nicrophorus* may carry at any one time up to 500 mites of various species and 1000s of nematodes. See Symbiosis.

Photosynthesis An important biological process where autotrophs such as green plants absorb electromagnetic radiation using carbon dioxide and water to accumulate energy stored as organic compounds (carbohydrates). A byproduct of this process is dioxygen which of course is important to human respiration. See Chlorophyll and Electromagnetic Spectrum.

Phyllosphere The above ground surfaces of a plant and according to some authorities the interior.

Phylum Traditionally 'division' is sometimes used instead of 'phylum' in botany although it is not strictly correct. See Taxonomic Rank.

Phytotoxin This term is used in two ways. Firstly to describe a substance that is toxic to a plant and secondly to describe a toxic substance that is produced by a plant. Phytotoxins can be found in unfinished compost but this is a natural part of the process. Along with the level of aerobic respiration, phytotoxins have been used in the past as a parameter of compost stability. Phytotoxins in compost can inhibit plant growth, delay seed germination or have other undesirable effects when applied to plants. When compost is properly made and cured, phytotoxins are not a problem. See Stabilisation and Maturity. In the first category are herbicides and bacterial phytotoxins. Regarding the second category these products of plants can function in plant defence or are apparently just byproducts of the plant's primary metabolism. Examples include alkaloids, terpenes and phenolics. See Allelopathy and Anti-Feedant.

Pigments Or biological pigments, are used by plants and other organisms to reflect certain wavelengths of light and absorb others which may then be used in biochemical reactions such as a part of photosynthesis. See Chlorophyll.

Pollen The male sex cells of a plant.

Polyculture Is a holistic style of growing often used by indigenous cultures where multiple crops and food organisms are grown in the same area in a style that integrates with natural systems. Trees with edible nuts, and indigenous food animals often play an important part in such systems. It beats contemporary agriculture in terms of productivity, quality, calories per square foot and sustainability but is more labour intensive. Such a system existed in the Philippines before the Green Revolution but the application of pesticides to protect the new rice cultivars destroyed it and devastated the natural population of frogs and fish that had previously been nurtured by the local people for sustenance. See Permaculture.

Polysaccharide These are carbohydrate molecules such as starch and glycogen that are composed of monosaccharides.

Post-Enlightenment This social phenomena is emerging from disillusionment with what the Enlightenment has developed into and its limitations. Counter-Enlightenment trends were around from the very start of the Enlightenment. They gained in traction as the rigidity and blinkered stances associated with the Enlightenment became more obvious. Michel Foucault (1926-1984) was a key critic against much of the dogma and self-interest of Enlightenment based institutions. The Muslims took a different course and critiqued the Enlightenment because of its lack of a spiritual dimension. To achieve the advances in science during the Enlightenment and continuing into the scientific revolution much of what it was to be human was sacrificed leaving today, if the logic of materialism is followed through, a nihilistic void in an ultimately pointless existence. In today's era when the most fundamental aspects of science have proved uncertain and the direction of science is ethically questionable, the Post-Enlightenment discourses are gaining considerable credence. See Materialism and Post-Materialism.

Post-Materialism Contradicts materialism and makes the claim based on robust, peer reviewed studies that matter is affected by mind. These findings undermine scientific rationalism and its assertion that the observer and the observed can be completely separate. Post-materialists also pose alternative theories that challenge the materialist conception that the mind originates from the physical mechanisms in the brain. Regarding theories of consciousness both sides of the debate are still a long way from being either refuted or proven. See Quantum Mechanics, Materialism.

Potager Originating from medieval France was written of as early as 1626 in La Quintinye's *Instruction pour les Jardins fruitiers et potagers*. The potager is closely related to the cottage garden where there is an intensive close planting of ornamentals, beneficials, fruit, vegetables and herbs together in raised beds for recreation, health, medicine, self sufficiency or aesthetic effect.

Precession of the Equinoxes Or general precession, is a gradual change in orientation of the Earth's orbital plane with respect to the equator. In effect this means that every year the tropical calendar goes back very slightly in relation to the equator which changes the traditional zodiac's alignment with the constellations. As celebrated in the popular song, and because of this phenomena, we are on the dawning of the Age of Aquarius. Currently the vernal equinox is in Pisces. See Zodiac (Tropical), Zodiac (Sidereal) and Vernal Equinox.

Primary Metabolism These are vital processes that plants need for survival such as photosynthesis and respiration. See Secondary Metabolism.

Producer Producers or autotrophs (strictly speaking, photosynthesising autotrophs) mainly obtain their energy from the sun. They are at the base of food chains and are used by consumers for food.

Prokaryotes Are organisms that lack a cell nucleus as distinct from Eukaryotes. Prokaryotes are mostly unicellular.

Propagule Any vegetative structure that can become detached from a plant, fungus or bacteria and produce a new organism. With regards to fungi a spore would be a standard example. See Vegetative Reproduction.

Protist Domain Eukarya. This taxon has been used in the past for the misfits of the Eukaryotes that do not really fit into the other groupings. Mostly unicellular organisms with a relatively simple structure. They can also be multicellular when without specialised tissues. The simple cell structure is the main feature that distinguishes them from other groups which indicates that it is a poorly conceived taxon and has been challenged by modern taxonomists. See Eukaryotes.

Pupae (Singular Pupa) Is a life stage in the lives of some insects where they create an immobile casing which envelops them while they undergo metamorphosis. The pupa is the third of four immature life stages: embryo > larva > pupa > imago. Pupa is only used to refer to insects that undergo a complete metamorphosis such as lepidopterans (Chrysalis) or mosquitoes (Tumbler).

Pyrethrin A naturally occurring insecticide extracted from the flowers of pyrethrum daisies (*Chrysanthemum cinerariaefolium* and *C.coccineum*). See Botanical Control.

Pyrethroid Is a synthesised relative of naturally occurring pyrethrin. It is more potent and persistent in the environment and is used as an insecticide which can penetrate the porous exoskeletons of arthropods. Generally considered safe for vertebrates except cats. Toxic to beneficials such as dragonflies and bees. Also toxic to mayflies, gadflies, fish and invertebrates. Most of these organisms are at the lower level of the food web so disruption at this level would have reverberations throughout the whole network.

Pyrethrum A number of plants of the genus *Chrysanthemum*, family Asteraceae. Important to agriculture because of its insecticidal properties. The flower heads of

Chrysanthemum cinerariaefolium and *C.coccineum* and are the most useful for this botanical pesticide. See Insecticide and Botanical Control.

Quanta (plural of Quantum) In physics quanta are the smallest amount of energy into which something can be divided and are restricted to discrete values. In this context it is a more accurate term than referring to particles and sub-atomic particles because the latter makes assumptions regarding their essential nature (i.e. neglecting their wavelike aspect). See Quantum Mechanics.

Quantum Mechanics Is the most fundamental branch of physics that investigates phenomena at the very small energy scales of atomic and sub-atomic particles. Developed in the early 19th century it was established that objects have both wavelike and particle-like properties. Either the position or the momentum only of quanta (particles) can be known with any precision. Its varied interpretations have led to a great deal of unresolved debate regarding the fundamental nature of the universe and our experience of it. See Ontology, Scientific Realism, Materialism and Post-Materialism.

Redox Reduction-Oxidation reaction. All reactions where atoms have their oxidation states changed. When reduction or oxidation are described separately they are half reactions. When there is reduction there is always oxidation and vice versa. See Reduction, Oxidation and Oxidation State.

Reduction A gain in electrons by an ion, atom or molecule. Equivalent to a decrease in the oxidation state. One half of a Redox Reaction. See Reduction, Oxidation and Redox.

Reductionism In the context of science, reductionism is the theory that a whole system can be described by the mechanisms of its individual parts. The counter point of reductionism is holism although holism acquires broader connotations in other contexts. Used a great deal in conventional science. See Scientific Method.

Reliability One of the cornerstones of scientific method. If an experiment or trial is 100% reliable it will give the same result every time it is done. If a test consistently gives the same result which is incorrect then the test is reliable but not valid. See Validity.

Retreating Moon After the perigee has been passed and leading up to the apogee. See Perigee and Approaching Moon.

Rhizosphere This is an area of high microbial activity that is in the soil closest to plant roots. In this area the bacteria feed on discarded plant cells and root secretions, the protozoa and nematodes graze on the bacteria and the cycle continues. Nutrient cycling and disease suppression are strongly linked to the rhizosphere. See Food Web and Bulk Soil.

Rodenticide Kills rodents e.g. rats and mice. See Pesticide.

Scape An erect and leafless flower stalk or peduncle as found on garlic plants.

Scientific Method According to the *Oxford English Dictionary*: 'A method or procedure that has characterised natural science since the 17th century, consisting in systematic observation, measurement, and experiment, and the formulation, testing, and modification of hypotheses.' See Validity, Reliability and Enlightenment.

Scientific Realism The belief that Nature exists independently of human consciousness. It is still the dominant viewpoint in science today. This notion was brought into question with the development of quantum mechanics when it was observed that consciousness appeared to affect the results of experiments with quanta. For instance in the original double slit experiment by Davisson and Germer (1927). To hold onto the latter belief increasingly bizarre explanations are now being presented. In fairness when the results of robust, peer reviewed studies in quanta are interpreted there is no explanation that does not appear, in our current milieu, to offer an extremely outlandish ontological picture. See Ontology, Materialism and Post-Materialism.

Secondary Metabolism In plants secondary metabolism aids development, growth and supports the primary metabolism, but is not required for survival. For example secondary metabolism takes care of adapting to the environment and defence against enemies.

Semiochemicals From the Greek 'semeon' meaning signal. These chemicals can be made industrially or occur in an organic form from plant and animal sources. Such chemicals can pass a message from one organism to the next by odour or other means. Some plants use semiochemicals to disrupt herbivores and attract beneficial predators.

Senescence Is the age-dependent process of deterioration that ultimately leads to death. See Leaf Senescence.

SEM An acronym for Scanning Electron Microscope. Images are made by scanning with a focussed beam of electrons which has a range of up 500,000 times magnification. Can produce finely detailed images in subjects such as pollen grains, snow flakes and microscopic organisms. See Microorganism.

Sessile Used in two main contexts:

1. For animals or animal-like microorganisms it means that they are either permanently fixed in one spot throughout their lives or mostly except for certain phases e.g. rotifers and sessile nematodes such as *Cacopaurus* spp.
2. Regarding plants it means that leaves or flowers grow directly from the stem.

Sextile (60°. Astrology/Astronomy) An aspect of 60° between two planets and the reference planet. A sextile between the Moon and Sun is considered favourable in lunar gardening. Sextiles between Venus and the Moon are considered good periods for flower work. See Aspect.

Sidereal Month The average time it takes the Moon to revolve around the Earth. Approximately 27.322 days.

Sidereal Zodiac. See Zodiac (Sidereal).

Soft Water Water such as rainwater and distilled water that contains almost no calcium, magnesium or other metal cations. For soft water, calcium must be less than 85.5 ppm. See Hard Water.

Soil Dressing A liquid or solid amendment that is applied to the soil. Such amendments may improve the soil structure, combat pathogens, retain soil moisture, add important nutrients, add beneficial organisms or otherwise improve the soil's fertility. See Soil Fertility.

Soil Fertility The potential of a soil to facilitate plant growth and provide sustained, high quality consistent yields. Important factors of soil fertility include retention of soil water and the availability of essential plus beneficial elements. Other related factors may include soil depth, drainage, soil space, aggregation, organic matter, the presence of microorganisms and certain macroorganisms. When fertility is referred to in the book it is with regards to this context. See Soil Dressing, Biodynamics, Cation

Exchange Capacity and Soil Productivity.

Soil Horizon A layer of soil that is roughly parallel with the land surface. It has different properties to other adjacent layers in the soil profile which can vary by structure, texture, consistency, chemistry or biology.

Soil Productivity The capacity of a soil in its usual environment to support the growth of plants. It is closely linked to soil fertility and its associated factors such as the capacity of a soil to hold water, soil texture, organic matter, Cation Exchange Capacity, Essential Elements, Beneficial Elements and general nutrient availability in the soil.

Soil Profile A vertical succession of soil horizons. Such horizons are often labelled alphabetically in texts.

Solanaceae Or Nightshade family. Has a number of agriculturally important species such as potatoes, chillies, eggplant and tomatillo.

Solar Eclipse A conjunction where the Moon is directly in front of the Sun and casts a shadow on the Earth which is directly behind. Considered extremely unfavourable in lunar gardening and bad for composting because many food web organisms find it disruptive. See Lunar Eclipse.

Species The lowest level of the standard taxonomic tree, see Taxonomic Ranking. Generally a group of organisms that is capable of interbreeding and producing offspring that can reproduce.

All species are given a two part (binomial) name. The first part is the genus and is given a capital and the second part is the species in lower case. To be absolutely correct this is then placed in italics e.g. the herb yarrow is *Achillea millefolium*.

'Species' is a disputed term and there are many alternative definitions based on different trends in taxonomy such as DNA comparison, ecological niche or morphology. This disunity is well recognised and is called the species problem.

Spp. and Sp. In the literature and this book the plural 'spp.' is used to denote certain species that belong to a named Genus, e.g. *Tagetes* spp. means a number of marigolds in the *Tagetes* genus. The singular is sp.

Square (90°. Astrology/Astronomy) An aspect of 90° between two planets and the reference planet. A 'square' between the Moon and Sun is considered unfavourable in lunar gardening. See Aspect.

Stabilisation (Alt. spelling Stabilization) In the context of composting could be described as the process of decomposition towards maturity, at which point the compost is ready to use. Greater stability proceeds in geological time spans after most of the biological activity has subsided.

Maturity can take from three months to two years depending on conditions. The level of aerobic respiration has been used in the past as a parameter of stability because it correlates with a reduction in biological activity. However such a parameter needs to be considered with other factors since this parameter does not take into account the level of activity going down for other reasons such as fluctuations in oxygen levels, an insufficient supply of essential nutrients or anoxic conditions. See Phytotoxin.

Subphylum An additional division amongst numerous other divisions that have been extended from the standard taxonomic rankings. Below Phylum (and above Infraphylum and Microphylum if these divisions are being used) and above Class (and Superclass if this division is being used).

Subtend To be close to, below or extend under. See Involucral Bract.

Superfamily An additional division amongst numerous other divisions that have been extended from the standard taxonomic rankings. Below Order (and below Parvoder if this division is being used) and above Family.

Superorganism According to Wilson and Sober (1989): 'A collection of single creatures that together possess the functional organization implicit in the formal definition of organism.' Wilson and Sober (1989) 'Reviving the Superorganism'. *J. theor. Biol.* (1989) 136. University of New York. 13901, U.S.A. p.338

The line between an organism and a superorganism is disputed. Evolutionary biology does not acknowledge that superorganisms exist because the concept interferes with their theories of individual selection (i.e. Natural selection, individual competition). Will not acknowledge a beehive as a superorganism even though the hive mind has a longer memory span than any individual bee.

Broader and altered conceptions of the superorganism have been suggested with farms as superorganisms (Steiner) and the Earth itself as a superorganism (Lovelock). See Homeostasis, Organism, Biotic, Teleology, Holon and Guild.

Suspension A mixture of usually a liquid and a solid that has particles large enough (approx. > 1µm) to settle out in time e.g. milk. See Colloids, Emulsion and Heterogenous.

Surfactant Also known as a wetting agent. This is added to a spray to help it flow more easily and stick better to plant leaves on application. e.g. soft soap.

Sustainable Agriculture Looks to create a system or systems of agriculture that can perpetuate without certain factors degenerating and potentially destroying the whole system. Such factors could include soil, manpower or productivity. Sustainable agriculture attempts to conserve resources and reduce environmental impact plus 'waste' related to agricultural practice. Agroecosystems are used in the terminology of sustainable agriculture which is an acknowledgement of the significance of relating with a clearer understanding towards natural processes in contrast with conventional agricultural practice.

Symbiont An organism in a symbiotic relationship. See Symbiosis.

Symbiosis A close and often lifelong interaction between at least two different species which can be: Mutualism – Mutually beneficial, Commensalism – Just one organism benefits, Paratism – One benefits, the other is harmed. Synnecrosis – Both organisms are damaged (rare), Amensalism – One organism is destroyed, the other is unaffected. Some authorities assert a narrower definition of symbiosis which states that to be symbiotic the interaction has to be physically close, biochemical and long term (e.g. legumes and nitrogen fixing bacteria). By this definition mutualism for example would have to be distinguished between symbiotic mutualism and non-symbiotic mutualism.

Synergy The combined action of two or more things that create phenomena that do not happen independently. See Emergence. This could involve chemicals, organisms or ideas, concepts and behaviours. As a biological example consider the behaviour of some species of social amoebae that forage independently but come together into a multicellular sluglike form which moves a certain distance and then generates a tower in order to disperse.

Synodic Month The period between one new Moon and the next. On average 29.531 days. This is the most noticeable cycle of the moon because of the Moon's reflection of sunlight during waxing and waning. During full Moon, the Moon is in opposition to the Sun and rises as the Sun sets. See Lunar Day.

Synthetic A substance that is made via human mediated chemical synthesis. e.g. synthetic fertilisers. See Nitrogen Fixing.

Syntrophy One organism feeding off the products of another organism. These interactions are important in all living communities. For example microorganisms mineralising leaf litter. A special kind of syntrophy called metabolic symbiosis is where two populations of organisms consume the other's waste in a mutualistic relationship. See Mutualism and Symbiosis.

Systemic This term applies to the use of pesticides and means that the substance will penetrate the plant rather than just staying on the surface. See Non-Systemic.

Systemic Resistance Refers to disease resistance in plants and has two broad categories. The first are defences that exist by default in the plant and the second are resistances that are latent, only manifesting in response to infections. Attempts to aid plants in resisting certain diseases can involve inoculating a plant to induce a defence response in the plant's immune system and thereby increase resistance.

Syzygy A configuration of at least three planets in a straight line. On Earth syzygy occurs when the Moon is either in opposition or conjunction with the Sun. Solar and lunar eclipses can only occur during syzygy. The highest tides occur from the gravitational pulls of both the Sun and Moon combined during this period.

Target On the application of a pesticide only a specific pest is affected. This can be an advantage when minimising the impact of a spray on beneficials and other organisms. See Non-Target.

Taxon A taxonomic unit such as 'Species'. Such a unit encompasses all organisms below that rank. For instance the taxon 'Order' would include 'Family', 'Genus' and 'Species' taxa.

Taxonomic Rankings This is the generally accepted hierarchy in biology:

Life > Domain > Kingdom > Phylum > Class > Order > Family > *Genus* > *species*

Traditionally 'Phylum' was used to refer to animals and 'Division' replaced phylum for plants and fungi. Today there are a variety of competing codes with claims to being the most correct such as Phylocode and ICZN.

Teleology From the Greek telos meaning 'end, purpose'. The notion that something is moving towards a final goal or state (e.g. succession in natural systems). It is difficult for evolutionary biologists to navigate partly because it infers intent, awareness and intelligence in a non-human agent. See Superorganism and Homeostasis.

Temperate Areas across the globe that are not too hot and not too cold including Britain, most of Europe, areas in North East and North West America, Eastern Asia and New Zealand. The scientific definition is The North Temperate Zone which extends from the edge of the Arctic Circle (66.5° North Latitude) to the Tropic of Cancer (23.5° North Latitude) and the South Temperate Zone extends from the Antarctic Circle (66.5° South Latitude) to the Tropic of Capricorn 23.5° (South Latitude).

Temperateness is also influenced by proximity to the sea (warmer in winter), altitude (colder as getting higher), winds (can warm and cool) and aspect/microclimate (e.g. south facing slopes in the Northern Hemisphere are much warmer than North facing). Other areas and pockets may be temperate because of these additional factors.

Terroir A term originating from viniculture that embraces the total environment of the vine and the uniqueness of a growing area. Using the same vines but different approaches and location, grapes with considerably different qualities can be produced. A term not recognised within science, it is a topic of passionate debate within professional wine growing circles. Its existence is anathema to the culture of mass production and homogenisation that is attempting to overwhelm the industry. The expression of terroir is made by the decisions of the grower and ultimately is experienced in drinking the wine. See Genus Loci.

Test In biology 'test' as derived from the Latin testa is used in a more specialist context as distinct from the general homonym for an exam. It denotes the siliceous shell of sea urchins and various microorganisms such as the siliceous tests of the testate amoeba in the order Euglyphid which are common to soils and rich organic matter.

Testate In a biological context any organism that has a test. e.g. testate amoebae of the order Arcellinid common to leaf litter, soils and around fresh water.

Thrips Order Thysanoptera. A genus of small (approx. <1mm) delicate, fringed winged insects. Many species are considered pests because of their preference for commercial crops. However certain thrips are also considered beneficial because they feed on mites which are more formidable pests. They can reproduce at exponential rates given the right conditions and travel great distances on the wind.

Top Predator Or Apex Predator. The predator at the top of a food chain that is not predated but predates other consumers at lower trophic levels.

Tracheophytes See Higher Plants.

Transit The passage of a celestial body in front of another celestial body that is apparently larger. See Occultation.

Trigons Or Sky Triangles. The 12 signs of the zodiac divided into four metaphysical qualities (Earth, Water, Air and Warmth), three signs for each quality, hence Trigon. Earth – Aquarius, Gemini, Libra. Water – Pisces, Cancer, Scorpio. Air – Aquarius, Gemini, Libra. Warmth/Fire – Aries, Leo, Sagittarius. See Formative Forces and Metaphysics.

Trine (120° Astrology/Astronomy) An aspect of 120° between two planets and the reference planet. A trine between the Moon and Sun is considered very favourable in lunar gardening. Trines between Venus and the Moon are considered very good periods for flower work. See Aspect.

Trophic Cascade In a food chain when a high level or apex predator suppresses the abundance of their prey a level down, this subsequently has an impact at the next trophic level down where these organisms are under less pressure from predation and become more abundant. The existence of trophic cascades in terrestrial ecosystems is well established but not as strong as in marine ecosystems due to factors that perturb this process such as omnivory and cannibalism. Top predators generally have a beneficial effect on biodiversity and the vigour of the food web. Their removal often has serious and unpredictable consequences.

Trophic Level A category of classification comprising of organisms. The trophic level is the position in a food chain where an organism functions. The organisms in each trophic level get most of their energy/nutrients from the next level down. The number of links an organism is placed from the bottom of a chain can be counted as a measure of the trophic level. It can be used as a category of classification for organisms that transform organic and inorganic material in similar ways.

Tropical Zodiac See Zodiac (Tropical).

μm The symbol used to denote micrometres. 1μm= 1 millionth of a metre (1x10-6m).

UV Radiation Or Ultra-violet radiation, is electromagnetic radiation in the range of 10nm to 400nm that is found in sunlight and specialised artificial lighting. It cannot be seen by human eyes but is responsible for sunburn and can be felt as heat. It is the part of the electromagnetic spectrum that can burn plant foliage when over exposed and can also be an agent in the destruction of crop damaging mildews. This part of the spectrum is highly visible to certain birds and insects. Some pollinators such as honey bees identify the UV light that plants reflect which aides nectar location.

Vaccination Exposing an organism to dead or weakened pathogens in an attempt to cause it to develop defences and ultimately achieve immunity. See Inoculation.

Vacuole A fluid filled space within a cell that is bound by a membrane. In some bacteria the vacuole contains gas.

Validity One of the cornerstones of scientific method. When a trial or experiment is internally valid it accurately reflects the cause and effect relationship under investigation. When the findings of an experiment are expanded to be significant to larger groups (i.e. other than the test group), we are considering the experiment's external validity e.g. if element X was supplied to a crop of beetroot and the beetroot grew better than a sample without element X then how much does this apply to all beetroot? See Reliability and Scientific Method.

Vascular Plants See Higher Plants.

Vegetative Reproduction (or **Vegetative Propagation**) This is a process where plants reproduce asexually to produce new plants without spores or seeds. The resulting clone is not as genetically diverse as plants produced sexually, although some variances to the original plant can exist. Almost all plants can reproduce in this way in theory but some are more disposed towards the process. e.g. strawberries (use runners), alliums (use bulbs), potatoes (use tubers) dandelions and raspberries (use suckers). Cuttings from plants are often treated with a rooting liquid or powder containing hormones to stimulate this kind of growth such as auxin. Seaweed is a rich source of these hormones. See Propagule.

Vermicasts Also known as Vermicompost. The solid fraction of worm excreta. See Vermiwash.

Verminephrobacter A symbiont found in the nephridia of an earthworm's gut. It is hypothesised as being beneficial to the earthworm, relating to the internal cycling of nitrogen.

Vermiwash The liquid fraction of worm excreta. Can be used as a liquid manure and an aid to seed germination. See Vermicasts.

Vernal Equinox The beginning of spring in the Northern Hemisphere when night and day are roughly equal. The moment during the apparent passage of the Sun when it intersects the ecliptic plane and the equator. See Zodiac (Tropical), Zodiac (Sidereal), Constellation.

Vertebrate Animals in the subphylum vertebrata that have a backbone such as amphibians, fish, mammals, reptiles and birds. Vertebrates account for approximately 3% of all animals. The historic distinction between vertebrate and invertebrate is considered to overemphasise the significance of vertebrates since they account for such a small branch in the tree of life. See Invertebrate.

Viniculture The growing of vines for making of wine.

Viscous A substance with a semi-liquid consistency e.g. treacle and honeydew.

Viscosity The degree of resistance to flowing which comes about from internal friction e.g. treacle has greater viscosity than water.

Vitalism The belief in a life force. This belief underlies biodynamics and to some extent organic growing. It also is the basis today of many oriental philosophies and practices e.g. Feng shui, acupuncture, sushi. There is no solid evidence for its existence according to Western scientific criteria. See Biocrystallisation, Biodynamics, Zero Point Energy and Scientific Method.

Vortex A kind of flow form often seen in nature that rotates around an axis line that can be straight or curved in any number of ways. In liquids vortices can be used for mixing or separating microorganisms depending on the flow. Where mixing is required a vortex is established and then flow is reversed which initiates counter vortices which helps to bring in regions on different levels in the system that may not be fully engaged with the mixing process.

Watershed Also called a catchment basin or drainage basin, is an area that forms a natural or man-made boundary which water drains into. A watershed ecosystem is bounded by this area and can be an important element in agroecosystems, sustainable water management and bioregional governance.

Weed In agriculture a plant that is a threat to crops either from processes such as shading out crop plants and depriving them of nutrients, or otherwise interfering with their growth such as in allelopathy. Weeds have a tendency to be extremely vigorous and overwhelm other plants. From a holistic perspective the concept of a weed does not exist, it is just a plant that has become overly dominant and may be an indication of other imbalances.

Wetting Agent Also known as a surfactant. This is added to a spray to help it flow more easily and stick better to plant leaves on application e.g. soft soap.

Whorl The conjunction on a plant of petals, leaves, sepals or branches at a single point. See Involucral Bract.

Windrow A type of composting where compostable material is built up in rows with the newest material being added to one end, gradually extending the heap. This system is sometimes just used in the curing phase in combination with other systems where migration of pathogens into matured compost is a major preoccupation. The system used alone has the advantage of enhanced migration of larger decomposers, together with the natural expansion of beneficial fungi and bacteria from older segments via phoresy and enhanced flows of moisture.

Zero Point Free Energy or Zero Point Energy (ZPE) In quantum mechanics is the lowest level of energy possible that a quantum system can have. It is an energy that exists even in a vacuum at absolute zero and is thought to be all around, ubiquitous in the universe. Richard Feynman (Nobel Laureate) stated that with regards to its potential there was 'more than enough energy in the volume of a coffee cup to evaporate all the world's oceans'. It is a phenomenon that has been proven to exist but is not well understood and has not been successfully applied in society. Arguably ZPE as understood by Nikola Tesla is not compatible with the conventional understanding of energy transfer because ZPE has features that are not local and do not fit into any classically based description. Some suggest it can be likened to Dark Matter and even the mythical chi. See Quantum Mechanics, Scientific Realism, Materialism and Vitalism.

Zodiac (Sidereal) This zodiac is influenced by Eastern astrology and bears a closer relationship to the actual position of the constellations in the sky throughout the year. It is the zodiac of choice for lunar gardeners. In some versions the sidereal zodiac is divided into 12 equal divisions like the Tropical Zodiac, however biodynamic versions divide the constellations unequally to take account of their different sizes. See Zodiac (Tropical), Vernal Equinox and Sidereal Month.

Zodiac (Tropical) The Tropical Zodiac or 'The Circle of Animals' is the traditional zodiac of Western astrologers and follows the sequence – Aries, Taurus, Gemini, Cancer, Leo, Virgo, Libra, Scorpio, Sagittarius, Capricorn, Aquarius, Pisces. This zodiac was divided into 12 equal divisions of 30°. Originally this began in the house of Aries at the vernal equinox which marked the beginning of spring. However because of the precession of the equinoxes it is out of alignment with the constellations. See Zodiac (Sidereal), Constellation and Vernal Equinox.

APPENDIX TWO

Steiner's Preparations

To follow is a description of the method and application of Steiner's famous preparations interpreted from his original 'Agriculture Course' lectures[1] made in 1924. The intention of this appendix is to provide a simple account that can be applied in your own preparations whilst staying within the original frame of reference, which would certainly include a strong emphasis on intuition and personal exploration. The preparation instructions found from various biodynamic authorities tended to vary widely in the level of detail and often went well beyond the original author's intentions. Sometimes details were refused completely for financial reasons, which raises a degree of suspicion.

From the lectures the preparations were not numbered, only acquiring these later. The numbering system used by Dr Ehrenfried Pfeiffer, an early collaborator, and biodynamic authorities such as IFOAM,[2] has been adopted solely for the sake of clarity and scholarly comparison. Extracts of Steiner's work have been taken from 'The Rudolf Steiner Archive'[3] which is the most extensive and reliable source found for his works and the only one quoted from in this account.

Steiner was essentially a vitalist, someone who believes in a life force similar to that followed today in the Orient. In Taoism and many related systems there is the conception of chi flowing through the body and the universe. The Zapotecs of Mexico also hold to a vitalist philosophy where the chilli has a special place in their culture as a vessel or channel where this energy is believed to be particularly strong. As introduced in Chapter 10, the Maoris of New Zealand attach great significance to Mauri, which is sometimes described as an all pervading energy awareness, in living water and sometimes also as separate entities in the flows of different streams. There is no scientific proof of an existence of life force.

In Steiner's cosmology different planets and constellations had an influence on plant growth and the chemical elements. Plants, animals, even the soil was thought to have awareness or the potential for it to develop.

Although extremely outlandish by today's standards Steiner's work often offers food for thought that can be overlooked in more rational and conventional modes of thinking. Holistic thinkers such as Steiner can often bring out truths or half-truths that are difficult to pin down in a rigorous way, slipping through more rigid and strictly more 'correct' approaches. An example of this is illustrated in the first quote overleaf.

Horns in Spring containing BD 500, just unearthed. With kind permission of Bleu Grijalava, New Urban Farmers.[4]

To apply directly to the soil

BD 500. Horn manure[5]

> And so, throughout the winter — in the season when the Earth is most alive — the entire content of the horn becomes inwardly alive. For the Earth is most inwardly alive in winter-time. All that is living is stored up in this manure. Thus in the content of the horn we get a highly concentrated, life-giving manuring force.[5]

In winter things have the appearance of quietening down but there is still a great deal of activity from organisms retreating into the relative warmth of the soil to hibernate and so forth, rather in the way humans tend to spend more time inside and away from the cold.

Preparation method

Obtain some fresh cow manure and some cow horns. Stuff the manure into the horns and bury it 46cm-61cm (18ft to 30ft) in well-balanced soil, not too much clay or sand. Leave this over winter and dig it up in the spring. Thoroughly mix the contents of the horn with half a bucket of lukewarm water; this could be taken from a rainwater butt that has been heated during the day in the Sun. The resulting solution is used sparingly for up to 1/3 of an acre. Best used on soil that has just been dug.

Stirring method

> Stir quickly, at the very edge of the pail, so that a crater is formed reaching very nearly to the bottom of the pail, and the entire contents are rapidly rotating. Then quickly reverse the direction, so that it now seethes round in the opposite direction. Do this for an hour and you will get a thorough penetration.[5]

Benefits

Enhances soil fertility, preserves nitrogen and allows it to be used in a form suitable for plants.

BD 501. Horn silica[5]

> So you dig out what has been exposed to the summery life within the earth...you will presently see how well this supplements the influence which is coming from the other side, out of the earth itself, by virtue of the cow-horn manure.[5]

Preparation method

Obtain some ground quartz, silica, orthoclase or feldspar. With a pestle and mortar mix with a small amount of water to form a thin dough. Stuff this into the horn and bury as in BD 500 but this time in summertime. Leave until late autumn and then dig up. Store the contents out of the ground until the spring.

Use a small amount, between a pinch and the size of a pea. Use a full bucket of water and mix the same way as in BD 500.

Benefits

Intended for use as foliar spray to complement BD 500. Beneficial to vegetables and other crops.

BD 508. Horsetail (Equisetum arvense)[6]

> Look at the equisetum plant. It has this peculiarity: it draws the cosmic nature to itself; it permeates itself with the silicious nature... yet in such a way that it does not go upward and reveal itself in the flower but betrays its presence in the growth of the lower parts.[7]
> This aristocratic gentleman, silica, lives either in the ramparts of his castle – as in the equisetum plant – or else distributed in very fine degree, sometimes indeed in highly homeopathic doses. And he contrives to tear away what must be torn away from the limestone.[8]

Preparation method

Obtain some *Equisetum*. Make a decoction[9] and dilute. Sprinkle it on the fields in small quantities wherever needed.

Benefits

Reduces Moon influences. Prevents rust, mildew, blight, other fungal growth, parasites, and over vigorous weeds, since according to Steiner this comes about when there is too much Moon influence such as an overly wet winter followed by an overly wet spring.

Compost, manure and liquid manure additions

Preparations BD 502-BD 506 are intended to be applied in conjunction to different areas of the same compost heap. The idea behind this is linked to seeing the compost heap and the area intended for its application as an entity. Each preparation draws a different quality to culminate in a metaphysical-physical creature which is then expanded on application to encompass the entire farm. The intent is to facilitate life and vitality.

BD 502. Yarrow flowers (Achillea millefolium)[10]

> One would fain say, 'In no other plant do the Nature-spirits attain such perfection in the use of sulphur as they do in yarrow'.[10]

Preparation method

Begin in spring. Pick fresh flowers, which are best, and let dry for a short amount of time. Otherwise use dried yarrow mixed with a small amount of water and squeezed onto more dried yarrow. Press the yarrow and put it into a stag's bladder and bind it up. Hang it up during the summer in as sunny place as possible. In autumn take it down and bury it shallowly in the ground then leave it there throughout winter. It can be left there until needed. Insert a small amount into the compost heap, manure pile or liquid manure.

Benefits

Soil tonic and an aid to soil fertility and vitality. According to Steiner, yarrow was good for drawing cosmic influences into the soil and making sulphur and potash useful to surrounding plants.

BD 503. Chamomile flowers (Matricaria recutita)[10]

> ... living vitality connected as nearly as possible with the earthy nature must be allowed to work upon the substance. Therefore you should take these precious little sausages – for they are truly precious – and expose them to the earth throughout the winter.[10]

Preparation method

Begin the preparation in spring. Prepare in a similar way to yarrow. Pick the yellow-white flower heads and stuff into bovine intestines. Bury shallowly in the ground throughout the winter in an area with the best topsoil. Ideally find a spot where the snow will remain for a considerable time and there is plenty of sun. A South facing frost pocket would be a suitable location. Dig out in the spring and insert into the compost heap, manure pile or liquid manure.

Benefits

Draws earthly energies. Stabilises nitrogen in the heap. It will stimulate plant growth and improve plant health.

BD 504. Stinging nettle (Urtica dioica)[10]

> This 'condiment' will make the manure intelligent, nay, you will give it the faculty to make the earth itself intelligent – the earth into which the manure is worked.[10]

Preparation method

Pick some nettles and leave them in the open for a few days. Dig a shallow hole lined with peat-moss (ethical moss substitutes are widely available). Press the nettles together lightly and place them in the hole (no bladder or intestines needed). Cover with peat-moss and mark the spot. In this way it can be identified later when it will be difficult to distinguish it from the surrounding soil. Leave for the following winter and summer. It must stay in the ground for at least a year.

Benefits

Prevents denitrification and anaerobic decomposition. Encourages synergy of living organisms to create favourable living conditions.

BD 505. Oak bark (Quercus spp.)[10]

> I refer to the oak – notably the rind of the oak, which represents an intermediate product between plant-nature and the living earthy nature.[10]

Preparation method

Take oak bark then crush it up and pulverise it to a crumblike texture. Obtain the skull of any domestic animal and place the oak bark inside. Close off the skull with any other bones. Dig a shallow hole then cover this over with peat-moss. Set up some way to get a considerable amount of rainwater running through the skull, such as re-directing the flow of a water butt fed from a drain pipe. Put into the skull some vegetable matter that will readily supply a continued presence of vegetable slime. Steiner also suggests an alternative method using a full barrel that has a flow of rainwater where vegetable matter is put inside and the skull put inside, instead of within the ground. Whichever method is chosen it must then be left through autumn and winter and added to manure heap, compost heap or liquid manure.

Benefits

Protection against disease.

BD 506. Dandelion flowers (Taraxacum officinale)[10]

> Your plant will then benefit not only by what is in the tilled field itself, whereon it grows, but also by that which is in the soil of the adjacent meadow, or of the neighbouring wood or forest. That is what happens, once it has thus become inwardly sensitive. We can bring about a wonderful interplay in Nature, by giving the plants the forces which tend to come to them through the dandelion in this way.[10]

Preparation method

Gather dandelion heads and leave them for a few days uncovered. Press them together and sew them into bovine mesentery.[11] Bury it shallowly in the soil and leave over winter. Remove the dandelion and keep until needed. Add to manure and compost or liquid manure immediately prior to application.

Benefits

A cosmic channel and when added to the earth around plants, enables them to become more self aware and more able to draw to themselves what they need from beyond their immediate area.

BD 507. Valerian flowers (Valeriana officinalis)[10]

> Add this diluted juice of the valerian flower to the manure in very fine proportions. There you will stimulate it to behave in the right way in relation to what we call the 'phosphoric' substance.[10]

Preparation method

Take flowers of valerian, mash up and mix up with warm water then strain. Add a small amount to manure pile, compost heap or liquid manure.

Benefits

Stimulate the manure to function with phosphates.

References

Chapter One

1. Edwards A. (2005) *Celtic Knot Tree*. The Ugly Goat, PO BOX 84, Keiraville 2500, NSW Australia. Email: Amyedwards@toeknuckles.com
2. Brehaut E. (1933) *Cato the Censor On Farming*. Columbia University Press, NY. p.156.

3.0 Scheuerell S. and Mahafee W. (2002) 'Compost tea: principles and prospects for plant disease control'. *Compost Science and Utilization* 10, 313-338.

3.1. Scheuerell S. and Mahafee W. (2002) 'Assessing aerated and nonaerated watery fermented compost tea and *Trichoderma harzianum* T-22 for control of powdery mildew (*Sphaerotheca pannos* var. *rosae*) of rose in the Willamette valley, Oregon'. *Phytopathology*, 90, 69. pp. Abstract.

3.2. Tränkner A. (1992) 'Use of agricultural and municipal organic wastes to develop suppressiveness to plant pathogens'. In E.S. Tjamos, G.C. Papavizas and R.J. Cook (Eds.) *Biological control of plant diseases*. Plenum Press, New York, USA. pp.35-42.

3.3. Ma L.P., Qiao X.W., Gao F. and Hao B.Q. (2001) 'Control of sweet pepper fusarium wilt with compost extracts and its mechanism'. *Chinese Journal of Applied and Environmental Biology* 7, pp.84-87.

3.4. Weltzein H.C. and Ketterer N. (1986) 'Control of Phytophthora infestans on tomato leaves and potato tubers through water extracts of composted organic wastes'. *Phytopathology* 76, 1104 (abstract 372).

3.5. Nataranjan N., Cork A., Boomathi N., Pandi R., Velavan S. and Dhakshnamoorthy G. (2006) 'Cold aqueous extracts of African marigold, *Tagetes erecta* for control'. *Crop Protection* 25 (2006), pp.1210-1213.

4.0. IAASTD (2008) 'International Assessment of Agricultural Knowledge, Science and Technology for Development Global Report'. In LaSalle T, Hepperly P. and Amadou D. (2008) *The Organic Green Revolution*. The Rodale Institute. [Re: Ecologically Degenerate, p.6.]

4.1. Xu H-I., Parr J. and Umermura H. (Eds.). 2000. *Nature Farming and Microbial Applications*. London: Food Products Press. In Taurayi S. (2011). *An investigation of natuurboerdery (natural farming) approach: a ZZ2 case study*. Faculty of Economic and Management Sciences School of Public Management and Planning. Stellenbosch University. [Re: Ecologically Degenerate/Unsustainable, p.2.]

4.2. Badgley C., Moghtader J., Quintero E., Zakem E., Chappell M.J., Avile's-Va'zquez K., Samulon. A. and Perfecto I. (2006) *Organic Agriculture and the Global Food Supply. Renewable Agriculture and Food Systems*: 22(2). Cambridge University Press, UK. pp.86-108. In Taurayi S. (2011) *An investigation of natuurboerdery (natural farming) approach: a ZZ2 case study.* Faculty of Economic and Management Sciences School of Public Management and Planning. Stellenbosch University. [Re: Ecologically Degenerate/Unsustainable, p.2.]

4.3. Bavec F. and Bavec M. (2007) *Organic Production and Use of Alternative Crops*. Taylor and Francis; Boca Raton. In Taurayi S. (2011) *An investigation of natuurboerdery (natural farming) approach: a ZZ2 case study*. Faculty of Economic and Management Sciences School of Public Management and Planning. Stellenbosch University. [Re: Ecologically Degenerate/Unsustainable, p.2.]

4.4. Altieri M. (2007) 'Agroecology: Principles and Strategies for Designing Sustainable Farming Systems. Agroecology in Action'. Berkeley. University of California. In Taurayi S. (2011) *An investigation of natuurboerdery (natural farming) approach: a ZZ2 case study.* Faculty of Economic and Management Sciences School of Public Management and Planning. Stellenbosch University. [Re: Ecologically Degenerate/Unsustainable, p.2.]

4.5. Gliessman S. (2007) *Agroecology: The Ecology of Sustainable Food Systems*, 2nd Ed. CRC Press. New York. In Taurayi S. (2011) *An investigation of natuurboerdery (natural farming) approach: a ZZ2 case study*. Faculty of Economic and Management Sciences School of Public Management and Planning. Stellenbosch University. [Re: Ecologically Degenerate, p.24 and 29.]

4.6. LaSalle T., Hepperly P. and Amadou D. (2008) *The Organic Green Revolution*. The Rodale Institute. [Re: Ecologically Degenerate, p.3. Re: Unsustainable, p.2.]

4.7. Pimentel D. and Pimentel M. (Eds.)(1996) *Food, Energy and Society*. Niwot, CO. University of Colorado Press. In Horrigan L. Robert S., Walker L. and Walker P. (1996) 'How Sustainable Agriculture Can Address the Environmental and Human Health Harms of Industrial Agriculture'. Center for a Livable Future, Johns Hopkins Bloomberg School of Public Health, Baltimore, Maryland, USA. [Re: Ecologically Degenerate/Toxic/Unsustainable, pp.445-447.]

4.8. Fookes C. et al (2001). *Organic food and farming: Myth and Reality*. The Soil Association. Bristol House. 40-56 Victoria Street. Bristol BS1 6BY. [Re: Ecologically Degenerate, p.23.]

4.9. Hamaker J.D. and Weaver (2002) *The Survival of Civilization*. Publishers Hamaker-Weaver. Don Weaver Earth Health Regeneration P. O. Box 620478, Woodside, CA 94062. [Re: Ecologically Degenerate, p.167.]

4.10. Esther D. et al (2010) *Organic Pastoral Resource Guide*. 2nd Ed.. Soil & Health Association and Bio Dynamic Association. [Re: Toxic (Fungicides and mycorrhiza), p.23.]

4.11. Horrigan L., Robert S. Walker, L and Walker P. 'How Sustainable Agriculture Can Address the Environmental and Human Health Harms of Industrial Agriculture'. Center for a Livable Future, Johns Hopkins Bloomberg School of Public Health, Baltimore, Maryland, USA. [Re: Toxic (Human Health), p.450.]

4.12. Committee on the Role of Alternative Farming Methods in Modern Production Agriculture, National Research Council (1989) *Alternative Agriculture*. National Academy Press. [Re: Toxic, p.125. Human Health, p.185.]

4.13. Germ M., Stibilj V. and Kreft I. (2007) 'Metabolic Importance of Selenium for Plants'. *The European Journal of Plant Science and Biotechnology* 1(1). Global Science Books. [Re: Toxic (Bioaccumulation of Se from Agriculture), p.91.]

4.14. Taurayi S. (2011) *An investigation of natuurboerdery (natural farming) approach: a ZZ2 case study*. Faculty of Economic and Management Sciences School of Public Management and Planning. Stellenbosch University. [Re: Toxic (Humans, wildlife, livestock), p.36.]

4.15. Noffke J. (2010) General Manager Natuurboerdery Products. Personal Interview. Mooketsi. South Africa. In Taurayi S. (2011) *An investigation of natuurboerdery (natural farming) approach: a ZZ2 case study*. Faculty of Economic and Management Sciences School of Public Management and Planning. Stellenbosch University. [Re: Toxic (Pest and Disease Management), p.77.]

4.16. Taurayi S. (2011) *An investigation of natuurboerdery (natural farming) approach: a ZZ2 case study*. Faculty of Economic and Management Sciences School of Public Management and Planning. Stellenbosch University. [Re: Toxic (Agricultural Workers), p.146.]

4.17. Van Zwieten M., Stovold G. and Van Zwieten L. (2007) *Alternatives to Copper for Disease Control in the Australian Organic Industry*. Environmental Centre of Excellence, NSW Agriculture, Wollongbar. [Re: Toxic (Non Target Effects), p.1.]

4.18. Hamaker J.D. and Weaver (2002) *The Survival of Civilisation*. Publishers Hamaker-Weaver. Don Weaver Earth Health Regeneration P. O. Box 620478, Woodside, CA 94062. [Re: Toxic, p.41.]

4.19. Anderson A., 'Calculating biodegradation rates'. *Journal of Environmental Science and Health*, Part B, 1986. 21(1): pp.41-56. In Coolong T., John S. and Smigell C. *2010 Fruit and Vegetable Research Report*. Agricultural Experiment Station. University of Kentucky. Lexington, KY, 40546, 40546. [Re: Toxic (Agricultural Workers), p.53.]

4.20. Antonious G., et al. 'Pyrethrins and piperonyl butoxide adsorption to soil organic matter'. *Journal of Environmental Science and Health*, Part B, 2005. 39(1): p. 19-32. In Coolong T. and John S. *2010 Fruit and Vegetable Research Report*. Agricultural Experiment Station. University of Kentucky. Lexington, KY, 40546, 40546. [Re: Toxic (Agricultural Workers), p.53.]

4.21. Higa T. and Parr J.F. (1994) *Beneficial and Effective Microorganisms for a Sustainable Agriculture and Environment*. International Nature Farming Research Center, Atami, Japan. [Re: Toxic, p.1.]

4.22. Weber C.(2005). In Bissdorf J. (2005) *Field Guide to Non-chemical Pest Management in Tomato Production*.

Pesticide Action Network (PAN) Germany Nernstweg 32, 22765 Hamburg, Germany. [Re: Toxic, p.7.]

4.23. Cayuela M.L., Millner P.D., Meyer S.L.F and Roig A. (2008) 'Potential of olive mill waste and compost as biobased pesticides against weeds, fungi, and nematodes'. *Science of the Total Environment 399.* [Re: Toxic, p.12.]

4.24. Parrott N. and Marsden T. (2002) *The Real Green Revolution.* Department of City and Regional Planning, Cardiff University. [Re: Unsustainable, p.7.]

4.25. Dixon G.R and Tilston E.L. (Eds.) (2010) *Soil Microbiology and Sustainable Crop Production.* Springer. [Re: Unsustainable, p.197.]

4.26. Committee on the Role of Alternative Farming Methods in Modern Production Agriculture, National Research Council. (1989) *Alternative Agriculture.* National Academy Press. [Re: Unsustainable, p.10.] www.nap.edu/catalog/1208/alternative-agriculture

4.27. Stapper M. (2006) 'Soil Fertility Management – Towards Sustainable Farming Systems and Landscapes'. CSIRO Sustainability Network. [Re: Unsustainable, p.1.] www.bml.csiro.au/susnetnl/netwl61E.pdf

5. USDA Study Team on Organic Farming (1980) *Report and Recommendations on Organic Farming.* Washington DC: USDA, 1980. National Agricultural Library. In Dan Rigby D. and Caceres D. (1997) 'The Sustainability of Agricultural Systems'. School of Economic Studies, University of Manchester & Department of Rural Development at the National University of Cordoba, Argentina. Working Paper No.10: p.1.

6. Steiner R. (1924) *Geisteswissenshaftliche Grundlagen zum Gedeihen der Landwirtschaft.* 5th Ed. 8 Vortrage 1924. Steiner Verlag, Dormnach, 1975.

7. Northbourne Lord (1940) *Look to the Land.* J.M. Dent. London. In Scofield A. (1986). 'Organic Farming – The Origin of the Name'. *Biological Agriculture and Horticulture.* 4: 1-5.

8. Taurayi S. (2011) *An investigation of natuurboerdery (natural farming) approach: a ZZ2 case study.* Faculty of Economic and Management Sciences School of Public Management and Planning. Stellenbosch University. p.23-24.

8.1. The Food and Agriculture Organisation (FAO) (1998) *The State of Food and Agriculture 1998.* In Taurayi S. (2011) *An investigation of natuurboerdery (natural farming) approach: a ZZ2 case study.* Faculty of Economic and Management Sciences School of Public Management and Planning. Stellenbosch University. p.23.

8.2. Hawken P., Amory Lovins A., and Lovins L.H. (1999) *Natural Capitalism: Creating the Next Industrial Revolution.* Boston: Little Brown and Co. p.196. In Stein M. (2000) *When Technology Fails: A Manual for Self-Reliance & Planetary Survival.* Clear Light Publishers Santa Fe, New Mexico. p.94.

9. Xu H-I., Parr J. and Umermura, H. (Eds.). (2000) *Nature Farming and Microbial Applications.* London: Food Products Press. In Taurayi S. (2011) *An investigation of natuurboerdery (natural farming) approach: a ZZ2 case study.* Faculty of Economic and Management Sciences School of Public Management and Planning. Stellenbosch University. p.2.

9.1. Badgley C., Moghtader J., Quintero, E., Zakem E. Chappell M, J., Avile's-Va'zquez K., Samulon. A. and Perfecto I. (2006) 'Organic Agriculture and the Global Food Supply'. *Renewable Agriculture and Food Systems*: 22(2); Cambridge University Press, UK. pp.86-108. In Taurayi S. (2011) *An investigation of natuurboerdery (natural farming) approach: a ZZ2 case study.* Faculty of Economic and Management Sciences School of Public Management and Planning. Stellenbosch University. p.2.

9.2. Bavec F. and Bavec M. (2007) *Organic Production and Use of Alternative Crops.* Taylor and Francis; Boca Raton. In Taurayi S. (2011) *An investigation of natuurboerdery (natural farming) approach: a ZZ2 case study.* Faculty of Economic and Management Sciences School of Public Management and Planning. Stellenbosch University. p.2.

9.3. Altieri M. (2007) *Agroecology: Principles and Strategies for Designing Sustainable Farming Systems.* Agroecology in Action. Berkeley: University of California. In Taurayi S. (2011) *An investigation of natuurboerdery (natural farming) approach: a ZZ2 case study.* Faculty of Economic and Management Sciences School of Public Management and Planning. Stellenbosch University. p.2.

9.4. Alexander G. and the Ambassador College Ag. Res. Dept. (1974) *World Crisis in Agriculture.* Ambassador College Press, Pasadena. In Hamaker J.D. and Weaver (2002) *The Survival of Civilisation.* Publishers Hamaker-Weaver. Don Weaver Earth Health Regeneration P. O. Box 620478 Woodside, CA 94062. p.116.

9.5. Fookes C. et al. (2001) *Organic food and farming: Myth and Reality.* The Soil Association. Bristol House. 40-56 Victoria Street. Bristol BS1 6BY. p.7.

9.6. Harvey C.H. (2007) *Biological Farming: A Practical Guide.* Nuffield Australia. "Tauwitchere" Narrung SA 5259. p.11.

10. Parrott N. and Marsden T. (2002) *The Real Green Revolution.* Department and Regional Planning. Cardiff University. p.7.

10.1. Gliessman S. (2007) *Agroecology: The Ecology of Sustainable Food Systems,* 2nd Ed. CRC Press. New York. In Taurayi S. (2011) *An investigation of natuurboerdery (natural farming) approach: a ZZ2 case study.* Faculty of Economic and Management Sciences School of Public Management and Planning. Stellenbosch University. p.35 and p.138.

10.2. Ikerd J.E. (1996) 'Sustaining the Profitability of Agriculture'. Presented at the Extension Pre-Conference: The Economist's Role in the Agricultural Sustainability Paradigm, San Antonio, TX, 27th July (1996) Available: www.ssu.missouri.edu/faculty/JIkerd/papers/aae-sasa.htm [cited 5 February 2001]. In Horrigan, L. Robert S. Walker, L and Walker P. 'How Sustainable Agriculture Can Address the Environmental and Human Health Harms of Industrial Agriculture'. Center for a Livable Future, Johns Hopkins Bloomberg School of Public Health, Baltimore, Maryland, USA. p.452.

11. McGiffen M. (1999) 'Introduction to the Organic Horticulture Colloquium'. Department of Botany and Plant Sciences, University of California, Riverside, CA 92521-0124. In *Hort Science* Vol. 34(3) June 1999. p.557.

12. Doran J.W. (1980) 'Soil microbial and biochemical changes associated with reduced tillage'. *Soil Sci. Soc. Am. J.*, 44, 765. In 'Developing weed-suppressive soils through improved soil quality management'. *Soil & Tillage Research* 72 (2003) 193.

13.0. Kennedy, A.C., (1999) 'Soil microorganisms for weed management'. *J. Crop Prod.*, 2, 123-138.

13.1. Li J. and Kremer R.J., (2000) 'Rhizobacteria associated with weed seedlings in different cropping systems'. *Weed Sci.*, 48, 734-741.

14. Bragwen B. and Hopkins R. (2008), Transition Initiatives Primer, http://transitiontowns.org/TransitionNetwork/TransitionNetwork#primer

15. NHM (2012) National History Museum. Cromwell Road. London. SW7 5BD www.opalexplorenature.org/?q=Identification

16. Steiner R. (1924) 'Agriculture Course Lectures 1-8'. The Rudolf Steiner Archive. http://wn.rsarchive.org/Biodynamics/search=context?query=preparations

17.0. Raupp J. (2001) 'Manure Fertilization for Soil Organic Matter Maintenance and its Effects Upon Crops and the Environment, Evaluated in a Long-term Trial'. Institute of Biodynamic Reseach. Brandschneise 5, D-642965. Darmstadt, Germany. CAB International. Sustainable Management of Soil Organic Matter. [Re: Improved Yield, p.305.]

17.1. Reeve J.R., Carpenter-Boggs L., Reganold J.P., York A.L. and Brinton W.F. (2010) 'Influence of biodynamic preparations on compost development and resultant compost extracts on wheat seedling growth'. *Bioresource Technology* 101. [Re: Improved Yield, p.5658.]

18.0. Alföldi T., Mäder P., Niggli, U., Spiess, E., Dubois, D. and Besson, J.M. (1996) 'Quality investigations in the long-term DOC-trial'. Publications of the Institute for Biodynamic Research 9, 34-43. In Hathaway-Jenkins L.J. (2011) 'The Effect of Organic Farming On Soil Physical Properties, Infiltration and Workability'. Cranfield University. p.13.

18.1. Reganold J.P., Paimer A.S., Lockhart J.C. and Macgregor A.N. (1993) 'Soil quality and financial performance on biodynamic and conventional farms in New Zealand'. *Science* 260, 344-349. In Hathaway-Jenkins L.J. (2011) 'The Effect of Organic Farming On Soil Physical Properties, Infiltration and Workability'. Cranfield University. p.13.

19. Reginold J.P. (1995) 'Soil quality and profitability of biodynamic and conventional farming systems: A review'. Department of Crop and Soil Sciences, Washington State, University, Pullman, WA 99164-6420. In *American Journal of Alternative Agriculture.* Vol. 10. Issue 01. March. p.36.

20.0. Goode J. (2003) 'Biodynamic Wine, The wineanorak's guide, Part 5: An audience with Nicolas Joly'. www.wineanorak.com/biodynamic5.htm

20.1. Revve J. (2007) 'Taking a Scientific Look at Biodynamics'. Camphill Village, Kimberton Hills. p.4. www.stellanatura.com

21. Madder P., Pfiffner L., Niggli U., Plochberger K., Velimirow A., Balzer U., Balzer F. and Besson J-M (1993) 'Effect of

three farming systems (biodymanics, bio-organic, conventional) on yield and quality of beetroot in a seven year crop rotation'. *Acta Horticulturae*, 339. pp.10-31. In 'Organic farming, food quality and human health : a Review of the Evidence'. Soil Association. p.54.

22. Martin O. (2007) 'French Intensive Gardening: A Retrospective'. *The Cultivar*. The Center for Agroecology & Sustainable Food Systems. University of California, Santa Cruz. Spring/ Winter. Vol 25, No1 & 2. p.21.
23. Bortoft H. (2010) *The Wholeness of Nature: Goethe's Way of Science*. Lindisfarne books and Floris Press. 15 Harrison Gardens, Edinburgh. p.25.
24. Jeavons J. (2002) *How To Grow More Vegetables (fruits, nuts, berries, grains and other crops) Than You Ever Thought Possible On Less Land Than You Can Imagine*. 6th Ed. Ecology Action of the Midpeninsula, 5798 Ridgewood Road, Willits, CA 95490-9730. Ten Speed Press. Berkeley. Toronto. [Re: Goethe, p.61. Seminal publishing 1974, p.3. Moon Planting, pp.71-74. Watering, pp.74-79. Departure from Steiner's solutions, p.44. Life creating more life (preamble).]
25. Weathers J. P. (1909) *French Market-Gardening including Practical Details of "Intensive Cultivation" For English Growers*. John Murray, Albemarle Street, W. London. California Digital Library, University of California. http://archive.org/details/frenchmarketgard00weatrich

26.0. De Meester L., Gómez A., Okamura B. and Schwenk K. (2002) 'The Monopolization Hypothesis and the dispersal-gene flow paradox in aquatic organisms'. *Acta Oecol.* 23, 121-135. In Fontaneto D. and Hortal J.(2012) *Biogeography of microscopic organisms, is everything small everywhere?* Cambridge, UK: Systematics Association & Cambridge University Press. p.10.

26.1. Urban M.C. and De Meester L. (2009) 'Community monopolization: local adaptation enhances priority effects in an evolving metacommunity'. P. Roy. Soc. B – *Biol. Sci.* 276, 4129-4138. In Fontaneto D. and Hortal J.(2012) *Biogeography of microscopic organisms, is everything small everywhere?* Cambridge, UK: Systematics Association & Cambridge University Press. p.10.

26.2. Hortal, J. (2011) 'Geographical variation in the diversity of microbial communities: research directions and prospects for experimental biogeography'. In Fontaneto D. and Hortal J. (2012) *Biogeography of microscopic organisms, is everything small everywhere?* Cambridge, UK: Systematics Association & Cambridge University Press. p.11.

27. Fontaneto D. and Hortal J. (2012) *Biogeography of microscopic organisms, is everything small everywhere?* Cambridge, UK: Systematics Association & Cambridge University Press. pp.1-11.
28. Higa T. and Parr J.F. (1994) 'Beneficial and Effective Microorganisms for a Sustainable Agriculture and Environment'. International Nature Farming Research Center, Atami, Japan. [Re: Benefits, p.1.]
29. Ingham E.R. (2005) 5th Ed. *The Compost Tea Brewing Manual*. Soil Food Web Inc. 728 SW Wake Robin Ave. Corvallis, Oregon 97333. [Re: Food Web, pp.7-10.]
30. Walter C., Berridge M.V and Tribe D. (2001). 'Evidence of Rebuttal. Executive Summary. Genetically engineered Klebsiella planticola: A threat to terrestrial plant life?' *Royal Commission on Genetic Engineering* (2001). Academic Address: Department of Microbiology and Immunology, University of Melbourne. Parkville, Australia 3010.
31. Chalker-Scott L. (2007) 'Horticultural Myths: Compost tea: Examining the science behind the claims'. Winter Edition. *Master Gardener*. www.MasterGardenerOnline.com. pp.8-10.
32. Smith R.F. and Smith G.L. (1949) 'Supervised control of insects: Utilizes parasites and predators and makes chemical control more efficient'. *California Agriculture*. May 3 (5): 3–12.

33.0. D.C., Reid C.P.P. and Cole C.V. (1983) 'Biological strategies of nutrient cycling in soil systems'. *Advances in Ecological Research*, 13: 1-55. In Harris A. and Hill R. (2007) 'Carbon-Negative Primary Production: Role of Biocarbon and challenges for Organics in Aoteroa/New Zealand'. Tauranga, New Zealand, *Journal of Organic Systems*. Vol.2 No.2. Lincoln University, New Zealand. p.3. and p.7.

33.1. Gupta V.V.S.R. and Roget D.K. (2004) 'Understanding soil biota and biological functions: Management of soil biota for improved benefits to crop production and environmental health'. CSIRO Land and Water, Glen Osmond SA. In Lines-Kelly R. (Ed) (2004) *Soil biology in agriculture. Proceedings of a workshop on current research into soil biology in agriculture*. Tamworth Sustainable Farming Training Centre 11-12 August 2004. NSW Department of Primary Industries, Orange 2800. p.3.

34. Allison F.E. (1973) 'Soil Organic Matter and its Role in Crop Production'. Developments in Soil Science. Vol 3 (1973). pp. Preface.
35. Lindeman R.L. (1942) 'The trophic-dynamic aspects of ecology'. *Ecology* 23, 399-418.

36.0. Smith, S.E. and Read D.J. (1997) *Mycorrhizal Symbiosis*, 2nd Ed. Academic Press, London. In Huang H., Zhang S., Shan X., Chen B., Zhu Y., Nigel J. and Bell B. (2006) 'Effect of arbuscular mycorrhizal fungus (*Glomus caledonium*) on the accumulation and metabolism of atrazine in maize (*Zea mays* L.) and atrazine dissipation in soil'. Center for Environmental Policy, Imperial College, London, UK. p.452.

36.1. Perry D.A, Amaranthus M.P, Borchers S.L. and Brainerd R.E. (1989), 'Bootstrapping in ecosystems: Internal interactions largely determine productivity and stability in biological systems with strong positive feedback'. *Bioscience* 39, 230-237. In Rose P.S. (1997) 'Mutualistic biodiversity networks: the relationship between soil biodiversity and mutualism, and their importance to ecosystem function and structural organisation'. Life Sciences. University of Manchester, UK. p.3.

37.0. DeAngelis D.L. (1992) *Dynamics and nutrient cycling and food webs*. Chapman and Hall, New York. In Polis G.A. and Strong D.R. (1996) 'Food Web Complexity and Community Dynamics'. *The American Naturalist*, Vol. 147. No.5. May. p.819.

37.1. Fenchel T. (1988) 'Marine plankton food chains'. *Annual Review of Ecology and Systematics* 19, 19-38. In Polis G.A. and Strong D.R. (1996) 'Food Web Complexity and Community Dynamics'. *The American Naturalist*, Vol. 147. No.5. May. p.819.

37.2. Hairston N.G. (1993) 'Cause-effect relationships in energy flow, trophic structure and interspecific interactions'. *American Naturalist* 142, 379-411. In Polis G.A. and Strong D.R. (1996) 'Food Web Complexity and Community Dynamics'. *The American Naturalist*, Vol. 147. No.5. May. p.819.

37.3. Odum E.P. and Biever L.J. (1984) 'Resource quality, mutualism and energy partitioning in food chains'. *American Naturalist* 124, 360-376. In Polis G.A. and Strong D.R. (1996) 'Food Web Complexity and Community Dynamics'. *The American Naturalist*, Vol. 147. No.5. May. p.819.

37.4. Polis G.A. (1981) 'The evolution and dynamics of intraspecific predation'. *Annual Review of Ecology and Systematics* 12, 225-251. In Polis G.A. and Strong D.R. (1996) 'Food Web Complexity and Community Dynamics'. *The American Naturalist*, Vol. 147. No.5. May. p.819.

37.5. Wieger and Owen (1971) 'Trophic structure, available resources and population density in terrestrial vs. aquatic ecosystems'. *Journal of Theoretical Biology* 30, 69-81. In Polis G.A. and Strong D.R. (1996) 'Food Web Complexity and Community Dynamics'. *The American Naturalist*, Vol. 147. No.5. May. p.819.

38. Mooney K.A., Halitschke R., Andre K. and Agrawal A.A. (2010) 'Evolutionary Trade-Offs in Plants Mediate the Strength of Trophic Cascades'. 26 March. *Science*, Vol. 327. p.1642.
39. Lavelle P. (1988) 'Earthworms and the soil system'. *Biol. Fertil. Soil.* 6, 237-251. In Zirbes L., Thonart P. and Haubruge E. (2012) 'Microscale interactions between earthworms and microorganisms: a review'. *Biotechnol. Agron. Soc. Environ.* 16(1). p.125.

40.0. Amador J.A. & Görres J.H. (2007) 'Microbiological characterization of the structure built by earthworms and ants in an agricultural field'. *Soil Biol. Biochem.* 39, 2070-2077. In Zirbes L., Thonart P. and Haubruge E. (2012) 'Microscale interactions between earthworms and microorganisms: a review'. *Biotechnol. Agron. Soc. Environ.* 16(1). p.128.

40.1. Brown G.G. (1995) 'How do earthworms affect microfloral and faunal community diversity?' *Plant Soil* 170, 209-231. In Zirbes L., Thonart P. and Haubruge E. (2012) 'Microscale interactions between earthworms and microorganisms: a review'. *Biotechnol. Agron. Soc. Environ.* 16(1). p.126.

40.2. Brown G.G. and Doube B.M. (2004) 'Functional interactions between earthworms, microorganisms, organic matter, and plants'. In Zirbes L., Thonart P. and Haubruge E. (2012) 'Microscale interactions between earthworms and microorganisms: a review'. *Biotechnol. Agron. Soc. Environ.* 16(1). p.126.

40.3. Graff O. (1971) 'Beeinflussen Regenwurmröhren die Pflanzenernährung'. *Landbauforschung Volkenrode* 21. 303-320. In Zirbes L., Thonart P. and Haubruge E. (2012) 'Microscale interactions between earthworms and microorganisms: a review'. *Biotechnol. Agron. Soc. Environ.* 16(1). p.128.

40.4. Jayasinghe B.A.T.D. and Parkinson D. (2009) 'Earthworms as the vectors of actinomycetes antagonistic to litter decomposer fungi'. *Appl. Soil Ecol.* 43, 1-10. In Zirbes L., Thonart P. and Haubruge E.(2012) 'Microscale interactions between earthworms and microorganisms: a review'. *Biotechnol. Agron. Soc. Environ.* 16(1). p.128.

40.5. Jégou D., Schrader S., Diestel H. and Cluzeau D. (2001) 'Morphological, physical and biochemical characteristics of burrow walls formed by earthworms'. *Appl. Soil Ecol.* 17, 165-174. In Zirbes L., Thonart P. and Haubruge E.(2012) 'Microscale interactions between earthworms and microorganisms: a review.' *Biotechnol. Agron. Soc. Environ.* 16(1). p.128.

40.6. Maraun M. et al. (1999) 'Middens of the earthworm *Lumbricus terrestris* (Lumbricidae): microhabitats for micro- and mesofauna in forest soil'. *Pedobiologia* 43, 276-287. In Zirbes L., Thonart P. and Haubruge E.(2012) 'Microscale interactions between earthworms and microorganisms: a review'. *Biotechnol. Agron. Soc. Environ.* 16(1). p.126.

40.7. Shuster W.D., Subler S. and McCoy E.L. (2001) 'Deep-burrowing earthworm additions changed the distribution of soil organic carbon in a chisel-tilled soil'. *Soil Biol. Biochem.* 33, 983-996. In Zirbes L., Thonart P. and Haubruge E.(2012) 'Microscale interactions between earthworms and microorganisms: a review'. *Biotechnol. Agron. Soc. Environ.* 16(1). p.126.

40.8. Tiunov A.V. and Scheu S., (1999) 'Microbial respiration, biomass, biovolume and nutrient status in burrow walls of *Lumbricus terrestris* L. (Lumbricidae)'. *Soil Biol. Biochem.* 31, 2039-2048. In Zirbes L., Thonart P. and Haubruge E.(2012) 'Microscale interactions between earthworms and microorganisms: a review'. *Biotechnol. Agron. Soc. Environ.* 16(1). p.128.

40.9. Tiunov A.V. and Sheu S., (2000) 'Microfungal communities in soil, litter and casts of *Lumbricus terrestris* L. (Lumbricidae): a laboratory experiment'. *Appl. Soil Ecol.* 14. 17-26. In Zirbes L., Thonart P. and Haubruge E. (2012) 'Microscale interactions between earthworms and microorganisms: a review.' *Biotechnol. Agron. Soc. Environ.* 16(1). p.126.

41.0. Bonkowski M., Griffin B.S. and Ritz K. (2000) 'Food preferences of earthworms for soil fungi'. *Pedobiologia* 44, 666-676. In Zirbes L., Thonart P. and Haubruge E. (2012) 'Microscale interactions between earthworms and microorganisms: a review'. *Biotechnol. Agron. Soc. Environ.* 16(1). p.127.

41.1. Doube B.M., Schmidt O., Killham K. and Correll R. (1997) 'Influence of mineral soil on the palatability of organic matter for lumbricid earthworms: a simple food preference study'. *Soil Biol. Biochem.* 29, 569-575. In Zirbes L., Thonart P. and Haubruge E. (2012) 'Microscale interactions between earthworms and microorganisms: a review.' *Biotechnol. Agron. Soc. Environ.* 16(1). p.127.

41.2. Doube B.M. and Brown G.G. (1998) 'Life in a complex community: functional interaction between earthworms, organic matter, microorganisms and plants'. In Zirbes L., Thonart P. and Haubruge E. (2012) 'Microscale interactions between earthworms and microorganisms: a review.' *Biotechnol. Agron. Soc. Environ.* 16(1). p.127.

41.3. Neilson R. and Boag B. (2003) 'Feeding preferences of some earthworm species common to upland pastures in Scotland'. *Pedobiologia* 47, 1-8. In Zirbes L., Thonart P. and Haubruge E. (2012) 'Microscale interactions between earthworms and microorganisms: a review'. *Biotechnol. Agron. Soc. Environ.* 16(1). p.127.

41.4. Satchell J.E. (1967) 'Lumbricidae'. In Zirbes L., Thonart P. and Haubruge E. (2012) 'Microscale interactions between earthworms and microorganisms: a review'. *Biotechnol. Agron. Soc. Environ.* 16(1). p.127.

42.0. Cooke A. and Luxton M. (1980) 'Effect of microbes on food selection by *Lumbricus terrestris* L'. *Rev. Ecol. Biol. Sol.* 17, 365-370. In Zirbes L., Thonart P. and Haubruge E. (2012) 'Microscale interactions between earthworms and microorganisms: a review'. *Biotechnol. Agron. Soc. Environ.* 16(1). p.127.

42.1. Wright M.A. (1972) 'Factors governing ingestion by the earthworm *Lumbricus terrestris* (L.), with special reference to apple leaves'. *Ann. Appl. Biol.* 70, 175-188. In Zirbes L., Thonart P. and Haubruge E. (2012) 'Microscale interactions between earthworms and microorganisms: a review'. *Biotechnol. Agron. Soc. Environ.* 16(1). p.127.

43.0. Sinha R.K., Chauhan K., Valani D., Chandran V., Soni B.K. and Patel V. (2010) 'Earthworms: Charles Darwin's "Unheralded Soldiers of Mankind": Protective & Productive for Man & Environment'. Griffith School of Engineering (Environment), Griffith University, Brisbane, Australia. *Journal of Environmental Protection* 1. p.252.

43.1. Edward C.A. (2000) 'Potential of vermicomposting for processing and upgrading organic waste', Ohio State University. Ohio. U.S. In Sinha R.K., Herat S., Valani D.and Chauhan K. (2010) 'Earthworms - The Environmental Engineers: Review of Vermiculture Technologies for Environmental Management & Resource Development'. Griffith School of Engineering, Griffith University, Brisbane, Australia. QLD-4111. p.7.

44.0. Horn M.A., Schramm A. and Drake H.L. (2003) 'The earthworm gut: an ideal habitat for ingested N2O-producing microorganisms'. *Appl. Environ. Microbiol.* 69, 1662-1669. In Zirbes L., Thonart P. and Haubruge E.(2012) 'Microscale interactions between earthworms and microorganisms: a review'. *Biotechnol. Agron. Soc. Environ.* 16(1). p.128.

44.1. Ihssen J. et al. (2003) 'N2O-producing microorganisms in the gut of the earthworm *Aporrectodea caliginosa* are indicative of ingested soil bacteria'. *Appl. Environ. Microbiol.* 69, 1655-1661. In Zirbes L., Thonart P. and Haubruge E. (2012) 'Microscale interactions between earthworms and microorganisms: a review'. *Biotechnol. Agron. Soc. Environ.* 16(1). p.128.

44.2. Aira M., Sampedro L., Monroy F. and Domínguez J. (2008) 'Detritivorous earthworms directly modify the structure, thus altering the functioning of a microdecomposer food web'. *Soil Biol Biochem.* 40, 2511-2516. In Dixon G.R and Tilston E.L. (Eds.) (2010) *Soil Microbiology and Sustainable Crop Production*. Springer. p.96.

45. Davidson S.K., Powell R.J. and Stahl D.A. (2010). 'Transmission of a bacterial consortium in *Eisenia fetida* egg capsules'. *Environ. Microbiol.* 12, 2277-2288. In Zirbes L., Thonart P. and Haubruge E. (2012) 'Microscale interactions between earthworms and microorganisms: a review'. *Biotechnol. Agron. Soc. Environ.* 16(1). p.127.

46. Post-materialism. See p.157 and p.197. E. Fisher (2019) *Compost Teas for the Organic Grower*. Permanent Publications.

47. Radin D. , Michel L., Johnston J., Delorme A. (2012) 'Psychophysical interactions with a double-slit interference pattern'. *Physics Essays*, 26, (2013). p.157. www.deanradin.com/evidence/RadinPhysicsEssays2013.pdf

47.1. PEAR (2010) 'Princeton Engineering Anomalies Research. Scientific Study of Consciousness-Related Physical Phenomena: Implications and Applications'. http://pearlab.icrl.org/implications.html

48. PEAR (2010) 'Scientific Study of Consciousness-Related Physical Phenomena: Experimental Research. Princeton Engineering Anomalies Research'. http://pearlab.icrl.org/experiments.html

Chapter Two

1. Jeavons J. (2002) *How To Grow More Vegetables (fruits, nuts, berries, grains and other crops) Than You Ever Thought Possible On Less Land Than You Can Imagine.* 6th Ed. Ecology Action of the Midpeninsula, 5798 Ridgewood Road, Willits, CA 95490-9730. Ten Speed Press. Berkeley. Toronto. p.61.

2.0. Arnon D. I. (1943) 'Mineral Nutrition of Plants'. *Annual Review of Biochemistry.* 12, pp.493-528.

2.1. Arnon D.I. and Stout P.R. (1939) 'The essentiality of certain elements in minute quantity for plants with special reference to copper'. *Plant Physiol.* 14, 371-375. In Barker A.V. and Pilbeam D.J. (Eds.) (2007) *Handbook of Plant Nutrition*. CRC Press. Taylor and Francis group. p.3.

3. Barker A.V. and Pilbeam D.J. (Eds.) (2007) *Handbook of Plant Nutrition*. CRC Press. Taylor and Francis group. [Re: Essential Elements, p.4. Dry mass volumes, p.5.]

4. von Liebig J.F. (1855) *Principles of Agricultural Chemistry, With Special Reference to the Late Researches Made in England*. London: Walton & Maberly. p.136. In Barker A.V. and Pilbeam D.J. (Eds.) (2007) *Handbook of Plant Nutrition*. CRC Press. Taylor and Francis group. p.12.

5.0. Epstein E. (2009) 'Silicon: its manifold roles in plants'. *Ann. Appl. Biol.* Department of Land, Air and Water Resources – Soils and Biogeochemistry, University of California, Davis, CA, USA. p.155.

5.1. Coder K.D. (2011) *Essential Elements of Tree Health*. Warnell School of Forestry & Natural Resources University of Georgia. p.69.

6. Barker A.V. and Pilbeam D.J. (Eds.) (2007) *Handbook of Plant Nutrition*. CRC Press. Taylor and Francis group. [Re: Sodium, p.571. Boron, p.244.]

7. Barker A.V. and Pilbeam D.J. (Eds.) (2007) *Handbook of Plant Nutrition*. CRC Press. Taylor and Francis group. [Re: Fe, p.330. Mn, p.352. Zn, p.412. B, p.242. Ni, p.396. Mg & K

interactions, p.150. Mo, p.376.]

8. Dijkstra E. (2010) *Organic Pastoral Resource Guide*. 2nd Ed. EcoAgriLogic Ltd. Published by the Soil & Health Association and Bio Dynamic Association. [Re: Root Metabolism and Copper and Root Metabolism, p.45. Sodium and White Clover, p.70.]
9. Leo M., Walsh D. and Beaton J.D. (Eds.) (1973) *Soil Testing and Plant Analysis*. Soil Science Society of America. Madison, Wisconsin.
10. Loneragan J. F. (2011) Feature Essay 16.1. 'A brief history of plant nutrition'. http://sci-wikibook.bacs.uq.edu.au
11. Nicholas D.J.D. (1961) 'Minor Mineral Elements'. *Annual Review of Plant Physiology* 12. pp.63-90.
12. Soetan et al. (2010) *African Journal of Food Science* Vol. 4(5) pp.200-222, May 2010. p.217.
13. Unnamed (2012) Holon (philosophy.) Wikipedia. [Re: Example of Holons being misrepresented] http://en.wikipedia.org/wiki/Holon_%28philosophy%29
14. Koestler A. (1967) *The Ghost in the Machine*. London: Hutchinson. 1990 reprint edition, Penguin Group. p.48.
15. Mollison B. (2001) *Introduction to Permaculture*. 9th Ed. Barking Frogs Permaculture Center. Publisher and Distributor of Permaculture Publications. POB 52, Sparr FL 32192-0052 USA. pp.12-13. Email: YankeePerm@aol.com.

16.0. Combs G.F. (2001) 'Selenium in global food systems'. *British Journal of Nutrition* 85, 517-547. In Zhao F.J., Su Y.H., Dunham S.J., Rakszegi M., Bedo Z., McGrath S.P., Shewry P.R. (2009) 'Variation in mineral micronutrient concentrations in grain of wheat lines of diverse origin'. *Journal of Cereal Science* 49 (2009). p.290.

16.1. Rayman M.P. (2002) 'The argument for increasing selenium intake'. *Proceedings of the Nutrition Society* 61, 203-215. In Zhao F.J., Su Y.H., Dunham S.J., Rakszegi M., Bedo Z., McGrath S.P., Shewry P.R. (2009) 'Variation in mineral micronutrient concentrations in grain of wheat lines of diverse origin'. *Journal of Cereal Science* 49 (2009). p.290.

17. Fairweather-Tait S.J., Bao Y., Broadley M.R., Collings R., Ford D., Hesketh J.E., and Hurst R. (2011) 'Selenium in Human Health and Disease'. *Antioxidant and Redox Signalling* Volume 14, Number 7. pp.1345-6.
18. Salisbury F.B. and Ross C.W. (1992) *Plant Physiology*. 4th Ed. Belmont, CA, Wadsworth Publishers. pp.124-125. In Barker A.V. and Pilbeam D.J. (Eds.) (2007) *Handbook of Plant Nutrition*. CRC Press. Taylor and Francis group. p.500.
19. Barker A.V. and Pilbeam D.J. (Eds.) (2007) *Handbook of Plant Nutrition*. CRC Press. Taylor and Francis group. p.501.
20. Jebb S.A. and Moore R. (2000) *MRC Human Nutrition Research*. Downhams Lane. Milton Rd. Cambridge. CB41XJ. p.55.
21. Barker A.V. and Pilbeam D.J. (Eds.) (2007) *Handbook of Plant Nutrition*. CRC Press. Taylor and Francis group. p.8.

22.0. USDA (1999) *Soil Quality Test Kit Guide*. United States Department of Agriculture. Agricultural Research Service. www.nrcs.usda.gov/Internet/FSE_DOCUMENTS/stelprdb1044790.pdf

22.1. USDA (prev new)(1999), Soil Quality Indicator Sheets, United States Department of Agriculture, Natural Resources Conservation Service: Soils. www.nrcs.usda.gov/wps/portal/nrcs/detail/soils/health/assessment

23. USDA (2009), Soil testing: Small Scale Solutions for your Farm, United States Department of Agriculture. p.1. www.nrcs.usda.gov/Internet/FSE_DOCUMENTS/stelprdb1167377.pdf
24. Pitchcare (2016). www.pitchcare.com/shop/pitchcare-soil-analysis-and-soil-testing-kits/
25. USDA (2008), Soil Quality Indicators: Bulk Density, United States Department of Agriculture. www.nrcs.usda.gov/wps/PA_NRCSConsumption/download?cid=nrcs142p2_051591&ext=pdf
26. Gershuny G. and Smillie J. (2010) *The Soul of the Soil, A Guide to Ecological Soil Management*. 2nd Ed. In Singh A. (2010). 'Signals from the Soil'. *The Canadian Organic Grower*, Summer, p.41.

27.0. Alexander S. (2014) *Illustrated Guide to Tillage Weeds*. Teagasc, Co. Dublin [Re: Emergence in Tilled Soil, p.2.] www.teagasc.ie/media/website/publications/2014/Guide_to_identifying_Tillage_Weeds_2014.pdf

27.1. Bond W., Davies G., Turner R. (2007) 'The biology and non-chemical control of Small Nettle (*Urtica urens* L.)'. HDRA, Ryton Organic Gardens, Coventry, CV8, 3LG, UK. [Re: Annual Nettle and Low Lime, p.1.]

28.0. Birkett D.A., Maggs C.A., Dring M.J. and Boaden P.J.S. and R.Seed (1998) *Infralittoral Reef Biotopes with Kelp Species An overview of dynamic and sensitivity characteristics for conservation management of marine SACs*. Scottish Association of Marine Science (UK Marine SACs Project). p.15.

28.1. DETR (1994) *CLR Report No 02 Volume Two (Of Two) – Guidance on Preliminary Site Inspection of Contaminated Land*. Department of the Environment, Transport and the Regions. p.77.

29. Hamilton N. (2007) *Grow Organic*. New Holland publishers (UK) Ltd. [Re: Tomato, p.244. Potato, p.230. Chilli, p.222. Leek, Garlic, Onion, p.204. Chives, p.260. Carrot, p.139. Brussel Spouts, p.125. Cabbage, p.129. Swede, p.238. Beetroot, p.112. Jerusalem Artichoke, p.90. Sweetcorn, p.241.]
30. Jett J.W.(2005) *Horticulture: Plant pH Preferences*. West Virginia University. p.4.
31. Patterson L. (2009) *Vegetable Families*. OSU/Lane County Extension Service. p.17. [Re: Rhubarb pH, p.17. Peas pH, p.12. Runner Beans pH, p.13.] http://extension.oregonstate.edu/lane/sites/default/files/VegFam.pdf
32. Backer H. (1992) *Growing Fruit*. The Royal Horticultural Society. Octopus Publishing Group Ltd. [Re: Blueberry, p.82. Cobnut, p.166.]
33. Tallec T., Diquélou S., Lemauviel S., Cliquet J.B., Lesuffleur F. and Ourry A. (2008) 'Nitrogen:sulfur ratio alters competition between *Trifolium repens* and *Lolium perenne* under cutting: production and competitive abilities'. *European Journal of Agronomy* 29, 94-101. In Varin S., Cliquet J., Personeni E., Avice J. and Lemauviel-Lavenant S. (2010) 'How does sulphur availability modify N acquisition of white clover (*Trifolium repens* L.)?' *Journal of Experimental Botany* Vol. 61, No. 1. p.226.
34. Thomas B. (2007) *Maritime Pasture Manual*. AgraPoint International Inc. Nova Scotia Dept. of Agriculture. p.54. Email: info@agrapoint.ca
35. Williams U.S. (1992) *A Textbook of Biology*. 3rd Ed. In Soetan K.O., Olaiya C.O. and Oyewole O.E. (2010) 'The importance of mineral elements for humans, domestic animals and plants: A review'. *African Journal of Food Science* Vol. 4(5). May. p.203.
36. Mason J. (1951) *Sustainable Agriculture*, 2nd Ed. National Library of Australia Cataloguing-in-Publication. Landlinks Press. PO Box 1139. Collingwood Vic. 3066. Australia. p.94.
37. Davies K. and Ballingall M. (2009) 'Technical Note TN615'. SAC Consultancy Services Division. Sandpiper House. Ruthvenfield Rd, Perth, PH1 3EE.

38.0. Wilman D. & Riley J. A. (1993) 'Potential nutritive value of a wide range of grassland species'. *Journal of Agricultural Science* 120, 43-49. In Bond W., Davies G. and Turner R. (2007) 'The biology and non-chemical control of Common Nettle (*Urtica dioica* L.)'. HDRA, Ryton Organic Gardens, Coventry, CV8, 3LG, UK. [Re: N, Ca, Mg, p.1.]

38.1. Grime J.P., Hodgson J.G. and Hunt R. (1988) *Comparative Plant Ecology*, Unwin Nt.2 Hyman Ltd, London, UK. In Bond W., Davies G. and Turner R. (2007) 'The biology and non-chemical control of Common Nettle (*Urtica dioica* L.)'. HDRA, Ryton Organic Gardens, Coventry, CV8, 3LG, UK. [Re: N, Ca, Mg, p.1.]

38.2. Salisbury E. (1962) 'The biology of garden weeds. Part II'. *Journal of the Royal Horticultural Society* 87, 458-470 & 497-508. In Bond W., Davies G. and Turner R. (2007) 'The biology and non-chemical control of Common Nettle (*Urtica dioica* L.)'. HDRA, Ryton Organic Gardens, Coventry, CV8, 3LG, UK. [Re: Iron, p.1.]

39. Kourik R. (2005) *Designing And Maintaining Your Edible Landscape Naturally*. Chelsea Green Publications. p.265.
40. Lawson L.D. (1998) 'Garlic: a review of its medicinal effects and indicated active compounds'. In Bongiorno P.B, Fratellone P.M. and LoGiudice P. (2008) 'Potential Health Benefits of Garlic (*Allium Sativum*): A Narrative Review'. The Berkeley Electronic Press. p.1.
41. (Eds) (2003) *Phytoremediation: Transformation and Control of Contaminants*. New Jersey, Wiley-Interscience Inc. [Re: Ni and Indian Mustard, Zn and Indian Mustard, Cu and Indian Mustard, p.19 and p.898. Cu and Sunflower, p.898. Mn and Indian Mustard, Zn and Sunflower, Ni and Sunflower, Mn and Sunflower, p.19.]
42. Hodson M.J., White P.J., Mead A. and Broadley M.R. (2005) 'Phylogenetic Variation in the Silicon Composition of Plants'. *Annals of Botany* 96. School of Biological and Molecular Sciences, Oxford Brookes University, UK. p.1028.

www.aob.oxfordjournals.org
43. Marschner H. (1995) *Mineral Nutrition of Higher Plants.* 2nd Ed. New York: Academic Press. pp.369-379. In Barker A.V. and Pilbeam D.J. (Eds.) (2007) *Handbook of Plant Nutrition.* CRC Press. Taylor and Francis group. p.381.
44. Whittenberger, R. T. (1945) 'Silicon absorption by rye and sunflower'. *Am. J. of Bot.* 32(9), 539-549. In Hogendorp B.K. (2008) 'Effects of silicon-based fertiliser applications on the development and reproduction of insect pests associated with greenhouse-grown crops'. University of Illinois at Urbana-Champaign. p.8.
45. Ragland K.W. and Aerts D.J. (1991) 'Properties of Wood for Combustion Analysis'. *Bioresource Technology* 37, p.166.
46.0. Zell. H. [Photo Credits]. GNU Free Licence, version 1.2. 'Shepherd's Purse' (2009): https://commons.wikimedia.org/wiki/File: Capsella_bursa-pastoris_002.JPG Annual Nettle (2010) : https://en.wikipedia.org/wiki/Urtica_urens#/media/File:Urtica_urens_002.JPG
46.1. O'Connor P. (2016) [Photo Credit]. 'Pineapple mayweed' (Matricaria discoidea). July 16. Cropped. Heartwood Forest, Sandridge, Herts.(https://creativecommons.org/licenses/by-sa/2.0/) www.flickr.com/photos/anemoneprojectors/28757464942

Chapter Three

1. Edmonds K. (2012) [Photo Credit]. 'Sockeye salmon, Adams River, British Columbia, Canada'. WWF Blog. World Wildlife Federation. http://blog.wwf.ca/blog/2012/04/30/fishy-business-budget-bill-rolls-back-protection-for-fish/sockeye-salmon-adams-river-british-columbia-canada-2/
2. Molinero L. (2005) 'The Meaning of Salmon in the Northwest: A Historical, Scientific and Sociological Study'. Friends of Thoreau Environmental Program Research Institute for North American Studies, University of Alcalá, Madrid. p.4.
3. Dixon G.R. and Tilston E.L (Eds.) (2010) *Soil Microbiology and Sustainable Crop Production.* Springer. [Re: Facultative anaerobes, p.75. Humus, p.84.]
4. Epstein E. and Bloom A. (2004) [Photo Credit]. *Plant Nutrition.* Sinauer Associates, Sunderland, MA. In Berry W. (2010) 'Symptoms of Deficiency in Essential Minerals'. Topic 5.1.
5.0. King F. H. *Farmers of Forty Centuries: Organic Farming in China, Korea, and Japan*, London, 1916. In Howard, Sir Albert (1943) *An Agricultural Testament.* Oxford University Press, Inc. p.53.
5.1. Jenkins J. (2013) *The Humanure Handbook.* Jenkins Publishing, 143 Forest Lane, Grove City, PA 16127 USA. [Re: Further Reading] http://humanurehandbook.com/about.html
6. Hodges S.C. (2000) *Soil Fertility Basics.* Soil Science Extension. North Carolina State University. p.19.
7. Barker A.V. and Pilbeam D.J. (Eds.) (2007) *Handbook of Plant Nutrition.* CRC Press. Taylor and Francis group. [Re: Phosphorus and Deficiency, p.54-55. Potassium and Essentiality, p.92. Potassium and Structure, p.92. Potassium and Deficiency, p.99.]
8. Soetan K.O., Olaiya C.O. and Oyewole O.E. (2010) 'The importance of mineral elements for humans, domestic animals and plants: A review'. *African Journal of Food Science* Vol. 4(5). May. p.207.
9. Sharma S., Mahotra P. and Bhattacharyya A.K. (2008) 'Effect of electroplating industrial waste on "available phosphorus" of soil in relation to other physico-chemical properties'. *Afr. J. Environ. Sci. Technol.*, 2(9): 257-264. In Soetan K.O., Olaiya C.O. and Oyewole O.E. (2010) 'The importance of mineral elements for humans, domestic animals and plants: A review'. *African Journal of Food Science* Vol. 4(5). May. p.207.
10. O'Neil P. (1998) *Environmental Chemistry.* 3rd ed. Blackie Academic and Professional. [Re: Phosphorus, p.123. Nitrogen, p.89. Shatter Rocks, p.51. Algal Blooms, p.125.]
11. Mullins G. (2011) 'Chapter 4. Basic Soil Fertility'. In Goatley M. Jr and Hensler K. (Eds.) (2011) *Urban Nutrient Management Handbook.* Virginia Cooperative Extension. p.50.
12. Dijkstra E. (2010) *Organic Pastoral Resource Guide.* 2nd Ed. EcoAgriLogic Ltd. Published by the Soil & Health Association and Bio Dynamic Association. p.39.
13. Mullins G. (2009) *Phosphorus, Agriculture & the Environment.* Department of Crop and Soil Environmental Sciences. Virginia State University. p.7.
14. Pajuelo E. and Stougaard J. (2005) 'Lotus Japonicus as a Model System'. In Marquez A.L. (Ed) (2005) *Lotus Japonicus Handbook.* Published by Springer. P.O. Box 17, 3300 AA, Dordrecht. The Netherlands. Chapter 1.1. p.4.
15. Mikkelsen R. (2008) 'Managing Potassium for Organic Crop Production'. In *Better Crops.* Vol 92. No. 2. International Plant Nutrition Institute (IPNI). 3500 Parkway Lane, Suite 550, Norcross, GA 30092-2844 USA. [Re: Metabolic processes, p.26. Crop Yield and Quality, p.29.]
16. Imas P. and Magen H. (2000) 'Potash Facts in Brief'. Internationa Potash Institute. c/o DSW, Potash House, P.O.B. 75. Beer-Sheva, 84100, Israel. p.2.
17. Beringer H. and Nothdurft F. (1985) 'Effects of Potassium on Plant and Cellular Structures'. In Munson R.D. (Ed) (1985) *Potassium in Agriculture.* American Society of Agronomy, Crop Science Society of America, Soil Science Society of America. p.354.
18. Boodley J.W. (1975) *Recognizing Nutrient Deficiency Symptoms.* Cornell University. 107 Nott Terrace, Suite 301, Schenectady, NY 12308. Ithaca, New York. April. p.2.
19. Arocena J. M., Glowa K. R., Massicotte H. B. and Lavkulich L. (1999) 'Chemical and mineral composition of ectomycorrhizosphere soils of subalpine fir (*Abies lasiocarpa* (Hook.) Nutt.) in the AE horizon of a Luvisol'. *Can. J. Soil Sci.* 79. p.25. www.nrcresearchpress.com/doi/pdf/10.4141/S98-037
20.0. Egamberdiyeva D. and Höflich G. (2004) 'Effect of plant growth-promoting bacteria on growth and nutrient uptake of cotton and pea in a semi-arid region of Uzbekistan'. *J. Arid Environ.* 56, 293-301. In Adesemoye A.O. and Kloepper J.W. (2009) 'Plant–microbes interactions in enhanced fertilizer-use efficiency'. *Appl. Microbiol Biotechnol.* 85. [Re: Maize, Growth, Uptake and Soil Properties, p.4.]
20.1. Saharan B.S. and Nehra V. (2011) 'Plant Growth Promoting Rhizobacteria: A Critical Review'. *Life Sciences and Medicine Research.* April 30.LSMR-21. Department of Microbiology, Kurukshetra University, Kurukshetra, Haryana 136 119, India. [Re: Mineral Availability, Uptake and Growth in Pepper and Cucumber, p.5.]
20.2. Basak B.B. and Biswas D.R. (2009) 'Influence of potassium solubilizing microorganism (*Bacillus mucilaginosus*) and waste mica on potassium uptake dynamics by sudan grass (*Sorghum vulgare* Pers.) grown under two Alfisols'. *Plant and Soil – Plant Soil.* Vol. 317, no. 1. [Re: Sudan Grass Yield and Uptake using KSB and Mica, p.235.]
20.3. Archana D.S. (2007) 'Studies on Potassium Solubilising Bacteria'. Department of Agricultural Microbiology. College of Agriculture, Dharwad. University of Agricultural Sciences. [Re: Study Review. Positive results for KSB with Greengram, Brinjal, Sorgum, Maize, Yam, Egg Plant, Pepper, Cucumber and Chilli , pp.14-16]

Chapter Four

1. Darwin C.R. (1881) *The formation of vegetable mould through the action of worms, with observations on their habits.* John Murray. London. p.328.
2. Gillings M. (1998) 'Annual Report'. Centre for Biodiversity and Bioresources, Department of Biological Sciences, Macquarie University, NSW 2109 Australia. p.3.
3. Carpenter D. (2012) 'Earthworms – their role in soil ecosystems'. Soil Biodiversity Group, Natural History Museum, London.
4. Taylor H. and Lowe C.N. (2012) 'Key to Common British earthworms'. In The OPAL Soil and Earthworm Survey. FSC publications. www.field-studies-council.org
5. Blakemore R.J. (2000) 'Vermicology I – Ecological considerations of the earthworm species used in vermiculture'. Vermillennium – International conference on vermiculture and vermicomposting, Kalamazoo, MI. In Mainoo N.O.K. (2007) *Feasibility of low cost vermicompost production in Accra, Ghana.* Department of Bioresource Engineering, McGill University, Montreal. p.36.
6. Fisher E. (2016) [Photo Credit]. '*Eisenia fetida*'. Newland Avenue Allotments, Hull.
7.0. Lee K.E. (1985) *Earthworms. Their Ecology and Relationships with Soils and Land Use.* Academic Press, Sydney. In Mainoo N.O.K. (2007) *Feasibility of low cost vermicompost production in Accra, Ghana.* Department of Bioresource Engineering, McGill University, Montreal. p.3.
7.1. Vetter S., Fox O., Elkschmitt K. and Wolters V. (2004) 'Limitations of faunal effects on soil carbon flow: biotic density dependence, biotic regulation mutual inhibition'. *Soil Biology and Biochemistry* 36, pp.387-397. In Mainoo N.O.K. (2007) *Feasibility of low cost vermicompost production in Accra,*

Ghana. Department of Bioresource Engineering, McGill University, Montreal. p.3.

7.2. Jones C.G., Lawton J.H. and Shachak M. (1994) 'Organisms as ecosystem engineers'. *Oikos* 69, pp.371-386. In Mainoo N.O.K. (2007) *Feasibility of low cost vermicompost production in Accra, Ghana*. Department of Bioresource Engineering, McGill University, Montreal. p.36.

8. Allison F.E. (1973) 'Soil Organic Matter and its Role in Crop Production'. *Developments in Soil Science*. Vol.3 (1973). p.*322*.

9. Edwards, C.A. and Bater, J.A., (1992) 'The use of earthworms in environmental management'. *Soil Biology and Biochemistry*, Volume 24, pp.1683-1689. In Mainoo N.K. (2007) *Feasibility of low cost vermicompost production in Accra, Ghana*. Department of Bioresource Engineering McGill University, Montreal. p.3.

10. Okafor F.C. (2009) 'The Varied Roles of Snails (Gastropod Molluscs) in the Dynamics of Human Existence'. Issue 47 of *Inaugural Lecture Series*. University of Nigeria, Senate Ceremonials Committee. p.51.

11. Finke M. and Winn D. (2004) 'Insects and Related Arthropods: A Nutritional Primer for Rehabilitators'. *J. Wildlife Rehab* 27(3-4). p.14.

12. Kompak K. (2005) [Photo Credit]. '*Geotrupes stercorarius*'. http://commons.wikimedia.org/wiki/File:Geotrupes_stercorarius2.jpg

13. Step E. (1919) *Insect Artizans and Their Work*. Dodd, Mead and company. New York. p.53.

14. Balog A., Marko V. and Ferencz L.(2008) 'Patterns in distribution, abundance and prey preferences of parasitoid rove beetles *Aleochara bipustulata* (L.) (Coleoptera: Staphylinidae, Aleocharinae) in Hungarian agroecosystems'. *North-Western Journal of Zoology* Vol. 4. No. 1. p.6.

15. Enkegaard A. (2012) 'Development of IPM-strategies for field vegetables (onion, lettuce)'. DFFE-project 2009-2012. AU-DJF. p.3.

16. 'Biocontrol of the cabbage root fly by release of predators'. Horticulture Research International Wellesbourne, Warwick. CV35 9EF. Defra Project Code: HH1830SFV. pp.1-3.

17. Shelton A. (2012) '*Aleochara bilineata* (Coleoptera: Staphylinidae)'. Biological Control. Cornell University.

18. Lee J.C. and Landis D.A. (2000) *Natural Enemies in Your Garden: A Homeowner's Guide to Biological Control.Extension Bulletin 2719*. October. Department of Entomology. Michigan State University. p.32.

19. Cloudsley-Thompson, J. L. (1952) 'Studies in diurnal rhythms, II. Changes in the physiological responses of the woodlouse *Oniscus asellus* to environmental stimuli'. *Exp. Biol.*, 29, 295-303. In Cloudsley-Thompson J.L. (1958) *Spiders, scorpions, centipedes, and mites; the ecology and natural history of woodlice, myriapods, and arachnids*. New York, Pergamon Press. p.6.

20. ©Phichak Limprasutr (2018) [Photo Credit] – stock.adobe.com

21. Compton J. (1954) *The Life of the Spider*. New American Library. p.55.

22.0. Riechert, S.E. and L. Bishop. (1990) 'Prey control by an assemblage of generalist predators: spiders in garden test systems'. *Ecology* 71:1441.1450. In Maloney D., Drummond F.A. and Alford R. (2003) 'Spider Predation in Agroecosystems: Can Spiders Effectively Control Pest Populations?' Department of Biological Sciences, University of Maine Orono ME 04469. pp.5-6.

22.1. Young O.P., and Edwards G.B. (1990) 'Spiders in United States field crops and their potential effect on crop pests'. *J. Arachnol*, 18:1.27. In Maloney D., Drummond F.A. and Alford R. (2003) 'Spider Predation in Agroecosystems: Can Spiders Effectively Control Pest Populations?'. Department of Biological Sciences. The University of Maine Orono ME 04469. pp.5-6.

22.2. Nyffeler M., Sterling W.L. and Dean. D.A. (1994) 'How spiders make a living'. *Environ. Entomol.*, 23:1357.1367. In Maloney D., Drummond F.A. and Alford R. (2003) 'Spider Predation in Agroecosystems: Can Spiders Effectively Control Pest Populations?'. Department of Biological Sciences. The University of Maine Orono ME 04469. pp.5-6.

23.0 Nentwig W. and Wissel. C. (1986) 'A comparison of prey lengths among spiders'. *Oecologia*, 68: 595-600. In Maloney D., Drummond F.A. and Alford R. (2003) 'Spider Predation in Agroecosystems: Can Spiders Effectively Control Pest Populations?'. Department of Biological Sciences. The University of Maine. Orono ME 04469. p.15.

23.1. Nyffeler M., Sterling W.L. and Dean. D.A. (1994) 'How spiders make a living'. *Environ. Entomol.* 23, 1357.1367. In Maloney D., Drummond F.A. and Alford R. (2003) 'Spider Predation in Agroecosystems: Can Spiders Effectively Control Pest Populations?'. Department of Biological Sciences. The University of Maine Orono ME 04469. p.15.

23.2. Marc P., Canard A. and Ysnel F.(1999) 'Spiders (Araneae) useful for pest limitation and bioindication'. *Agric., Ecosyst. Environ.* 74, 229-273. In Maloney D., Drummond F.A. and Alford R. (2003) 'Spider Predation in Agroecosystems: Can Spiders Effectively Control Pest Populations?'. Department of Biological Sciences. The University of Maine Orono ME 04469. p.15.

24.0. Fagan W.F., Hakim A.L., Ariawan H., and Yuliyantiningsih S. (1998) 'Interactions between biological control efforts and insecticide applications in tropical rice agroecosystems: the potential role of intraguild predation'. *Biological Control: Theory and Applications in Pest Management* 13, 121-126. In Maloney D., Drummond F.A. and Alford R. (2003) 'Spider Predation in Agroecosystems: Can Spiders Effectively Control Pest Populations?'. Department of Biological Sciences. The University of Maine Orono ME 04469. p.21.

24.1. Marc P., Canard A., and F. Ysnel F. (1999) 'Spiders (Araneae) useful for pest limitation and bioindication'. *Agric., Ecosyst. Environ.* 74, 229-273. In Maloney D., Drummond F.A. and Alford R. (2003) 'Spider Predation in Agroecosystems: Can Spiders Effectively Control Pest Populations?'. Department of Biological Sciences. The University of Maine Orono ME 04469. p.21.

24.2. Wise D.H., and Chen B. (1999) 'Impact of intraguild predators on survival of a forest-floor wolf spider'. *Oecologia* 121, 129-137. In Maloney D., Drummond F.A. and Alford R. (2003) 'Spider Predation in Agroecosystems: Can Spiders Effectively Control Pest Populations?'. Department of Biological Sciences. The University of Maine Orono ME 04469. p.21.

25. Fontaneto D. (2007) ©Phichak Limprasutr (2018) [Photo Credit] – stock.adobe.com *Gross, L. PLoS Biology* Vol. 5, No. 4, e99. https://commons.wikimedia.org/w/index.php?curid=2507344 In compliance with Creative Commons Attribution 2.5 Generic. Cropped and Resolution Altered.

26. Barron G.L. (2004) 'Chapter 19. Fungal Parasites and Predators of Rotifers, Nematodes and other Invertebrates'. In *Biodiversity of Fungi: Inventory and Monitoring Methods* by Mueller G.M, Bills G.F, Foster M.S. Elsevier, Academic Press. p.436.

27. Witty M.(2009) 'Bdelloid Rotifer Feeding and Locomotory Behavior'. American Society for Microbiology. Microbe Library. pp.Abstract.

28. Gibson C.H. (2011) 'Turbulence and fossil turbulence lead to life in the universe'. Turbulent Mixing and Beyond. August 2011, Abdus Salam International Centre for Theoretical Physics, Trieste, Italy. p.21.

29. Wallace R.L. (2001) 'Rotifera'. Ripon College, Ripon, Wisconsin, USA. In *Encyclopaedia of the Life Sciences*. John Wiley & Sons, Ltd. p.2.

30. Ashraf M., Ullah S., Rashid T., Ayub M., Bhatti E.M., Naqvi S.A, Javaid M. (2010) 'Optimization of Indoor Production of Fresh Water Rotifer, *Brachionus calyciflorus*, b: Feeding Studies'. *Pakistan Journal of Nutrition* 9 (6). Asian Network for Scientific Information. p.582.

31. Trautmann N. (1995) 'Invertebrates of the Compost Pile'. Cornell Composting Resources. Cornell University.

32. JSB Marines (2012) 'Rotifer culture kit grow your own rotifers not coral includes air pump'. In eBay. Category: Pet Supplies > Fish > Food. www.ebay.co.uk/itm/Rotifer-culture-kit-grow-your-own-rotifers-not-coral-includes-air-pump-/271024721955?pt=UK_Pet_Supplies_Fish&hash=item3f1a551023

33. Hoorman J.J. (2011) 'The Role of Soil Protozoa and Nematodes'. Ohio State University. SAG-15-11. pp.1. https://studylib.net/doc/18560755.

34. Stockdale E.A, Watson C.A., Black H.I.J. and Philipps L. (2006) 'Do farm management practices alter below-ground biodiversity and ecosystem function? Implications for sustainable management'. JNNC report No. 364. p.17.

35. Bonkowski M. (2004) 'Tansley review. Protozoa and plant growth: the microbial loop in soil revisited'. *New Phytologist* (2004) 162. p.617.

36.0. Kuikman P.J., Jansen A.G., van Veen J.A. (1991) '^{15}N-nitrogen mineralization from bacteria by protozoan grazing at different soil moisture regimes'. *Soil Biology and Biochemistry*, 23: 193–200. In Bonkowski M. (2004) 'Tansley review. Protozoa and plant growth:the microbial loop in soil revisited'. *New Phytologist* (2004) 162, p.619.

36.1. Alphei J., Bonkowski M. and Scheu S. (1996) 'Protozoa, Nematoda and Lumbricidae in the rhizosphere of *Hordelymus europaeus* (Poaceae): Faunal interactions, response of microorganisms and effects on plant growth'. *Oecologia* 106, 111-126. In Bonkowski M. (2004) 'Tansley review. Protozoa and plant growth : the microbial loop in soil revisited'. *New Phytologist* (2004) 162, p.619.

37.0. Darbyshire J.F. and Greaves M.P. (1973) 'Bacteria and protozoa in the rhizosphere'. *Pesticide Science* 4, 349-360. In Bonkowski M. (2004) 'Tansley review. Protozoa and plant growth: the microbial loop in soil revisited'. *New Phytologist* (2004) 162, p.619.

37.1. Foster R.C. and Dormaar J.F. (1991) 'Bacteria-grazing amoebae in situ in the rhizosphere'. *Biology and Fertility of Soils* 11, 83-87. In Bonkowski M. (2004) 'Tansley review. Protozoa and plant growth: the microbial loop in soil revisited'. *New Phytologist* (2004) 162. p.619.

38. Foissner W. (2001) [Photo Credit]. National Science & Technology Center. P.O. Box 25047 Bldg. 50, Denver Federal Center. Denver, Colorado 80225-0047.

39. Harvey C. (2007) *Biological Farming: A Practical Guide.* Nuffield Australia. "Tauwitchere" Narrung SA 5259. p.18.

40. Foissner W. (1999) 'Soil protozoa as bioindicators: pros and cons, methods, diversity, representative examples'. *Agriculture, Ecosystems & Environment* Volume 74, Issues 1-3, June 1.

41. Treonis A.M. and Diana H.W. (2005) 'Soil Nematodes and Desiccation Survival in the Extreme Arid Environment of the Antarctic Dry Valleys'. *Integr. Comp Biol.* 45, p.741.

42. Barron G.L. and Allin N. (2008) [Photo Credit]. University of Guelph. ON N1G 2W1. Canada.

43. Hopkins L. (2004) *Migratory Nematodes.* Bayer Crop Science Ltd. Hauxton. Cambridge. CB2 5HU. p.4.

44. Tarjan A.C., Esser R.P. and Chang S.L. (1977) [Photo Credit]. 'Diagnostic Key to Plant Parasitic, Freeliving and Predaceous Nematodes'. *J. Water Pollution Cont. Fed.* 1, Vol 49, 2318-2337.

45. R.H.C. Curtis (2007) 'Plant parasitic nematode proteins and the host parasite interaction'. *Briefings in Functional Genomics and Proteomics* Vol. 6. No.1. p.50.

46. Melakeberhan H., Mennan S., Darby B. and Dudek T. (2007) 'Integrated approaches to understanding and managing *Meloidogyne hapla* populations' parasitic variability'. *Crop Protection* 26, p.894.

47. Kruger R. and McSorley R. (2008) *Nematode Management in Organic Agriculture.* University of Florida IFAS Extension. pp.6-7. http://edis.ifas.ufl.edu/ng047

48. Al Kader M.A. (2008) 'In vitro Studies on Nematode Interactions with their Antagonistic Fungi in the Rhizosphere of Various Plants'. Faculty of Forest and Environmental Sciences. Alber-Ludwigs-Universitat. Freigburg im Breisgau, Germany. p.11.

49. Okore C.C., Ononuju C.C, and Okwujiako I.A. (2011) 'Management of *Meloidogyne incognita* with *Pleurotus ostreatus* and *P. tuberregium* in Soybean'. *Int. J. Agric. Biol.* Vol. 13, No.3. p.401.

50. Georgis R. (2002) 'The Biosys experience: an insider's perspective'. In Divya K. and Sankar M. (2009) 'Entomopathogenic nematodes in pest management'. *Indian Journal of Science and Technology*, Vol. 2 No. 7. Multiplex Biotech Pvt. Ltd., Andarasanahalli, Tumkur, Karnataka-570 106, India. p.53.

51. Poinar G.O. (1990) 'Taxonomy and biology of Steinernematidae and Heterorhabditidae'. In Divya K. and Sankar M. (2009) 'Entomopathogenic nematodes in pest management.' *Indian Journal of Science and Technology* Vol. 2, No. 7. Multiplex Biotech Pvt. Ltd., Andarasanahalli, Tumkur, Karnataka-570 106, India. p.53.

52. Ettema C.H. (1998) 'Soil Nematode Diversity: Species Coexistence and Ecosystem Function'. *Journal of Nematology* 30(2), p.166.

53. Fleming V. (2010) [Photo Credit]. 'Oyster Mushroom'. B250/0959. Science Photo Library. www.aboutallonline.com/oyster-mushrooms-in-nature

54.0. Smith S.E. and Read D.J. (1997) *Mycorrhizal Symbiosis.* 2nd Ed. Academic Press, London. In Huang H., Zhang S., Shan X., Chen B., Zhu Y., Nigel J. and Bell B. (2006) 'Effect of arbuscular mycorrhizal fungus (*Glomus caledonium*) on the accumulation and metabolism of atrazine in maize (*Zea mays* L.) and atrazine dissipation in soil'. Center for Environmental Policy, Imperial College, London, UK. p.452.

54.1. Perry D.A., Amaranthus M.P., Borchers S.L. and Brainerd R.E. (1989) 'Bootstrapping in ecosystems: Internal interactions largely determine productivity and stability in biological systems with strong positive feedback.' *Bioscience* 39, 230-237. In Rose P.S. (1997) 'Mutualistic biodiversity networks: the relationship between soil biodiversity and mutualism, and their importance to ecosystem function and structural organisation'. Life Sciences, University of Manchester, UK. p.3.

55. Rose P.S. (1997) 'Mutualistic biodiversity networks: the relationship between soil biodiversity and mutualism, and their importance to ecosystem function and structural organisation'. Life Sciences, University of Manchester, UK. p.3. https://pdfs.semanticscholar.org/b27d/0a1834c-5d564342a290cad21b1a1a01d7e97.pdf

56. Barea J.M., Ferrol N., Azcón-Aguilar C., Azcón R. (2008) 'Mycorrhizal symbioses'. In Richardson A.E., Barea J., McNeill A.M. and Prigent-Combaret C. (2009) 'Acquisition of phosphorus and nitrogen in the rhizosphere and plant growth promotion by microorganisms'. *Plant Soil* (2009) 321, p.316.

57.0. Manjunath A. and M. Habte. (1988) 'The development of vesicular-arbuscular mycorrhizal infection and the uptake of immobile nutrients in *Leucaena leucocephala*'. *Plant and Soil* 106, 97–103. In Habte M. and Osorio N.W. (2010) 'Arbuscular Mycorrhizas: Producing and Applying Arbuscular Mycorrhizal Inoculum'. Department of Tropical Plant and Soil Sciences. p.4.

57.1. Pacovsky R.S. (1986) 'Micronutrient uptake and distribution in mycorrhizal or phosphorus-fertilized soybeans'. *Plant and Soil* 95, 379-388. In Habte M. and Osorio N.W. (2010) 'Arbuscular Mycorrhizas: Producing and Applying Arbuscular Mycorrhizal Inoculum'. Department of Tropical Plant and Soil Sciences. p.4.

58.0. Bolan N.S. (1991) 'A critical review on the role of mycorrhizal fungi in the uptake of phosphorus by plants'. *Plant and Soil* 134, 189-207. In Huang H., Zhang S., Shan X., Chen B., Zhu Y., Nigel J. and Bell B. (2006) 'Effect of arbuscular mycorrhizal fungus (*Glomus caledonium*) on the accumulation and metabolism of atrazine in maize (*Zea mays* L.) and atrazine dissipation in soil'. Center for Environmental Policy, Imperial College, London, UK. p.452.

58.1. Marschner H. and Romheld D.J. (1998) 'Role of root growth, arbuscular mycorrhiza and root exudates for the efficiency in nutrient acquisition'. *Field Crops Research* 56, 203-207. In Huang H., Zhang S., Shan X., Chen B., Zhu Y., Nigel J. and Bell B. (2006) 'Effect of arbuscular mycorrhizal fungus (*Glomus caledonium*) on the accumulation and metabolism of atrazine in maize (*Zea mays* L.) and atrazine dissipation in soil'. Center for Environmental Policy, Imperial College, London, UK. p.452.

59.0. Chakravarty P. and Unestam T. (1987) 'Mycorrhizal fungi prevent disease in stressed pine seedlings'. *J. Phytopath* 118, 335-340. In Brundrett M. (2009) 'Mycorrhizas in Natural Ecosystems'. *Advances in Ecological Research* Vol. 21. Academic Press Limited. p.225.

59.1. Dehne W.H. (1986) 'Influence of va mycorrhizae on host plant physiology'. In Brundrett M. (2009) 'Mycorrhizas in Natural Ecosystems'. *Advances in Ecological Research* Vol. 21. Academic Press Limited. p.225.

59.2. Duchesne L.C., Peterson R. L. and Ellis B.E. (1987) 'The accumulation of plant produced antimicrobial compounds in response to ectomycorrhizal fungi: a review'. *Phytoprotection* 68, 17-27. In Brundrett M. (2009) 'Mycorrhizas in Natural Ecosystems'. *Advances in Ecological Research* Vol. 21. Academic Press Limited. p.225.

60. Marx D.H. and Bryan W.C. (1971) 'Influence of ectomycorrhizae on survival and growth of aseptic seedlings of loblolly pine at high temperature.' *For. Sd.* 17, 37-41. In Chanipawat R.S. (1990) 'Effect of different YAM fungi under varying levels of phosphorus on growth and nutrition uptake of pigeon pea (*Cajanus cajan*)'. Department of Plant Pathology, Rajasthan Agricultural University S. K. N. College of Agriculture. p.141.

61. Molina R., Massicotte H. and Trappe J.M. (1992) 'Specificity phenomena in mycorrhizal symbioses: community-ecological consequences and practical implications'. In Dixon G.R. Dixon and Tilston E.L. (Eds.) (2010) *Soil Microbiology and Sustainable Crop Production*. Springer. p.124.

62. Al Kader M.A. (2008) 'In vitro Studies on Nematode Interactions with their Antagonistic Fungi in the Rhizosphere of Various Plants'. Faculty of Forest and Environmental Sciences. Alber-Ludwigs-Universitat. Freigburg im Breisgau,

Germany. p.13.
63. Li Y., Hyde K.D., Jeewon R., Cai L., Vijaykrishna D. and Zhang K. (2005) 'Phylogenetics and evolution of nematode-trapping fungi (Orbiliales) estimated from nuclear and protein coding genes'. *Mycologia*, September/October. Vol. 97. p.1034 and p. Abstract.
64. Davies K.G. (2005) 'Interactions between nematodes and microorganisms: bridging ecological and molecular approaches'. *Adv. Appl. Microbiol.* 57, 53-78. In Dixon G.R. Dixon and Tilston E.L.(Eds.) (2010) *Soil Microbiology and Sustainable Crop Production*. Springer. p.326.
65. Unnamed (2014) [Photocredit] In *Looking At The Amazing Mycorrhiza Fungi*. Vineyard Son Alegre, Mallorca. p.1.
66. Smith S.E. and Read D.J. (2008) *Mycorrhizal Symbiosis*. 3rd Ed. Academic Press, Elsevier, Amsterdam. In Richardson A.E., José-Miguel Barea J. McNeill A.M.and Prigent-Combaret C. (2009) 'Acquisition of phosphorus and nitrogen in the rhizosphere and plant growth promotion by microorganisms'. *Plant Soil* 321, pp.315-316.
67. Richardson A.E., Barea J., McNeill A.M. and Prigent-Combaret C.(2009) 'Acquisition of phosphorus and nitrogen in the rhizosphere and plant growth promotion by microorganisms'. *Plant Soil* (2009) 321, p.316.
68. Fitter A.H. (2005) 'Darkness visible: reflections on underground ecology'. *J. Ecol.* 93, 231-243. In Dixon G.R. Dixon and Tilston E.L. (Eds.) (2010) *Soil Microbiology and Sustainable Crop Production*. Springer Dordrecht Heidelberg, London, New York. p.123.
69.0. Gerdemann, J. W. (1968) 'Vesicular-arbuscular mycorrhizae and plant growth'. *Ann. Rev. Phytopath.* 6, 397-418. In Jalili B.L. and Chand H. (Eds.) (1990) Proceedings of the National Conference on Mycorrhiza. Haryaña Agricultural University, Hisar. February 14-16. p.180.
69.1. Harley J. L. and Smith S.E. (1983) *Mycorrhizal Symbiosis*. Academic Press London, N. Y. In Jalili B.L. and Chand H. (Eds.) (1990) Proceedings of the National Conference on Mycorrhiza. Haryaña Agricultural University, Hisar. February 14-16. p.180.
69.2. Hayman D. S. (1982) 'Influence of soil and fertility on activity and survival of vesicular-arbuscular mycorrhizal fungi'. *Phytopathology* 72, 1119-1125. In Jalili B.L. and Chand H. (Eds.) (1990) Proceedings of the National Conference on Mycorrhiza. Haryaña Agricultural University, Hisar. February 14-16. p.180.
69.3. Habte M. and Osorio N.W. (2001) *Arbuscular Mycorrhizas: Producing and Applying Arbuscular Mycorrhizal Inoculum*. Department of Tropical Plant and Soil Sciences. University of Hawaii, Manoa. p.3.
70.0. Allen (1992) *Mycorrhizal Functioning: An Integrative Plant-Fungal Process*. New York: Chapman and Hall. In He X., Xu M., Qiu G.Y, and Zhou J. 'Use of 15N stable isotope to quantify nitrogen transfer between mycorrhizal plants'. *Journal of Plant Ecology* 2(3). p.108.
70.1. Marschnerr H. and Dell B. (1994) 'Nutrient uptake in mycorrhizal symbiosis'. *Plant Soil* 159, 89-102. In He X., Xu M., Qiu G.Y, and Zhou J. 'Use of 15N stable isotope to quantify nitrogen transfer between mycorrhizal plants'. *Journal of Plant Ecology* 2(3). p.108.
70.2. Read D.J. (1991) 'Mycorrhizas in ecosytems'. *Experientia* 47, 376-91. In He X., Xu M., Qiu G.Y, and Zhou J. 'Use of 15N stable isotope to quantify nitrogen transfer between mycorrhizal plants'. *Journal of Plant Ecology* 2(3). p.108.
70.3. Smith S.E. and Read D.J. (2008) *Mycorrhizal Symbiosis*. 3rd Ed. San Diego, CA: Academic Press. In He X., Xu M., Qiu G.Y, and Zhou J. 'Use of 15N stable isotope to quantify nitrogen transfer between mycorrhizal plants'. *Journal of Plant Ecology* 2(3). p.108.
70.4. Smith S.E, Gianinazzi-Pearson V. and Koide R. et al. (1994) 'Nutrient transport in mycorrhizas: structure, physiology and consequences for efficiency of the symbiosis'. *Plant Soil* 159,103-13. In He X., Xu M., Qiu G.Y, and Zhou J. 'Use of 15N stable isotope to quantify nitrogen transfer between mycorrhizal plants'. *Journal of Plant Ecology* 2(3). p.108.
71. Foissner W., Agatha S.N. and Berger H. (2002) 'Soil ciliates (Protozoa, Ciliophora) from Namibia (Southwest Africa), with emphasis on two contrasting environments, the Etosha Region and the Namib Desert'. *Denisia* 5, 1-1459. In Foissner W. (2006) 'Biogeography and Dispersal of Micro-organisms: A Review Emphasizing Protists'. *Acta Protozool.* 45, 111-13. p.130.
72. Read D.J. (1996) 'The structure and function of the ericoid mycorrhizal root'. *Ann. Bot.* 77, 365-374. In Scagel C.F. (2005) 'Inoculation with Ericoid Mycorrhizal Fungi Alters Fertilizer Use of Highbush Blueberry Cultivars'. *Hortscience*. Vol. 40(3) June 2005. p.786.
73. Eck P., Gough R.E., Hall I.V. and Spears J.M. (1990) 'Blueberry management'. In Prodorutti D., et al. (2007) 'Highbush Blueberry: Cultivation, Protection, Breeding and Biotechnology'. *The European Journal of Plant Science and Biotechnology*. Global Science Books. p.46.
74.0. Goulart B.L., Schroeder M.L., Demchak K., Lynch J.P., Clark J.R, Darnell R.L. and Wilcox W.F. (1993) 'Blueberry mycorrhizae: current knowledge and directions', *Acta Horticulturae* 346, 230-239. In Prodorutti D., et al. (2007) 'Highbush Blueberry: Cultivation, Protection, Breeding and Biotechnology'. *The European Journal of Plant Science and Biotechnology*. Global Science Books. p.46.
74.1. Koron D. and Gogala N. (2005) 'The use of mycorrhizal fungi in the growing of blueberry plants (*Vaccinium corymbosum* L.)'. Acta *Horticulturae* 525, 101-106. In Prodorutti D., et al. (2007) 'Highbush Blueberry: Cultivation, Protection, Breeding and Biotechnology'. *The European Journal of Plant Science and Biotechnology*. Global Science Books. p.46.
75. Koron D. and Gogala N. (2005) 'The use of mycorrhizal fungi in the growing of blueberry plants (*Vaccinium corymbosum* L.)'. Acta *Horticulturae* 525, 101-106. In Prodorutti D., et al. (2007) 'Highbush Blueberry: Cultivation, Protection, Breeding and Biotechnology'. *The European Journal of Plant Science and Biotechnology*. Global Science Books. p.46.
76.0. Scagel, C.F., A. Wagner, and P. Winiarski. (2005a). 'Frequency and intensity of root colonization byericoid mycorrhizal fungi in nursery production of blueberry'. *Small Fruits Rev.* In Scagel C.F. (2005) 'Inoculation with Ericoid Mycorrhizal Fungi Alters Fertilizer Use of Highbush Blueberry Cultivars'. *Hortscience* Vol. 40(3) June 2005. p.786.
76.1. Scagel C.F., Wagner A., and Winiarski P. (2005) 'Inoculation with ericoid mycorrhizal fungi alters root colonization and growth in nursery production of blueberry plants from tissue culture and cuttings'. *Small Fruits Rev.* In Scagel C.F. (2005) 'Inoculation with Ericoid Mycorrhizal Fungi Alters Fertilizer Use of Highbush Blueberry Cultivars'. *Hortscience* Vol. 40(3) June 2005. p.786.
77. Allison F.E. (1973) 'Soil Organic Matter and its Role in Crop Production'. *Developments in Soil Science*. Vol.3 (1973). p.41.
78. Trautmann N. and Olynciw E. (1996) 'Compost Microorganisms'. Cornell University, Ithaca, NY 14853-5601 607-255-1187. p.1.
79.0. Requena N., Perez-Solis E., Azcon-Aguilar C., Jeffries P. and Barea J.M. (2001) 'Management of indigenous plant-microbe symbioses aids restoration of desertified ecosystems'. *Appl. Environ. Microbiol.* 67(2), 495-498. In Dixon G.R. Dixon and Tilston E.L. (Eds.) *Soil Microbiology and Sustainable Crop Production* (2010) p.124.
79.1. Celik I., Ortas I. and Kilic S. (2004) 'Effects of compost, mycorrhiza, manure and fertilizer on some physical properties of a Chromoxerert soil.' *Soil Tillage Res.* 78, 59-67. In Dixon G.R. Dixon and Tilston E.L. (Eds.) (2010) *Soil Microbiology and Sustainable Crop Production*. Springer. p.124.
79.2. Turnau K., Orlowska E., Ryszka P., Zubek S., Anielska T., Gawronski S. and Jurkiewicz A. (2007) 'Role of mycorrhizal fungi in phytoremediation and toxicity monitoring of heavy metal rich industrial wastes in southern Poland'. In Dixon G.R. Dixon and Tilston E.L. (Eds.) (2010) *Soil Microbiology and Sustainable Crop Production*. p.124.
80. Sandle T. (2016) '*Geobacillus stearothermophilus*' [Photo Credit]. In Sandle T. (2016) *Pharmaceutical Microbiology: Essentials for Quality Assurance and Quality Control*. www.prlog.org/12525651-geobacillus-stearothermophilus-1.jpg
81. Downie A. (Undated) [Photo Credit]. John Innes Centre, Norwich Research Park, Colney, Norwich NR4 7UH, UK.
82. Suneetha V. and Khan Z.A. (2011) 'Chapter 14. Actinomycetes : Sources for Soil Enzymes'. In Shukla G. and Varma A. (Eds.) (2011) *Soil Biology* Volume 22. p.259.
83. Wilkins K. (1996) 'Volatile metabolites from actinomycetes'. *Chemosphere*, 32, 1427-1434. In George M., Anjumol A., George G. and Hatha A.A.M (2012) 'Distribution and bioactive potential of soil actinomycetes from different ecological habitats'. *African Journal of Microbiology Research* Vol. 6(10). p.2265.
84. Finstein M.S. and Morris M.L. (1975) 'Microbiology of municipal solid waste composting'. *Adv. Appl. Microbiol.* 19,

113-151. In Ashraf R., Shahid F. and Ali T.A. (2007) 'Association of Fungi, Bacteria and Actinomycetes with Different Composts'. *Pak. J. Bot.* 39(6). p.2142.

85. Arnold P. (2011) 'Actinomycetes: The Sign of Composting Success'. Acadia University/Bio-Logic Environmental Systems.

86.0. Epstein E. (1997) *The Science of Composting*. Technomic Publishing AG, Basel, Switzerland. p.483.

86.1. Wilkins K. (1996) 'Volatile metabolites from actinomycetes'. *Chemosphere* 32, 1427-1434. In George M., Anjumol A., George G. and Hatha A.A.M (2012) 'Distribution and bioactive potential of soil actinomycetes from different ecological habitats'. *African Journal of Microbiology Research* Vol. 6(10). p.2265.

87.0. Kloepper J. and Ryu C. (2004) *Phytopathology* 94, 1259-1266. In Fernando W.G.D. and Li R. (2011) 'Do cropping practices and systems influence fluctuations in PGPR populations or vice versa? *Is it a chicken or egg situation?*' University of Manitoba, Winnipeg, R3T 2N2, Canada. p.14.

87.1. Haas D. and Défago G. (2005) *Nat. Rev. Microbiol.* 3, 307-319. In Fernando W.G.D. and Li R. (2011) 'Do cropping practices and systems influence fluctuations in PGPR populations or vice versa? *Is it a chicken or egg situation?*' University of Manitoba, Winnipeg, R3T 2N2, Canada. p.14.

88. Sirota-Madi A., Olender T., Helman Y., Ingham C., Brainis I., Roth D., Hagi E., Brodsky L., Leshkowitz D., Galatenko V., Nikolaev V., Mugasimangalam R.C., Bransburg-Zabary S., Gutnick D.L., Lancet D. and Ben-Jacob E. (2010) [Photo Credit]. 'Genome sequence of the pattern forming Paenibacillus vortex bacterium reveals potential for thriving in complex environments'. *BMC Genomics* 11, 710. BioMed Central Ltd. p.3.

89. Bloemberg G.V. and Lugtenberg B.J. (2001) 'Molecular basis of plant growth promotion and biocontrol by rhizobacteria'. *Curr. Opin. Plant Biol. 2001* 4, 343-350. In Sirota-Madi A., et al. (2010) 'Genome sequence of the pattern forming Paenibacillus vortex bacterium reveals potential for thriving in complex environments'. *BMC Genomics* 11, 710. BioMed Central Ltd. p.11.

90. Sirota-Madi A., Olender T., Helman Y., Ingham C., Brainis I., Roth D., Hagi E., Brodsky L., Leshkowitz D., Galatenko V., Nikolaev V., Mugasimangalam R.C., Bransburg-Zabary S., Gutnick D.L., Lancet D. and Ben-Jacob E. (2010) [Photo Credit]. 'Genome sequence of the pattern forming Paenibacillus vortex bacterium reveals potential for thriving in complex environments'. *BMC Genomics* 11, 710. BioMed Central Ltd. p.4.

Chapter Five

1.0. Leveau J. (2009) [Quote]. 'Microbiology Life on Leaves'. *Nature* Vol 461. October 7th. Nature Publishing Group (NPG). p.7265, Abstract.

1.1. Howard L. (2010) [Photo Credit]. 'Scanning electron microscope image of *Nicotiana alata* upper leaf surface, showing tricomes and a few stomates'. Released to Public Domain. https://en.wikipedia.org/wiki/Leaf#/media/File:Leaf_epidermis_w_scale.jpg

2.0. Coleman D.C., Reid C.P.P. and Cole C.V.(1983) 'Biological strategies of nutrient cycling in soil systems'. *Advances in Ecological Research* 13, 1-55. In Harris A. and Hill R. (2007) 'Carbon-Negative Primary Production: Role of Biocarbon and challenges for Organics in Aoteroa/New Zealand'. Tauranga, New Zealand, *Journal of Organic Systems*, Vol.2 No.2. Lincoln University, New Zealand. p.3 and p.7.

2.1. Gupta V.V.S.R. and Roget D.K. (2004) 'Understanding soil biota and biological functions: Management of soil biota for improved benefits to crop production and environmental health'. CSIRO Land and Water, Glen Osmond SA. In Lines-Kelly R. (Ed) (2004) 'Soil biology in agriculture. Proceedings of a workshop on current research into soil biology in agriculture'. Tamworth Sustainable Farming Training Centre 11-12 August 2004. NSW Department of Primary Industries, Orange 2800. p.3.

3. Allison F.E. (1973) 'Soil Organic Matter and its Role in Crop Production'. *Developments in Soil Science*. Vol.3 (1973). p. Preface.

4. Bending G., Rodriguez-Cruz M. and Lincoln S.D. (2007) 'Fungicide impacts on microbial communities in soils with contrasting management histories'. *Chemosphere* 69, 82-88. In Dixon G.R. and Tilston E.L. (Eds.) (2010) *Soil Microbiology and Sustainable Crop Production*. Springer. p.281.

5. Clouds Hill Imaging. (2010) [Photo Credit]. 'Scanning Electron Micrograph (SEM): Radula of a Garden Snail, *Helix aspersa*'. Image ID: CHI0158. Last Refuge Ltd. Batch Farm, Panborough, Wells, Somerset, BA5 1PN, UK.

6. Speiser B. (2002) 'Moluscicides'. Research Institute of Organic Agriculture (FiBL), Frick, Switerland. In Pimental D. (Ed) *Encyclopedia of Pest Management*. CRC Press. p.506.

7. USDA (2008) 'New Pest Response Guidelines : Temperate Terrestrial Gastropods'. USDA. Inner Cover and Section 6-9.

8. San Martin G. (2010) [Photo Credit]. '*Tetranychus urticae* with silk threads'. http://en.wikipedia.org/wiki/File:Tetranychus_urticae_with_silk_threads.jpg

9. Hall K. (2006) [Photo Credit]. '*Phytoseiulus persimilis* eating a *T. urticae* egg'. http://photo.net/photodb/photo?photo_id=4306037

10. Gupta D. (2009) 'Pest Control: Insects and Other Arthropods'. *Agricultural Sciences* Vol 2. Encyclopedia of Life Support Systems (EOLSS). Department of Entomology, Dr. Y.S. Parmar University of Horticulture and Forestry, Nauni (Solan), India. p.298.

11. Dixon A. F. G. and Kindlmann P. (1998) 'Generation time ratio and the effectiveness of ladybirds as classical biological control agents'. In Stadler B. and Dixon T. (2008) *Mutualism: ants and their insect partners*. Cambridge University Press. p.68.

12. Soil Association (2009) *Organic Horticultural Production: An introductory guide*. Soil Association. South Plaza, Marlborough Street, Bristol BS1 3NX. p.11.

13. Wild A. (2003) [Photo Credit]. '*Acyrthosiphon pisum* (Pea Aphids)'. Item ID:6810. http://tolweb.org/treehouses/?treehouse_id=4421 Email: alwild@myrmecos.net

14. Savage S. (2011). [Photo Credit]. http://urbanwildlifejottings.blogspot.co.uk/2011/05/ladybird-larvae.html

15. Perrin R.M. (1976) 'The population dynamics of the stinging nettle aphid, *Microlophium carnosum*'. *Ecological Entomology* Vol 1, Issue 1. Wiley Online Library. pp. Abstract.

16.0. Ekesi S., Shah P.A., Clark S.J. and Pell J.K. (2005) 'Conservation biological control with the fungal pathogen Pandora neoaphidis: implications of aphid species, host plant and predator foraging'. *Agric. For Entomol.* 7, 21-30.

16.1. Alhmedi A., Haubruge E. and Francis F. (2009) 'Effects of stinging nettle habitats on aphidophagous predators and parasitoids in wheat and green pea fields with special attention to the invader *Harmonia axyridis* Pallas (Coleoptera: Coccinellidae)'. *Entomol. Sci.* 12, 349-358.

17. Bond W., Davies G. and Turner R. (2007) 'The biology and non-chemical control of Common Nettle (*Urtica dioica* L.)'. HDRA, Ryton Organic Gardens, Coventry, CV8, 3LG, UK. p.4.

18. Bedford I.D. (2004)[Photo Credit]. In *Crop Protection Compendium*. CAB international Publishing, Wallingford, UK. www.infonet-biovision.org/default/ct/77/pests

19. Hoddle M. (2011) [Photo Credit]. '*Encarsia formosa*, Beltsville strain'. Dept. of Entomology, University of California, Riverside, CA 92521. Email: mark.hoddle@ucr.edu

20. Martin N.A. (1998) 'Greenhouse Whitefly Life Cycle'. The Horticulture and Food Research Institute of New Zealand Ltd. www.hortnet.co.nz/publications/hortfacts/hf401041.htm

21. Augustin A.M. (2002) *Foraging behaviour and resource utilisation of Encarsia formosa (Hymenoptera: Aphelinidae), a parasitoid of the Greenhouse whitefly Trialeurodes vaporarium (Homoptera: Aleyrodidae) in greenhouses: the effect of host plant species and host plant structure*. PhD Thesis.University of Bayreuth. p.8.

22. Rice-Mahr S.E., Cloyd R.A., Mahr D.R. and Sadof C.S. (2001) *Biological Control of Pest and Insects and other pests of greenhouse crops*. University of Wisconsin. Cooperative Extension Publishing. 45 N, Charter Street, Madison, WI 53715. p.8.

23. Hoddle M.S., R.G. Van Driesche and Sanderson J.P. (1998) 'Biology and use of the whitefly parasitoid *Encarsia Formosa*'. *Annu. Rev. Entomol.* 1998. 43:645–69.

24. Vet L.E.M. and van Lenteren J.C. (1981) 'The parasite-host relationship between *Encarsia formosa* Gah. (Hymenoptera: Aphelinidae) and *Trialeurodes vaporiarorum* (West w.) (Homoptera: Aleyrodidae)'. *Zeitschrift fur Angewandte Entomologie* 91, 327-348. In Perdikis D., Kapaxidi E. and Papadoulis G. (2008) 'Biological Control of Insect and Mite Pests in Greenhouse Solanaceous Crops'. *The European Journal of Plant Science and Biotechnology*. Global Science Books. p.127.

25. Arakawa R. (1982) 'Reproductive capacity and amount of host feeding of *Encarsia formosa* Gahan (Hymenoptera, Aphelinidae)'. *Journal of Applied Entomology* 93, 175-185. In Perdikis D., Kapaxidi E. and Papadoulis G. (2008) 'Biological Control of Insect and Mite Pests in Greenhouse Solanaceous Crops'. *The European Journal of Plant Science and Biotechnology*. Global Science Books. p.127.

26. Smith L. (2009) 'Stingless wasp species discovered in Kent'. *The Guardian Online*. www.guardian.co.uk/environment/2009/oct/18/wasp-andrew-polaszek-natural-history-museum
27. Polaszek A. (2011) Unpublished Email to the Author.
28. Bukovinszky T. (2008) [Photo Credit]. '*Diadegma* sp.' Wageningen University. Bugsinthepicture.com. In Gershenzon J. (2012) 'Insects turn up their noses at sweating plants'. National Academy of Sciences. www.pnas.org/content/105/45/17211.full
29. Canola Council of Canada (Undated) [Photo Credit] 'dbmlarvalores'. www.canolacouncil.org/media/510962/dbmlarvalores.jpg
30. Schiffers B. and Wainwright H. (2011) [Photo Credit]. 'Biological Control and Integrated Crop Protection'. Training Manual 10. COLEACP. 130 rue du Trône, B-1050 Brussels, Belgium. p.84.
31. Zhang W., Ricketts T.H., Kremen C., Carney K. and Swinton S.M. (2007) 'Ecosystem services and dis-services to agriculture'. *Ecological Economics* 64(2) (2007). p.255. www.uvm.edu/rsenr/taylorricketts/documents/Zhang_2007_Ecological%20Economics.pdf
32. Choi H., Kim G. and Shin H. (2011) 'Biocontrol of Moth Pests in Apple Orchards: Preliminary Field Study of Application Potential for Mass Trapping'. *Biotechnology and Bioprocess Engineering* 16. The Korean Society for Biotechnology and Bioengineering and Springer. p.153.
33. Callaghan A. and Fellowes M. (2008) 'Garden Entomology'. *Introduction to Insect Science*. Royal Entomological Society. p.27.
34.0. Messenger C. (1997) 'The Sphinx Moths (Lepidoptera: Sphingidae) of Nebraska'. *Transactions of the Nebraska Academy of Sciences and Affiliated Societies*. Paper 72. pp.96-97. University of Nebraska – Lincoln. http://digitalcommons.unl.edu/tnas/72
34.1. Covell C.V. (1984) *A Field Guide to the Moths of Eastern North America*. Boston, Houghton Mifflin: 496. In Messenger C. (1997) 'The Sphinx Moths (Lepidoptera: Sphingidae) of Nebraska'. *Trans. Nebr. Acad. Sci. Affil. Soc.* Paper 72. p.103 and 104. University of Nebraska – Lincoln. http://digitalcommons.unl.edu/tnas/72
34.2. Heitzman J. R., and Heitzman J.E. (1987) *Butterflies and Moths of Missouri*. Jefferson City, Missouri Department of Conservation: 385 pp. In Messenger C. (1997) 'The Sphinx Moths (Lepidoptera: Sphingidae) of Nebraska'. *Trans. Nebr. Acad. Sci. Affil. Soc.* Paper 72. p.103 and 104. University of Nebraska – Lincoln. http://digitalcommons.unl.edu/tnas/72
34.3. Vandenborre G., Groten K., Smagg G., Lanno N., Baldwin I.T. and van Damme E.J. (2009) 'Nicotiana tabacum agglutinin is active against Lepidopteran pest insects'. *Journal of Experimental Botany*. Oxford University Press. p.1.
35. Sarfraz M., Keddie A.B. and Dosdall L.M. (2005) 'Biological control of the diamondback moth, *Plutella xylostella*: A review'. *Biocontrol Science and Technology*, December 2005; 15(8). p.763.
36. Tabashnik B.E., Cushing N.L., Finson N. and Johnson M.W. (1990) 'Field development of resistance to *Bacillus thuringiensis* in diamondback moth (Lepidoptera: Plutellidae)'. *Journal of Economic Entomology* 83: 1671-1676. In Sarfraz M., Keddie A.B. and Dosdall L.M. (2005) 'Biological control of the diamondback moth, *Plutella xylostella*: A review'. *Biocontrol Science and Technology*, December 2005; 15(8). p.776.
37.0. Talekar N.S., Liu S., Chen C. and Yiin Y. (1994) 'Characteristics of oviposition of diamondback moth (Lepidoptera: Yponomeutidae) on cabbage'. *Zoological Studies* 33, 72-77. In Sarfraz M., Keddie A.B. and Dosdall L.M. (2005) 'Biological control of the diamondback moth, *Plutella xylostella*: A review'. *Biocontrol Science and Technology*, December 2005; 15(8). p.765.
37.1. Justus K.A., Dosdall L.M., Mitchell B.K. (2000) 'Oviposition by *Plutella xylostella* (Lepidoptera: Plutellidae) and effect of phylloplane waxiness'. *Journal of Economic Entomology* 93, 1152-1159. In Sarfraz M., Keddie A.B. and Dosdall L.M. (2005) 'Biological control of the diamondback moth, *Plutella xylostella*: A review'. *Biocontrol Science and Technology*, December 2005; 15(8). p.765.
38. van Mele P. et al. (2002) *Cabbage Pest Management Manual*, CABI Bioscience, CAB International, Bakeham Lane, Egham, Surrey, TW20 9TY, UK. pp.47-48.
39. Schiffers B. and Wainwright H. (2011) 'Biological Control and Integrated Crop Protection'. Training Manual 10. COLEACP. 130 rue du Trône, B-1050 Brussels, Belgium. pp.83-84.
40. Harcourt D.G. (1957) 'Biology of the diamondback moth, *Plutella maculipennis* (Curt.) (Lepidoptera: Plutellidae), in Eastern Ontario. II. Life-history, behaviour, and host relationship'. *The Canadian Entomologist* 12, 554-564. In Sarfraz M., Keddie A.B. and Dosdall L.M. (2005) 'Biological control of the diamondback moth, *Plutella xylostella*: A review'. *Biocontrol Science and Technology*, December 2005; 15(8). p.765.
41. Lee J.C. and Landis D.A. (2000) 'Natural Enemies in Your Garden: A Homeowner's Guide to Biological Control'. Extension Bulletin 2719, New. Department of Entomology, Michigan State University. pp.45-46.
42. Harcourt D.G. (1986) 'Population dynamics of the diamondback moth in southern Ontario'. In Sarfraz M., Keddie A.B. and Dosdall L.M. (2005) 'Biological control of the diamondback moth, Plutella xylostella: A review'. *Biocontrol Science and Technology*, December 2005; 15(8). p.765.
43. Badenes-Perez F. R., Nault B.A. and Shelton A.M. (2006) 'Dynamics of diamondback moth oviposition in the presence of a highly preferred non-suitable host. Entomologia Experimentalis et Applicata'. 120. *Journal of Economic Entomology*. 83, 1671-1676. In Sarfraz M., Keddie A.B. and Dosdall L.M. (2005) 'Biological control of the diamondback moth, *Plutella xylostella*: A review'. *Biocontrol Science and Technology*, December 2005; 15(8). Taylor and Francis. Mortimer House, 37- 41 Mortimer Street, London W1T 3JH, UK. p.776.
44. PAN (2008) *How to Grow Crops Without Endosulphan*. Pesticides Action Network. Nernstweg. Hamburg. Germany. p.26. www.pan-germany.org.
45. Fitton M. and Walker A. (1992) 'Hymenopterous parasitoids associated with diamondback moth: the taxonomic dilemma'. pp.225-232. In Khatri D. (2011) 'Reproductive biology of *Diadegma semiclausum* Hellen (Hymenoptera: Ichneumonidae)'. Massey University, Palmerston North, New Zealand. p.7.
46.0. Soebe (2005) [Photo Credit: Left]. 'A Large White (*Pieris brassicae*)'. Northern Germany (Eckernförde).
46.1. Entomart (2009) [Photo Credit: Right]. '*Mamestra brassicae*'. In Wikipedia. http://en.wikipedia.org/wiki/File:Mamestra_brassicae01.jpg
47. Devetak M., Vidrih M. and Trdan S. (2010) 'Cabbage moth (*Mamestra brassicae* [L.]) and bright-line brown-eyes moth (*Mamestra oleracea* [L.]) – presentation of the species, their monitoring and control measures'. *Acta agriculturae* Slovenica, 95 – 2, julij 2010 str. p.149.
48. Rasmann. S., and Turlings T.C. (2007) 'Simultaneous feeding by aboveground and belowground herbivores attenuates plant-mediated attraction of their respective natural enemies'. *Ecol. Lett.* (2007) 10: 926-36.
49. Akin S. (2009) [Photo Credit]. 'Green Lacewing Adult'. Insect Pest Photo Gallery. University of Arkansas. Division of Agriculture. S.E.Research & Extension Center. P.O. Box 3508 1408 Scogin Drive. Monticello, AR 71656. http://insectphotos.uark.edu/beneficial/index.html
50. Akin S. (2009) [Photo Credit]. 'Lacewing Larva'. Insect Pest Photo Gallery. University of Arkansas. Division of Agriculture. S.E.Research & Extension Center. P.O. Box 3508 1408 Scogin Drive. Monticello, AR 71656. http://insectphotos.uark.edu/beneficial/index.html
51. Schmidt M.L. (2009) 'Lacewing pupa/cocoon – eyes become visible – Chrysoperla'. Delaware County, Pennsylvania, USA, July 19. In http://bugguide.net/node/view/308381
52. Akin S. (2009) [Photo Credit]. 'Green Lacewing Eggs'. Insect Pest Photo Gallery. University of Arkansas. Division of Agriculture. S.E.Research & Extension Center. P.O. Box 3508 1408 Scogin Drive. Monticello, AR 71656. http://insectphotos.uark.edu/beneficial/index.html
53. Chaney W.E. and Wunderlich L.R. (2000) 'Developing an IPM strategy for the new aphid pest, *Nasonovia ribisnigri* (Mosley), in head and leaf lettuce'. UC Cooperative Extension, Monterey County 1432 Abbott St. Salinas, Ca. 93901. p.6.
54. Oatman E.R., Badley M.E. and Platner G.R. (1985) 'Predators of the two spotted spider mite on strawberry'. *California Agriculture* Jan-Feb. University of California. 1301 S. 46th St., Bldg. 478. Richmond, CA 94804-4600. p.12.
55. Lee J.C. and Landis D.A (2000) 'Natural Enemies in Your Garden: A Homeowner's Guide to Biological Control'. Extension Bulletin 2719, New. Department of Entomology, Michigan State University. p.26.
56. Lind K., Lafer K., Schloffer K., Innerhofer G. and Meister H. (2003) *Organic Fruit Growing*. CABI Publishing. 44 Brattle Street. Wallingford 4th Floor, Oxon OX10 8DE. p.138.
57. Woolfolk S.W., Smith D.B., Martin R.A., Sumrall B.H., Nordlund D.A. and Smith R.A (2007) 'Multiple Orifice Distribution System for Placing Green Lacewing Eggs into Verticel Larval Rearing Units'. *J. Econ. Entomol.* 100(2). p.283.

58. Weinzierl R. and Henn T. (1991) *Alternatives in Insect Management. Biological and Biorational Approaches.* North Central Regional Extension Publication 401. Publications Office, University of Illinois, 69 Mumford Hall, 1301 W. Gregory Drive, Urbana, IL 61801. p.59.
59. Bissdorf J. (2005) *Field Guide to Non-chemical Pest Management in Tomato Production.* Pesticide Action Network (PAN) Germany. Nernstweg 32, 22765 Hamburg, Germany. p.37.
60. Mochizuki A., Naka H., Hamasaki K. and Mitsunaga T. (2006) 'Larval Cannibalism and Intraguild Predation Between the Introduced Green Lacewing, *Chrysoperla carnea*, and the Indigenous Trash-Carrying Green Lacewing, *Mallada desjardinsi* (Neuroptera: Chrysopidae), as a Case Study of Potential Nontarget Effect Assessment'. Entomology Group, Dept. of Biological Safety, National Institute for Agro-Environmental Sciences, 3-1-3 Kannondai, Tsukuba, Ibaraki 305-8604, Japan. In *Environmental Entomology* 35(5). p.1298.
61. Kunafin F. (1998) 'Commercialization of predators'. *American Entomol.* 4(1): 26-38. In Nadeem S. (2010) 'Improvement in Production and Storage of Trichogramma Chilonis (IshII), Chrysoperla Carnea (Stephens) and their Hosts for Effective Field Releases against Major Insect Pest of Cotton'. Department of Agri. Entomology. Faculty of Agriculture, Faisalabad, Pakistan. p.8.
62. Hodgson E.W. and Trina J. (2008) 'Beneficial insects: lacewings and antlions'. Ent-124-08. July. Utah Plant Pest Diagnostic Laboratory and Utah State University. UppDL, 5305 Old Main Hill, Logan UT 84322-5305. p.2.
63. Ruzicka Z. (1997) 'Protective role of the egg stalk in Chrysopidae (Neuroptera)'. Institute of Entomology, Czech Academy of Sciences, Branisovska 31, 370 05. Ceske Budejovice, Czech Republic. In *Eur. J. Entomol.* 94. p.111.
64. Mohammad T.H. (2008) 'Pistachio Psylla, Agonoscena pistasiae Burck. And Laut. (Hom.: Psyllidae) Stages Preference by Chrysoperla carnea Steph. (Neuro.: Chrysopidae)'. *Academic Journal of Entomology* 1 (1). Faculty of Agriculture, Shahid Bahonr University Kerman, Iran. p.7.
65. Bailey M.J., Lilley A.K. and Spencer-Phillips P.T.N. (Eds.) (2006) 'Microbial Ecology of Aerial Plant Surfaces'. CABI Head Office, Nosworthy Way, Wallingford, Oxfordshire, OX10 8DE. pp. Preface pp.viii-ix.
66. Lindow S.E. and Brandl M.T. (2003) 'Minireview: Microbiology of the Phyllosphere'. *Applied and Environmental Microbiology*, April 2003, pp. 1875-1883, Vol. 69, No. 4. Department of Plant and Microbial Biology, University of California, Berkeley, California 94720. p.1876.
67. Bailey M.J., Lilley A.K. and Spencer-Phillips P.T.N. (Eds.) (2006) *Microbial Ecology of Aerial Plant Surfaces.* CABI Head Office, Nosworthy Way, Wallingford, Oxfordshire, OX10 8DE. p.3.
68. Lan N.T.P. and Mew T.W. (2001) 'Suppression of sheath blight development in rice by rice-associated biocontrol agents'. In Mew T.W. and Cottyn B. (Eds.) (2001) 'Seed Health and Seed-Associated Microorganisms for Rice Disease Management'. *Limited Proceedings.* No.6. International Rice Research Institute (IRRI). DAPA Box 7777, Metro Manila, Philippines. pp.47-48.
69.0. Wilson M., Hirano S.S., and S.E. Lindow S.E. (1999) 'Location and survival of leaf-associated bacteria in relation to pathogenicity and potential for growth within the leaf'. *Appl. Environ. Microbiol.* 65, 1435-1443. In Lindow S.E. and Brandl M.T. (2003) 'Microbiology of the Phyllosphere'. *Applied and Environmental Microbiology.* April. Vol. 69, No. 4. American Society for Microbiology. p.1876.
69.1. Paul N. D., Rasanayagam S., Moody S.A., Hatcher P.E. and Ayres P.G. (1997) 'The role of interactions between trophic levels in determining the effects of UV-B on terrestrial ecosystems', *Plant Ecol.* 128, 296-308. In Jacobs J.L and Sunding G.W.(2001) 'Effect of Solar UV-B Radiation on a Phyllosphere Bacterial Community'. *Applied and Environmental Microbiology.* December. Vol. 67. No.12. American Society for Microbiology. p.5488.
70. Sterk G., Heuts F., Merck N. and Bock J. (2002) 'Sensitivity of non-target arthropods and beneficial fungal species to chemical and biological plant protection products: results of laboratory and semi-field trials'. 1st International Symposium on Biological Control of Arthropods. p.307.
71. Thungrabeab M. and Tongma S. (2007) 'Effect of entomopatheogenic Fungi, *Beauveria bassiana* (Balsam) and Metarhizium anisopliae (Metsch) on non target insects'. *KMITL Sci. Tech. J.* Vol. 7 No. S1 Nov. 2007. p.11.
72. Al mazra'awi M.S. (2007) 'Impact of the Entomopathogenic Fungus *Beauveria bassiana* on the Honey Bees, *Apis mellifera* (Hymenoptera: Apidae)'. *World Journal of Agricultural Sciences* 3(1), 07-11, 2007.
73. Meyling N.V, Judith K., Pell J.K and Eilenberg J. (2006) Dispersal of *Beauveria bassiana* by the activity of nettle insects. *Journal of Invertebrate Pathology* 93 (2006). p.121.
74. Hoffmann M.P. and Frodsham A.C. (1993) 'Natural Enemies of Vegetable Insect Pests'. Cooperative Extension, Cornell University, Ithaca, NY. p.63.
75. Lorito M., Woo S.L., D'Ambrosio M., Harman G.E., Hayes C.K., Kubicek C.P. and Scala F. (1996) 'Synergistic interaction between cell wall degrading enzymes and membrane affecting compounds'. *Molecular Plant–Microbe Interactions* 9, 206–213. In Barea J., Pozo M.J., Azcon R. and Azcon-Aguilar C. (2005) 'Microbial co-operation in the rhizosphere'. *Journal of Experimental Botany* Vol. 56, No. 417. July. p.1766.
76. Sharma D.D. et al. (2008) 'BAPI-001: Organic Production System'. Indira Gandhi National Open University. p.30.
77.0. Smith P. (1993) 'Control of *Bemisia tabaci* and the potential of *Paecilomyces fumosoroseus* as a biopesticide'. *Biocontrol News and Information* 14 (4), 71N-78N. In Skrobek A. (2001) *Investigations on the effect of entomopathogenic fungi on whiteflies.* Friedrich-Wilhelms-University, Bonn. pp. Abstract.
77.1. Panyasiri C., Attathom T. and Poehling H.M. (2007) 'Pathogenicity of entomopathogenic fungi-potential candidates to control insect pests on tomato under protected cultivation in Thailand'. *Journal of Plant Diseases and Protection* 114 (6), 2007. Eugen Ulmer KG, Stuttgart. p.279.
78. Kirk A.A., Mercadier G., Bordata D., Delvare G., Pichon A., Arvanitaki L., Goudégnon A.E. and Rinçon C. (2001) 'Variability in Plutella and its natural enemies: implications for biological control'. European Biological Control Laboratory, CILBA, Montferrier sur Lez 34988, St. Gely du Fesc, France. In Proceedings of the 4th International Workshop. Nov. 2001, Melbourne, Australia. p.71.

Chapter Six

1. Evins C. (1975) Technical Report TR8: 'The Toxicity Of Chlorine To Some Fresh Water Organisms'. Water Research Centre. Medmenham Lab. P.O Box 16. Medmenham, Marlow, Bucks, SL7 2HD. p.24.
2. Phillips J.H. (1966) 'The Discovery and Control of Live Organisms in Great Yarmouth Water Supply'. *J. Instr. Wat. Engns.* 20(3), p.207.
3. Brigano F.A., Ruhstorfer R. B., Gottlieb M., Trickle G., Harrison J. F., Verstrat S. J., Petty B. L. and Yoder R. (2004) 'Technical Application Bulletin: Chloramines'. Water Quality Association. National Headquarters & Laboratory. 4151 Naperville Road. Lisle, Illinois 60532. p.3.
4. Mason J. (1951) *Sustainable Agriculture.* 2nd Ed. National Library of Australia Cataloguing-in-Publication. Landlinks Press. PO Box 1139. Collingwood Vic. 3066. Australia. p.94.
5.0. Wilman D. & Riley J. A. (1993) 'Potential nutritive value of a wide range of grassland species'. *Journal of Agricultural Science* 120, 43-49. In Bond W., Davies G. and Turner R. (2007) 'The biology and non-chemical control of Common Nettle (*Urtica dioica* L.)'. HDRA, Ryton Organic Gardens, Coventry, CV8, 3LG, UK. [Re: N, Ca, Mg, p.1.]
5.1. Grime J.P., Hodgson J.G. and Hunt R. (1988) 'Comparative Plant Ecology'. Unwin Hyman Ltd, London, UK. Plant strategies in shade. In Bond W., Davies G. and Turner R. (2007) 'The biology and non-chemical control of Common Nettle (*Urtica dioica* L.)'. HDRA, Ryton Organic Gardens, Coventry, CV8, 3LG, UK. [Re: N, Ca, Mg, p.1.]
5.2. Salisbury E (1962) 'The biology of garden weeds. Part II'. *Journal of the Royal Horticultural Society* 87, 458-470 & 497-508. In Bond W., Davies G. and Turner R. (2007) 'The biology and non-chemical control of Common Nettle (*Urtica dioica* L.)'. HDRA, Ryton Organic Gardens, Coventry, CV8, 3LG, UK. [Re: Iron, p.1.]
5.3. Mason J. (1951) *Sustainable Agriculture.* 2nd Ed. National Library of Australia Cataloguing-in-Publication. Landlinks Press. PO Box 1139. Collingwood Vic. 3066. Australia. p.94.
6. Jeavons J. (2002) *How To Grow More Vegatables Than You Ever Thought Possible On Less Land Than You Can Imagine.* 6th Ed. Berkeley, CA. Ten Speed Press. p.144.
7.0. Kourik R. (2005) *Designing And Maintaining Your Edible Landscape Naturally.* Chelsea Green Publications. [Re:

Phosphorus, p.265.]
7.1. Mason J. (1951) *Sustainable Agriculture*. 2nd Ed. National Library of Australia Cataloguing-in-Publication. Landlinks Press. PO Box 1139. Collingwood Vic. 3066. Australia. p.94.
8.0. Tilley D. (2008) 'Yellow Sweetclover & White Sweetclover'. (USDA) United States Department of Agriculture and (NRCS) Natural Resources Conservation Service. [Re: Disuse and Green Manure, p.2. Weed Status, p.3.]
9. Rayburn E. (2007) 'Drought Management Before, During, and After the Drought'. West Virginia University. Extension Forage Agronomist. In 'National Grass-Fed Beef Conference: The Art and Science of Grass-Fed Production and Marketing'. Feb-Mar. The Pennsylanvania State University Department of Dairy and Animal Science. p.29.
10. Magdoff F. and van Es H. (2000) *Building Soils for Better Crops*. 2nd Ed. Sustainable Agriculture Network Handbook Series. Book 4. AFSIC, Room 304, National Agricultural Library, 10301 Baltimore Ave. Beltsville, MD 20705-2351. p.86.
11. Eisler R. (2000) *Handbook of Chemical Risk Assessment: Heath Hazards to Humans, Plants and Animals*. CRC Press. Vol 3. p.1616.
12. Marschner H. (1995) *Mineral Nutrition of Higher Plants*. 2nd Ed. New York: Academic Press. pp.369-379. In Barker A.V. and Pilbeam D.J. (Eds.) (2007) *Handbook of Plant Nutrition*. CRC Press. Taylor and Francis group. p.381.
13. Thomas B. (2001) 'Fertility for pastures'. In Thomas B. (2009) *Maritime Pasture Manual*. AgraPoint International Inc. Nova Scotia Dept of Agriculture. p.64. Email: info@agrapoint.ca
14. Thomas B. (2009) *Maritime Pasture Manual*. AgraPoint International Inc. Nova Scotia Dept of Agriculture. p.65. Email: info@agrapoint.ca
15. Jacob A. (1958) *Magnesium: The Fifth Major Plant Nutrient*. London: Staples Press Ltd London. In Barker A.V. and Pilbeam D.J. (Eds.) (2007) *Handbook of Plant Nutrition*. CRC Press. Taylor and Francis group. p.377.
16.0. D. Bertrand A. de Wolf. (1967) 'Nickel, a dynamic trace element for higher plants'. *C R Academic Sci.* 265, 1053-1055. In Barker A.V. and Pilbeam D.J. (Eds.) (2007) *Handbook of Plant Nutrition*. CRC Press. Taylor and Francis group. [Re: Nickel, pp.398-399.]
16.1. Cammack R. (1995) Splitting molecular hydrogen'. *Nature*, 373:556–557. In Barker A.V. and Pilbeam D.J. (Eds.) (2007) *Handbook of Plant Nutrition*. CRC Press. Taylor and Francis group. [Re: Nickel, p.399.]
16.2. Albrecht S.L., Maier R.J., Hanus F.J., Russel S.A., Emerich D.W. and H.J. Evans. (1979) 'Hydrogenase in *Rhizobium japonicum* increases nitrogen fixation by nodulated soybeans'. *Science* 203, 1255-1257. In Barker A.V. and Pilbeam D.J. (Eds.) (2007) *Handbook of Plant Nutrition*. CRC Press. Taylor and Francis group. [Re: Nickel, p.399.]
16.3. Salisbury F.B. and Ross C.W. (1992) *Plant Physiology*. 4th Ed. Belmont, CA: Wadsworth Publishers, pp.124-125. In Barker A.V. and Pilbeam D.J. Eds (2007) *Handbook of Plant Nutrition*. CRC Press. [Re: Cobalt, p.500.]
16.4. Davidova E.G., Belov A.P. and Zachinskii V.V. (1986) 'The accumulation of labelled cobalt in yeast cells'. *Izh. Timiryazev S-KH Akad.* Jul-Aug (4), pp.109-114, 1986. [Re: Cobalt, p.501.]
16.5. Venkateswerlu G. and Sivaramasastry K. (1970) 'The mechanism of uptake of cobalt ions by *Neurospora crassa*'. *Biochem. J.* 118, 497-503, 1970. In Barker A.V. and Pilbeam D.J. (Eds.) (2007) *Handbook of Plant Nutrition*. CRC Press. [Re: Cobalt, p.501.]
16.6. Marschner H. (1995) *Mineral Nutrition of Higher Plants*. 2nd Ed. New York: Academic Press, 1995, pp.369-379. In Barker A.V. and Pilbeam D.J. (Eds.) (2007) *Handbook of Plant Nutrition*. CRC Press. Taylor and Francis group. [Re: Molybdenum, p.378.]
17. Thomas B. (2007) *Maritime Pasture Manual*. AgraPoint International Inc. Nova Scotia Dept. of Agriculture. p.53. Email: info@agrapoint.ca
18. Hodson M.J., White P.J., Mead A. and Broadley M.R. (2005) 'Phylogenetic Variation in the Silicon Composition of Plants'. *Annals of Botany* 96. School of Biological and Molecular Sciences, Oxford Brookes University, UK. p.1028.
19. Epstein E. (2009) 'Silicon: its manifold roles in plant's. *Ann Appl Biol.* Department of Land, Air and Water Resources – Soils and Biogeochemistry, University of California, Davis, CA, USA. p.155.
20. Snyder G.H., Matichenkov V.V. and Datnoff L.E. (2007) 'Chapter 19: Silicon'. In Barker A.V. and Pilbeam D.J. (Eds.) (2007) *Handbook of Plant Nutrition*. CRC Press. Taylor and Francis group. [Re: Plant available forms, p.552.]
21.0. Adatia M.H. and R.T. Besford R.T. (1986) 'The effects of silicon on cucumber plants grown in recirculating nutrient solution'. *Ann. Bot.* 58, 343-351. In Barker A.V. and Pilbeam D.J. (Eds.) (2007) *Handbook of Plant Nutrition*. CRC Press. p.557.
21.1. Bocharnikova E.A. (1996) 'The study of direct silicon effect on root demographics of some cereals'. In *Root Demographics and Their Efficiencies in Sustainable Agriculture, Grasslands, and Forest Ecosystems*, Madrea Conference Conter-Clenson, South Carolina, 14-18 July. Proceedings of the Fifth Symposium of the International Society of Root Research. In Barker A.V. and Pilbeam D.J. (Eds.) (2007) *Handbook of Plant Nutrition*. CRC Press. p.557.
21.2. Kim H.M. (1987) 'The influence of nitrogen and soil conditioners on root development, root activity and yield of rice. 3. The effects of soil conditioners on development of root, rooting zone and rice yield'. Research Rep. Rural Development Admin., *Plant Environ., Mycol. & Farm Prod Util*, Korea Republic 29, 12-29. In Barker A.V. and Pilbeam D.J. (Eds.) (2007) *Handbook of Plant Nutrition*. CRC Press. p.557.
21.3. Kudinova L.I. (1975) 'The effect of silicon on growth, size of leaf area and sorbed surface of plant roots'. *Agrochemistry* 10, 117-120. In Barker A.V. and Pilbeam D.J. (Eds.) (2007) *Handbook of Plant Nutrition*. CRC Press. p.557.
21.4. Matichenkov V.V. (1996) 'The silicon fertilizer effect of root cell growth of barley'. Abstr. The Fifth Symposium of the International Society of Root Research, Clemson, SC, USA. p.110. In Barker A.V. and Pilbeam D.J. (Eds.) (2007) *Handbook of Plant Nutrition*. CRC Press. p.557.
22. Piekos R. and Paslawska S. (1976) 'Studies on the optimum conditions of extraction of silicon species from plants with water. I. *Equisetum arvense* L'. *Herb. Planta Med.* Jun.29(4), 351-6.
22.1. Bergner P. (2001) 'Equisetum: Silicon in horsetail and comfrey'. *Medical Herbalism* 10(4), 10.
23. Kelley W.P. (1956) 'Dennis Robert Hoagland (1884-1949) – *A Biographical Memoir*'. National Academy of Sciences, Washington D.C. p.124.
24. Martin D.L. and Gershunny G. (Eds.) (1992) *The Rodale Book of Composting: Easy methods for every gardener*. Rodale Press, Emmaus Pennsylvania. p.103.
25. Teas J., Pino S., Critchley A. and Braverman L.E. (2004) 'Variability of Iodine Content in Common Commercially Available Edible Seaweeds'. *Thyroid*, Volume 14, Number 10. Publishers. Mary Ann Liebert Inc. p.837.
26. Lightfoot D.G. (1991) 'Combined Fields (Electro-osmosis and Pressure) Dewatering of Kelp'. Department of Agricultural Engineering Macdonald Campus of McGill University Ste. Anne de Bellevue, Quebec. pp.1-2.
27. Temple W.D. and Bomke A.A. (1989). 'Effects of kelp (*Macrocystis integrifolia* and *Ecklonia maxima*) foliar applications on bean growth'. *Plant and Soil* 117, 85-92. In Lightfoot D.G (1991) 'Combined Fields (Electro-osmosis and Pressure) Dewatering of Kelp'. Department of Agricultural Engineering Macdonald Campus of McGill University Ste. Anne de Bellevue, Quebec. p.15.
28. Temple W.D. and Bomke A.A. (1989) 'Effects of kelp (*Macrocystis integrifolia* and *Ecklonia maxima*) foliar applications on bean growth'. *Plant and Soil* 117, 85-92. In Lightfoot D.G (1991) 'Combined Fields (Electro-osmosis and Pressure) Dewatering of Kelp'. Department of Agricultural Engineering Macdonald Campus of McGill University St. Anne de Bellevue, Quebec. p.13.
29.0. Blunden G. (1971) 'The effects of aqueous seaweed extracts as a fertilizer additive'. Proceedings of the 7th International Seaweed Symposium, Sopporo, Japan, August 8-12, 1971: 584-589. In Lightfoot D.G (1991) 'Combined Fields (Electro-osmosis and Pressure) Dewatering of Kelp'. Department of Agricultural Engineering Macdonald Campus of McGill University Ste. Anne de Bellevue, Quebec. p.12.
29.1. Blunhden G. and Wildgoose P.B. (1977) 'The effects of aqueous seaweed extract and kinetin on potato yields'. *Journal of the Science of Food and Agriculture* 28, 121-125. In Lightfoot D.G (1991) 'Combined Fields (Electro-osmosis and Pressure) Dewatering of Kelp'. Department of Agricultural Engineering Macdonald Campus of McGill University Ste. Anne de Bellevue, Quebec. p.13.
30. Booth E. (1965) 'Some properties of seaweed manures'. Proceedings of the 5th International Seaweed Symposium, Halifax, Nova Scotia, August 25-28. 349-357. In Lightfoot D.G (1991) 'Combined Fields (Electro-osmosis and Pressure) Dewatering of Kelp'. Department of Agricultural Engineering Macdonald Campus of McGill University Ste.

Anne de Bellevue, Quebec. p.64.

31. Chaudhary D.R., Bhandari S.C. and Shukla L.M. (2004) 'Role of Vermicompost in sustainable agriculture – a review'. *Agric. Rev.* 25 (1), p.29.
32. Sinha R.K., Valani D., Chauhan K. and Agarwal S.(2010) 'Embarking on a second green revolution for sustainable agriculture by vermiculture biotechnology using earthworms: Reviving the dreams of Sir Charles Darwin'. *Journal of Agricultural Biotechnology and Sustainable Development* Vol. 2(7), p.113.
33. Simsek-Ersahin Y. (2011) 'The Use of Vermicompost Products to Control Plant Diseases and Pests'. *Soil Biology* Volume 24, 191-213.
34. Pierre V., Phillip R. Margnerite L. and Pierrette C. (1982) 'Anti-bacterial activity of the haemolytic system from the earthworms *Eisinia foetida* Andrei'. *Invertebrate Pathology* 40(1), 21-27. In Sinha R.K., Agarwal S., Chauhan K. and Valani D. (2010) 'The wonders of earthworms & its vermicompost in farm production: Charles Darwin's 'friends of farmers', with potential to replace destructive chemical fertilizers from agriculture'. Vol.1, No.2, 76-94. Home Science, University of Rajasthan, Jaipur, India. p.91.
35. Harris (1978) In Multicote (2008) 'Nutritional recommendations for Potato Multicote Products'. Haifa. www.haifa-group.com/files/Guides/Potato.pdf [Re: NPK intake/comparison, pp.5-6. Tuber Bulking, p.6.]
36. Multicote (2008) 'Nutrional Recommendations for Potato Multicote Products.Haifa'. www.haifa-group.com/files/Guides/Potato.pdf [Re: Phosphorus, p.12. Potassium, p.13.]
37. Huett (1985) In Multicote (2008) 'Nutritional Recommendations for Tomato Multicote Products'. Haifa. p.8. www.haifa-group.com/files/Guides/tomato/Tomato.pdf
38. Multicote (2008) 'Nutritional recommendations for Cucumber Multicote Products.Haifa'. [Re: More K than N, p.31.] www.haifa-group.com/files/Guides/Cucumber.pdf
39. Laboski C.A.M. and Peters J.B. (2012) 'Nutrient Application Guidelines for field, vegetable, and fruit crops in Winsconsin (A2809)'. University of Wisconsin-Extension, Cooperative Extension Publishing at 432 N. Lake St., Rm. 227, Madison. WI 53706. p.81. Email: pubs@uwex.edu
40. Curley S. (Ed) (1994) 'Foliar Nutrition'. Midwest Laboratories Inc. 13611 B Street, Omaha, NE 68144. p.23. https://midwestlabs.com/wp-content/uploads/2017/01/foliar_nutrition.pdf
41. Engeland R. L. (1991) *Growing great garlic: the definitive guide for organic gardeners and small farmers.* Okanogan, WA: Filaree Productions. (HSA Library). In Meyers M. (2006) *Garlic: An Herb Society of America Guide*'. The Herb Society of America. 9019 Kirtland Chardon Rd. Kirtland, Ohio 44094. p.15.
42. Hegde D.M. (1986) 'Fruit development in sweet pepper (*Capsicum annuum* L.) in relation to soil moisture and nitrogen fertilization'. *Singapore Journal Primary Industries* 14, 64-75. In Hegde D.M. (2010) 'Nutrient Requirements of Solanaceous Vegetable Crops'. All India Coordinated Safflower Improvement Project Solapur, Maharashtra, India. p.4.
43. Hegde D.M. (2010) 'Nutrient Requirements of Solanaceous Vegetable Crops.' All India Coordinated Safflower Improvement Project Solapur, Maharashtra, India. p.1.

Chapter Seven

1. NRC (1989) [Photo Credit]. 'Agricultural Age'. In NRC (1989) *Alternative Agriculture.* Committee on the Role of Alternative Farming Methods in Modern Production Agriculture', National Research Council. p.123.
2. Alexenizer M. and Dorn A. (2007) 'Screening of medicinal and ornamental plants for insecticidal and growth regulating activity'. *J. Pest Sci.* (2007) 80. p.205.
3. Van Zwietan M., Stovold G. and Van Zwieten L. (2007) *Alternatives to Copper for Disease Control in the Australian Organic Industry.* 'Environmental Centre of Excellence', NSW Agriculture, Wollongbar. p.22.
4. Sculley F.X. (1989) 'God's gift, nature's soap'. *The Herb Quarterly* 1989. Spring. www.drugs.com/npp/soapwort.html#ixzz0uQNDgxms. p.7.
5. Güçlü-Üstündağ Ö. and Mazza G. (2007) 'Saponins: Properties, Applications and Processing'. *Critical Reviews in Food Science and Nutrition* Vol. 47. Issue 3. March 2007. Taylor and Francis group. p.238.
6. De Geyter E., Geelen D. and Smagghe G. (2007) 'First results on the insecticidal action of saponins'. *Commun. Agric. Appl. Biol. Sci.* 2007, 72(3), 645-8. Ghent University. p.645.

7.0. Ej Douda O., Zouhar M., Maza´kova´ J., Nova´kova´ E. and Pavela R. (2010) 'Using plant essences as alternative mean for northern root-knot nematode (*Meloidogyne hapla*) management'. *J. Pest. Sci.* p.1.

7.1. Bharadwaj A. and Sharma S. (2007) 'Effect of some plant extracts on the hatch of *Meloidogyne incognita* eggs'. *Int. J. Bot.* 3, p.312.

7.2. Sturz A.V. and Kimpinski J. (2004) 'Endoroot bacteria derived from marigolds (*Tagetes* spp.) can decrease soil population densities of root-lesion nematodes in the potato root zone'. *Plant and Soil* 262. Kluwer Academic Publishers. p.241.

8.0. HDRA (2000) 'Natural Pesticides No. TNP2. Mexican marigold, *Tagetes minuta*'. HDRA, Ryton Organic Gardens Coventry, CV8 3LG, UK.

8.1. HDRA (2000) 'Pest Control No. TPC2. Aphids, Family Aphididae'. HDRA, Ryton Organic Gardens Coventry, CV8 3LG, UK.

8.2. Mollison B. (1981) *Introduction to Permaculture.* The Rural Education Center, Wilton, NH USA. p.4.

8.3. Harris P., Jarratt J.H., Killebrew F. and Byrd J.D. et al. (2008) *Organic Vegetable IPM Guide.* Extension Service of Mississippi State University. p.5.

8.4. Little T. and Frost D. (Eds.) (2008) *Organic Farming Technical Guide: A farmer's guide to Organic fruit and vegetable production.* Organic Centre Wales, Institute of Biological, Environmental and Rural Sciences, Aberystwyth University, Aberystwyth, Ceredigion, SY23 3AL. p.10 and 26.

9. Linnaeus (2007) 'HOSTS – a Database of the World's Lepidopteran Hostplants'. The Natural History Museum. www.nhm.ac.uk/jdsml/research-curation/research/projects/hostplants/index.dsml.
10. McKenry M., Buzi T. and Kaku S. (1995) 'First-Year evaluations of tree and vine growth and nematode development following 17 pre-plant treatments'. Proc. Ann. Intl. Res. Conf. Methyl Bromide Alternatives and Emissions Reductions. November 6-8, 1995, San Diego, CA. p.37.

11.0. Vasudevan P., Kashyap S. and Sharma S. (1997) '*Tagetes*: A Multipurpose plant'. *Bioresources Technology.* p.29. Volume 62, Issues 1-2. Oct-Nov. pp.29-35.

11.1. Natarajana N., Corkb A., Boomathia N., Pandia R., Velavana S. and Dhakshnamoorthya G. (2006). 'Cold aqueous extracts of African marigold, *Tagetes erecta* for control'. p.1210. *Crop Protection* 25 (2006) 1210-1213.

12. Sasser J.N. (1998) 'A Perspective on Nematode Problems Worldwide'. In 'Nematode Parasitic to Cereals and Legumes in Temperate Semi Arid Regions'. Sasena M.C., Sikora R.A. and Srivastava (Eds.). ICARDA, Syria. pp.8-12.
13. Ruabete T.K. (2005). In S.R. Gowen S.R., Ruabete T.K and Wright J.G. (2005) 'Root Knot Nematodes'. Pest Advisory Leaflet No. 9. 3rd Ed. Secretariat of the Pacific Community. p.1.
14. PKF Consulting Ltd and International Research Network (2005) *Kenya's Pyrethrum Industry 2005.* Export Processing Zones Authority. EPZA Administration Building Athi River EPZ, Viwanda Road Off Nairobi - Namanga Highway. P.O Box 50563, 00200 Nairobi, Kenya. p.1.

15.0. Casida J.E. (1980) 'Pyrethrum Flowers and Pyrethroid Insecticides'. *Environmental Health Perspectives* Vol. 34, p.191.

15.1. Harris P., Jarratt J.H., Killebrew F. and Byrd J.D et al (2008) *Organic Vegetable IPM Guide.* Extension Service of Mississippi State University. p.14.

16.0. Ellis B. and Bradley F. (1996) *The Organic Gardener's Handbook of Natural Insect and Disease Control.* Rodale Press. Emmaus, Pensylvania. pp.480-481.

16.1. Stoll G. (2000) *Natural protection in the tropics.* Margraf Verlag Weikasheim. pp.150-152. In PAN (2008) *How to Grow Crops Without Endosulphan.* Pesticides Action Network. Nernstweg. Hamburg. Germany. p.53. www.pan-germany.org.

17.0. Hill N. (2008) 'A Novel Plant-Based Synergist For Pyrethrum and Pyrethroids Against Urban Public Health Pests'. In Robinson W.H. and Bajomi D. (Eds.) (2008) Proceedings of the Sixth International Conference on Urban Pests. OOK-Press Kft., H-8200 Veszprém, Pápai út 37/a, Hungary. p.235.

17.1. Casida J.E. (1980) 'Pyrethrum Flowers and Pyrethroid Insecticides'. *Environmental Health Perspectives* Vol. 34, pp191.

18. Mulaa M. and Maala E. (1995) Kenya Agricultural Research Institute and the Rockefeller Foundation. Kari-Kitale. P.O Box, Kitale. p.1.
19. Wauchope R. D., Butler T. M., Hornsby A.G., Augustijn-Beckers P.M. and Burt J. P. (1992) 'The SCS/ARS/CES pesticide properties database for environmental decision making'. In 'National Pesticide Information Center's Fact Sheet'. Oregon State University. Environmental and Molecu-

lar Toxicology. 333 Weniger Hall. Corvallis, OR 97331-6502.

20. Casida J.E. (1980) 'Pyrethrum Flowers and Pyrethroid Insecticides'. *Environmental Health Perspectives* Vol. 34, pp189.

21.0. Bernard C.B. and Philogne J.R. (1993) 'Insecticide synergists: Role, importance, and perspectives'. *Journal of Toxicology and Environmental Health*, Part A, Volume 38, Issue 2. February 1993. p.199.

21.1. Casida J.E. (1980) 'Pyrethrum Flowers and Pyrethroid Insecticides'. *Environmental Health Perspectives* Vol. 34, pp191.

22. Hill N. (2008) 'A Novel Plant-Based Synergist For Pyrethrum and Pyrethroids Against Urban Public Health Pests'. In Proceedings of the Sixth International Conference on Urban Pests. Robinson W.H. and Bajomi D. (Eds.) (2008). OOK-Press Kft., H-8200 Veszprém, Pápai út 37/a, Hungary. p.235.

23.0. Noetzel D., Ricard M. and Wiersma J. (1989). 'Effect of 3 pyrethroid and 4 suggested synergist combinations on pyrethroid-resistant Colorado Potato Beetle', 1987. *Insecticide and Acaricide Tests*. 14, 159-160.

23.1. Noetzel D. and Ricard M. (1989) 'Control of imported cabbageworm and cabbage looper using pyrethrin, rotenone, or Asana with of without one of four 'synergists'. *Insecticide and Acaricide Tests*. 14, 96-97.

24. Casida J.E. (1980) 'Pyrethrum Flowers and Pyrethroid Insecticides'. *Environmental Health Perspectives* Vol. 34, p.190.

25. CEMAS (2006) 'Final report to investigate any synergistic effects that may exist between pyrethrins and dill seed oil'. CEMR-3156. 1-19. In Proceedings of the Sixth International Conference on Urban Pests. Robinson W.H. and Bajomi D. (Eds.) (2008). OOK-Press Kft., H-8200 Veszprém, Pápai út 37/a, Hungary. p.235.

26. HDRA (2000) 'Natural Pesticides No. TNP1 Chillipepper, *Capiscum frutescens*'. HDRA – the organic organisation, Ryton Organic Gardens Coventry, CV8 3LG, UK.

27. Loevinsohn M., Meijerink G. and Salasya B. (1998) 'Integrated Pest Management in Smallholder Farming Systems in Kenya. Evaluation of a pilot project'. International Service for National Agriculture. In FAO Inter-Country Programme (2000). *Tomato Integrated Pest Management: An Ecological Guide*. FAO Inter-Country Programme for the Development and Application of Integrated Pest Management in Vegetable Growing in South and South-East Asia. p.74.

28. HDRA (2010) 'Natural Pesticides No. TNP3: Garlic, *Allium satvium*, natural pesticide'. HDRA - the organic organisation, Ryton Organic Gardens Coventry, CV8 3LG.

29. Dixon D. (Ed) (2003) *Appropriate Technology*, Vol 30, No 3. Research Information Ltd. 222 Maylands Avenue. Hemel Hempstead, Herts. HP2 7TD, UK. p.53.

30.0. Kazem M.G.T. and El-Shereif S.A.E.H.N. (2010) 'Toxic Effect of Capsicum and Garlic Xylene Extracts in Toxicity of Boiled Linseed Oil Formulations against Some Piercing Sucking Cotton Pests'. *American-Eurasian J. Agric. & Environ. Sci.* 8 (4), 390-396, 2010. IDOSI Publications. Central Agricultural Pesticides Laboratory, Dokki, Giza, Egypt. p.390.

30.1. Curtis H., Noll U.M., Stormann J. and Slusarenko A.J. (2004) 'Broad-spectrum activity of the volatile phytoanticipin allicin in extracts of garlic (*Allium sativum* L.) against plant pathogenic bacteria, fungi and oomycetres'. *Physiol. Mol. Plant Pathol.* 2004, 65, 79-89. In Barker A.V. and Pilbeam D.J. (Eds.) (2007) *Handbook of Plant Nutrition*. CRC Press. Taylor and Francis group. p.194.

30.2. Bongiorno P.B, Fratellone P.M. and LoGiudice P. (2008) 'Potential Health Benefits of Garlic (*Allium sativum*): A Narrative Review'. *Journal of Complementary and Integrative Medicine*, Vol. 5 (2008), Iss. 1, Art. 1. pp.8-9.

31.0. Bongiorno P.B, Fratellone P.M. and LoGiudice P. (2008) 'Potential Health Benefits of Garlic (*Allium sativum*): A Narrative Review'. *Journal of Complementary and Integrative Medicine*, Vol. 5, Iss. 1, Art. 1. p.2.

31.1. Lawson L.D. (1998) 'Garlic: A review of its medicinal effects and indicated active compounds'. In Pedraza-Chaverrí J., Gil-Ortiz M., Albarrán G., Barbachano-Esparza L., Menjívar M. and Medina-Campos O.N.(2004) 'Garlic's ability to prevent in vitro Cu^{2+}-induced lipoprotein oxidation in human serum is preserved in heated garlic: effect unrelated to Cu^{2+}-chelation'. Nutr J. 2004; 3: 10. www.ncbi.nlm.nih.gov/pmc/articles/PMC519022/

32. Evans M.R. (2008) 'Root Media play a role in plant health – Researchers work to design better disease suppressiveness'. June Edition. *Nursery Management & Production (NmPro)*. p.56.

33.0. Radulović N., Stojanović G. and Palić R. (2006) 'Composition and antimicrobial activity of *Equisetum arvense* L. essential oil'. *Phytother. Res.* 20. [Re: Antagonists of fungal diseases, p.85.]

33.1. Cwalina-Ambroziak B. and Nowak M.K. (2011) 'Fungi colonising the soil and roots of tomato (*Lycopersium esculentum* Mill.) Plants treated with biological control agents'. *Acta Agrobotanica* Vol. 64 (3). [Re: Equisetum and antagonism to fungal diseases, p.87. 'Antagonists of damping-off pathogens: *Fusarium equiseti, Fusarium oxysporum* and *Fusarium poae*', p.87.]

34. Willomitzer J. and Tomanek J. (1981) 'Larvicidal Efficiency of Some Inorganic Compounds and Plant Extracts Against the House Fly'. *Acta Vet. Brno.* 50. p.105.

35.0. Tilgner S. (1999) 'Biodynamic Farming'. p.4. www.herbal-transitions.com/Newsletters%20and%20other%20goodies/Biodynamic%20farming.pdf

35.1. Willis G. (2008) *The New Holistic Agriculture Field Broadcaster Handbook*. GW Agriculture. P.O. Box 3362. McKinney, USA, TX 75070. p.74.

36. Steiner R. (1924) 'Agriculture Course. Lecture 6'. The Rudolf Steiner Archive. http://wn.rsarchive.org/Lectures/Agri1958/19240614p01.html In Kristiansen P., Taji A. and Reganold J. (2006) *Agriculture – A Global Perspective*. CABI Publishing, Wallingford, Oxon OX10 8DE. p.143.

37. USDA (1997) 'Moderately Sensitive to Harvest'. USDA Forest Service Gen. Tech. Rep. PSW-GTR-162. p.38.

38. Oduor A. (2006) *Host searching of parasitoids: Does infochemically-complex habitat affect host searching efficiency of Cotesia wasps and could this affect indirect defense of plants.* Masters Thesis. Laboratory of Entomology. Wageningen University. Netherlands. p.4.

39. Botha D.D. (2010) *Melamine, from fertilizer to pasture to cow's milk*. Stellenbosch University. [Re: Melamine Scandal, Boosting Protein Value, p.1.]

40. Weinzierl R. and Henn T. (1998) *Alternatives in Insect Management: Biological and Biorational Approaches*. ENR Project IP13. Cooperative Extension Service of the University of Illinois. p.57.

41. Villenave J., Deutsch B., Lode T., and Rat-Morris E. (2006) 'Pollen preference of the Chrysoperla species (Neuroptera: Chrysopidae) occurring in the crop environment in western France'. *Eur. J. Entomol.* 103. [Re: Pollen preference in crop plants, p.772. Pollen of *Brassica napus*, p.775. Pollen consumption and availability, p.775.]

42. Flowerdew B. (2012) Quote. "The best time to..." In Graham N. (2012) 'Bob Flowerdew: I bartered jam for a photocopier'. Telegraph Media Group Ltd. www.telegraph.co.uk/finance/personalfinance/fameandfortune/9467155/Bob-Flowerdew-I-bartered-jam-for-a-photocopier.html

43. Royal Society of Chemistry (2015) 'Salicylic acid'. www.chemspider.com/Chemical-Structure.331.html

44. Larqué-Saavedra A., Martin-Mex R. (1998) 'Effects of Salicylic Acid on the Bioproductivity of Plants'. p.15. In Hayat S., Ahmad Aq. (Eds.) (2007) *Salicyclic Acid – A Plant Hormone*. Springer. www.springer.com/gp/book/9781402051838

45. Hayat S., Ahmad Aq. (Eds.)(2007) *Salicyclic Acid – A Plant Hormone*. Springer. www.springer.com/gp/book/9781402051838

46.0. Mallinger R.E., Hogg D.B., Gratton C. (2011) 'Methyl Salicylate Attracts Natural Enemies and Reduces Populations of Soybean Aphids (Hemiptera: Aphididae) in Soybean Agroecosystems'. *Journal of Economic Entomology* 104(1), p.115 Department of Entomology, University of Wisconsin-Madison, 1630 Linden Street, Madison, WI 53703. www.bioone.org/doi/abs/10.1603/EC10253?journalCode=ecen

46.1. Snoeren T.A.L, Mumm R., Poelman E.H.,Yang Y., Pichersky E., Dicke M. (2010) 'The Herbivore-Induced Plant Volatile Methyl Salicylate Negatively Affects Attraction of the Parasitoid Diadegma semiclausum'. *J. Chem. Ecol.* 36, p.479. http://link.springer.com/article/10.1007/s10886-010-9787-1

47. USDA (2011) 'Technical Evaluation report: Indole-3-Butyric Acid (IBA): Crop Production'. [Re: Uses: p.2, lines 40-43, Synergists: p.8 lines 347-351.] www.ams.usda.gov/sites/default/files/media/Indole%203%20Butyric%20Acid%20TR.pdf

48. USDA (2011) Technical Evaluation Report: 'Indole-3-Butyric Acid (IBA), Crop Production'. p.1, line 31. www.ams.usda.gov/sites/default/files/media/Indole%203%20Butyric%20Acid%20TR.pdf

49.0. Grunwald C. and Lockard R.G. (1973) 'Synergism between gibberellin and auxin in the growth of intact tomato'. *Physiol. Planta.* 29(1), p.61 and pp.64-67. www.researchgate.net/

publication/230055419

49.1. Haniyeh B., Shinjiro Y., Masashi A., Shinobu S.(2009) 'Effects of shoot-applied gibberellin/gibberellin-biosynthesis inhibitors on root growth and expression of gibberellin biosynthesis genes in *Arabidopsis thaliana*'. *Plant Root*. Jan. 2010. p.6. www.plantroot.org/PDFarchive/2010/4_4.pdf

50.0. Manoharan K. and Ganesamoorthi R. (2015) 'Nutrient Status and Plant Growth Promoting Potentiality of Vermicompost and Biocompost on Vigna Radiata'. *World Journal of Pharmacy and Pharmaceutical Sciences*. Vol 4, Issue 08. P.G. and Research Department of Microbiology, Sengamala Thayaar Educational Trust Women' College, Mannargudi, Tamilnadu, India. p.831. [Re: Gibberellins, Microorganisms and Auxins in vermicompost, p.831]. www.wjpps.com/download/article/1438357867.pdf

50.1. Talashilkar S.C. et al. (1999). *J. Indian Soc. Soil Sci.* 47, SO-53. In Chaudhary D.R., Bhandari S.C. and Shukla L.M. (2004) 'Role of vermicompost in sustainable agriculture – A Review' (2004). *Agric. Rev.* 25 (1), p.31. [Re: Carbohydrates (soluble) in vermicompost, p.31.] www.researchgate.net/publication/235545341

51. LeBannister M. (2011) *Maximum Yield*. September-October. [Re: Air in stems, p.30. Study, p.31]. www.joomag.com/magazine/maximum-yield-uk-2011-september-october/0924684001396475281?page=30

52. Gold L.S., Slone T.H., Ames B.N. and Manley N.B. (2001) *Handbook of Pesticide Toxicology*. 2nd Ed. San Diego, CA. Academic Press. p.823.

53. Çaliskan M. (1998) 'The Metabolism of Oxalic Acid'. *Turk. J. Zool.* 24 (2000). Mustafa Kemal University, Department of Biology 31040 Hatay-TURKEY. p.103.

54. PEI (1999) 'Agricultural Business Profile on Rhubarb'. PEI Department of Agriculture and Forestry. Agriculture Resources Team. Charlottetown Research Centre 440 University Ave. Charlottetown, PEI C1A 7N3. p.3.

55. Gold L.S., Slone T.H., Ames B.N. and Manley N.B. (2001) *Handbook of Pesticide Toxicology*. 2nd Ed. San Diego, CA. Academic Press. p.823.

56.0. Beier R.C., and Nigg H. N. (1994) 'Toxicology of naturally occurring chemicals in food'. In Gold L.S., Slone T.H., Ames B.N. and Manley N.B. (2001) *Handbook of Pesticide Toxicology*. 2nd Ed. San Diego, CA. Academic Press. p.823.

56.1. Hodgkinson A. (1977) 'Oxalic Acid in Biology and Medicine Academic Press' New York. In Gold L.S., Slone T.H., Ames B.N. and Manley N.B. (2001) *Handbook of Pesticide Toxicology*. 2nd Ed. San Diego, CA. Academic Press. p.823.

57.0. Sharma S., Vaidyanathan S., Nath R. and Thind S.K. (1991) 'Advances in pathophysiology of calcium oxalate stone disease'. *Ind. J. Urol.* 8, 25-37. In Çaliskan M. (1998) 'The Metabolism of Oxalic Acid'. *Turk. J. Zool.* 24 (2000). Mustafa Kemal University, Department of Biology 31040 Hatay-TURKEY. p.104.

57.1. Roth D.A. and Breitenfield R.V. (1977) 'Vitamin C and oxalate stones'. *J. Am Med. Ass.* 237, 768. In Çaliskan M. (1998) 'The Metabolism of Oxalic Acid'. *Turk. J. Zool.* 24 (2000). p.104.

57.2. Obzansky D.M. and Richardson K.E. (1983) 'Quantification of urinary oxalate with oxalate oxidase from beet stems'. *Clin. Chem.* 29, 1815-1819. In Çaliskan M. (1998) 'The Metabolism of Oxalic Acid'. *Turk. J. Zool.* 24 (2000). p.104.

58. Doubrava N.S., Dean R.A. and Kué J. (1988) 'Induction of systemic resistance to anthracnose caused by *Colletotrichum lagenarium* in cucumber by oxalate and extracts from spinach and rhubarb leaves'. *Physiological and Molecular Plant Pathology* (1988) 33. p.69.

59.0. Foerster P. et al. (2001) *Best Practices for the Introduction of Non-Synthetic Pesticides in Selected Cropping Systems*. Deutsche Gesellschaft für Technische Zusammenarbeit (GTZ) GmbH P.O.Box 5180, 65726 Eschborn, Germany. p.7.

59.1. Schmutterer H. (1995) *The Neem tree. Source of unique natural products for integrated pest management, medicine, industry and other purposes*. VCH Verlagsgesellschaft mbH. p.696. In Foerster P et al. (2001) 'Best Practices for the Introduction of Non-Synthetic Pesticides in Selected Cropping Systems'. Deutsche Gesellschaft für Technische Zusammenarbeit (GTZ) GmbH P.O.Box 5180, 65726 Eschborn, Germany. p.7.

60. Kleeberg H. and Zebitz P.W. (Eds.) (1993) *Proceedings of Practice-oriented Results on Use and Production of Neem Ingredients and Pheromones*, Workshop 1 + 3, Trifolio Gmbh, Lahnau, Germany. In Foerster P. et al. (2001) *Best Practices for the Introduction of Non-Synthetic Pesticides in Selected Cropping Systems*. Deutsche Gesellschaft für Technische Zusammenarbeit (GTZ) GmbH P.O.Box 5180, 65726 Eschborn, Germany. p.28.

61. Foerster P. et al. (2001) 'Best Practices for the Introduction of Non-Synthetic Pesticides in Selected Cropping Systems'. Deutsche Gesellschaft für Technische Zusammenarbeit (GTZ) GmbH P.O.Box 5180, 65726 Eschborn, Germany. p.100.

62. Extoxnet (1996) Rotenone Pesticide Information Profile. http://extoxnet.orst.edu/pips/rotenone.htm

63. Latha P., Anand T., Ragupathi N., Prakasam V. and Samiyappan R. (2009) 'Antimicrobial activity of plant extracts and induction of systemic resistance in tomato plants by mixtures of PGPR strains and Zimmu leaf extract against *Alternaria solani*'. *Biological Control* Volume 50, Issue 2, August 2009. p.85.

64.0. Karthikeyan M., Sandosskumar R., Radhajeyalakshmi R., Mathiyazhagan S., Khabbaz Salah E., Ganesamurthy K., Selvi B., and Velazhahan R. (2007) 'Effect of formulated zimmu (*Allium cepa* L. × *Allium sativum* L.) extract in the management of grain mold of sorghum'. *Journal of the Science of Food and Agriculture*, Volume 87, Number 13, October 2007. p.2495.

64.1. Sandosskumar R., Karthikeyan M., Mathiyazhagan S., Mohankumar M., Chandrasekar G. and Velazhahan R. (2006) 'Inhibition of *Aspergillus flavus* growth and detoxification of aflatoxin B_1 by the medicinal plant zimmu (*Allium sativum* L. × *Allium cepa* L.)'. *World Journal of Microbiology and Biotechnology* Volume 23, Number 7. p.1007.

64.2. Satya V.K., Radhajeyalakshmi R., Kavitha K., Paranidharan V., Bhaskaran R. and Velazhahan R. (2007) 'In vitro antimicrobial activity of zimmu (*Allium sativum* L.'*Allium cepa* L.) leaf extract'. *Archives Of Phytopathology And Plant Protection* Volume 38, Issue 3, 2005. p.185.

Chapter Eight

1. Golueke C.G. (1972) *Composting. A study of its processes and principles*. Rodale Press Inc. Emmaus, Pennsylvania. p.83.

2. Martin D.L. and Gershunny G.L. (Ed) (1992) *The Rodale Book of Composting*. The Rodale Press, Emmaus, Pennsylvania. p.34.

3. Sinha R.K., Herat S., Valani D. and Chauhan K. (2009) 'Earthworms: The "Unheralded Soldiers of Mankind" and "Farmer's Friend" Working Day and Night Under the Soil: Reviving the Dreams of Sir Charles Darwin for Promoting Sustainable Agriculture'. *American-Eurasian J. Agric. & Environ. Sci.* 5 (S). IDOSI Publications. Corresponding Author Dr. Rajiv K. Sinha, School of Engineering (Environment), (Vermiculture Unit), Griffith University, Nathan, Campus, Brisbane, QLD-4111, Australia. p.9.

4. ADSA Consulting Ltd (2005) *Assessment of options and requirements for Stability and Maturity Testing of Composts*. The Waste and Resources Action Programme. The Old Academy, 21 Horsefair, Banbury, Oxon OX16 0AH. p.10 and p.13.

5.0. Misra R.V., Roy R.N. and Hiraoka H. (2003) 'On-farm Composting Methods'. Sales and Marketing Group. Information Division. Food and Agriculture Organization of the United Nations. Viale delle Terme di Caracalla. 00100 Rome, Italy. p.27. E-mail: publications-sales@fao.org

5.1. Magdoff F. and van Es H. (2000) *Building Soil for Better Crops*. 2nd Ed. Sustainable Agriculture Network. SAN Coordinator c/o AFSIC, Room 304. National Agricultural Library. 10301 Baltimore Ave. Beltsville, MD 20705-2351. p.114.

6. Jenkins J. (2005) *The Humanure Handbook. A Guide To Composting Human Manure*. 3rd Ed. Publishers Joseph Jenkins, Inc. 143 Forest Lane, Grove City, PA 16127 USA. p.34.

7. Mason J. (2003) *Sustainable Agriculture*. 2nd Ed. Landlinks Press. PO Box 1139. Collingwood Vic. 3066. Australia. p.45.

8. Wurster C.M., Donald A. McFarlane D.A. and Bird M.I. (2007) 'Spatial and temporal expression of vegetation and atmospheric variability from stable carbon and nitrogen isotope analysis of bat guano in the southern United States'. *Geochimica et Cosmochimica Acta* 71 (2007) 3302-3310. p.3305.

9. USDA NRCS East National Technology Support Center and North Dakota NRCS (2011) 'Carbon To Nitrogen Ratios in Cropping Systems'. USDA Natural Resources Conservation Service. Greensboro NC. p.2. http://soils.usda.gov/sqi/management/files/C_N_ratios_cropping_systems.pdf

10. Rohlich G.A., Virginia W., Connor L.J., Golueke C.G., Hinesly T.D., Jones P.H., Herbert M.L., Raymond C.L., Lue-Hing C., John T.P., Prakasam T.B.S., Brown N.L., and Engel J. (1977) *Methane generation from human, and agricultural wastes. National Research Council*. National Academy of

Sciences, Washington D.C. p.67.

11.0. American Chemical Society (2007) 'Human Urine As A Safe, Inexpensive Fertilizer For Food Crops'. October 8th. *Science Daily.*

11.1. Pradhan S.K., Nerg A.M., Sjoblom A., Holopainen J.K. and Heinonen-Tanski H. (2007) 'Use of Human Urine Fertilizer in Cultivation of Cabbage (*Brassica oleracea*) –Impacts on Chemical, Microbial, and Flavor Quality'. *J. Agric. Food Chem.* 2007, 55, 8657-8663. p.8657.

12.0. Frederickson J., Howell G., Hobson A. and Lane J. (2006) 'Adding value to waste derived compost: performance and environmental impact of a combined in-vessel composting and vermicomposting system'. Integrated Waste Systems Research Group. The Open University, Walton Hall, Milton Keynes, MK7 6AA. p.52.

12.1. Munroe G. (2005) 'Manual of On-Farm Vermicomposting and Vermiculture'. Organic Agriculture Centre of Canada. p.8.

13. Unnamed (2013) 'Drapers Fence Post Auger', in Amazon www.amazon.co.uk/Draper-21304-Fence-Post-Auger/dp/B0049RG9VY

14.0. Rangarajan A. (2001) 'Outcome of Disease Suppressive Composts on Organic Vegetable Quality, Composition and Yield'. Organic Farming Research Foundation. https://ofrf.org/download/1782/97.08.01.Rangarajan.Spr97.IB9_.pdfredirect=node/991

14.1. Sullivan P. (2004) *Sustainable Management of Soil-Bourne Plant Diseases.* (ATTR) National Sustainable Agriculture Information Service . Postal Address: NCAT, Fayetteville, Arkansas (P.O. Box 3657, Fayetteville, AR 72702). p.10.

Chapter Nine

1. Ingham E.R. (2005) *The Compost Tea Brewing Manual.* 5th Ed. Soil Food Web Inc. 728 SW Wake Robin Ave. Corvallis, Oregon 97333. [Re: Putrefication, pp.4-5. Outcompeted, p.19. Few living organisms, p.67. Inhibition, p.63. Food Web, pp.7-10. Bacterial/Fungal Ratio, p.55.]
2. Ingram D.T. and Millner P.D. (2007) 'Factors Affecting Compost Tea as a Potential Source of *Escherichia coli* and Salmonella on Fresh Produce'. *Journal of Food Protection* Vol 70, No. 4. pp.832-833.
3. NOSB (2004) National Organic Standards Board. Compost Tea Task Force Report. USDA. In NOSB (2006) Formal Recommendation by the National Organic Standards Board (NOSB) to the National Organic Program (NOP). USDA. Addendum B, p.10. www.ams.usda.gov/sites/default/files/media/NOP%20Final%20Rec%20Guidance%20use%20of%20Compost.pdf
4. National Organic Standards Board (NOSB). www.ams.usda.gov/rules-regulations/organic/nosb
5. NOSB (2006) Formal Recommendation by the National Organic Standards Board (NOSB) to the National Organic Program. www.ams.usda.gov/sites/default/files/media/NOP%20Final%20Rec%20Guidance%20use%20of%20Compost.pdf
6. Sideman E. (2004) 'An Update on Compost Tea: Benefits, Risks, Regulation'. www.mofga.org/Publications/MaineOrganicFarmer-Gardener/Spring2004/CompostTea/tabid/1357/Default.aspx
7. Duffy B., Sarreal C., Subbarao R., Stanker L. (2004) 'Effect of molasses on Regrowth of *E. coli* O157:H7 and Salmonella in compost teas'. www.tandfonline.com/doi/abs/10.1080/1065657X.2004.10702163
8. NOSB (October 19, 2006) 'Crops Committee Recommendation for Guidance Use of Compost, Vermicompost, Processed Manure, and Compost teas'. NOSB. p.16.
9. Smith R.S., Bloomfield S.F. and Rook G.A. (2012) 'The Hygiene Hypothesis and its implications for home hygiene, lifestyle and public health'. *International Scientific Forum on Home Hygiene(IFH).* [Re: Certain Diseases, p.5. *E.Coli,* p.65.] www.ifh-homehygiene.org
10. Kelley S. (2004) 'Aerated compost tea and other alternative treatments for disease control in pumpkins'. Organic Farming Research Foundation. P.O. Box 4 40 Santa Cruz, California 95061. p.2. E-mail: info@ofrf.org

11.0. Fookes C. and Dalmeny D. (2001) *Organic Food and Farming. Myth and Reality Organic vs Non-Organic The Facts.* The Soil Association and Sustain. Bristol House 40-56 Victoria Street. Bristol BS1 6BY. pp.29-31.

11.1. Lotter D.W. (2003) 'Organic agriculture'. *J. Sustain. Agric.* 21(4), p.17.

12. Van Zyl T.D., Prinsloo P., Pieterse P.A., Morad-Khan R., Nzanza B.F. and Hern H. In Tourayi S. (2011) *An investigation of natuurboerdery (natural farming) approach: a ZZ2 case study.* MSc Thesis. University of Stellenbosch. Faculty of Economic and Management Sciences. School of Public Management and Planning. p.79.
13. Chalker-Scott L. (2008) 'The Myth of Compost Tea, Episode III'. Puyallup Research and Extension Center, Washington State University. pp.1-2. [At the time of writing her references to unpublished material from assorted Universities could not be located or confirmed.]
14. Chalker-Scott L. (2007) 'Compost Tea: Examining the science behind the claims'. WSU Puyallup Research and Extension Center 7612 Pioneer Way E. Puyallup, WA 98371. p.2.
15. Lishman G. and Fisher E. (2015) 'An Interview with Dr Gavin Lishman of Martin Lishman Ltd'. See downloadable interviews.
16. Bricknell C. (Ed.) (2008) *RHS A-Z Encyclopedia of Garden Plants.* Published by Dorling Kindersley, UK. p.349.
17. East of Eden Plants (2017) 'Rootgrow Mycorrhizal fungi ~ Feeding your Plants for Life'. [Note: There are many other sources available; this is quoted merely as an example.] www.eastofedenplants.co.uk/rootgrow.htm
18. Fodor Z.A.J. and Williams R.B. (2005) *Bentonite, Kaolin, and selected Clay Minerals.* World Health Organisation, Geneva. p.6. www.who.int/ipcs/publications/ehc/ehc_231.pdf
19. Maier R.M., Williamson R.C. (2016) 'Evaluation of Kaolin Clay as an Alternative Management Tactic for Japanese Beetle Feeding Damage in Grape Vineyards'. *Journal of Horticulture.* p.184. www.esciencecentral.org/journals/evaluation-of-kaolin-clay-as-an-alternative-management-tactic-forjapanese-beetle-feeding-damage-in-grape-vineyards-2376-0354-1000184.pdfm
20. Phillips M. (2005) *The Apple Grower: Guide for the Organic Orchardist.* 2nd Ed. Chelsea Green Publishing. p.235.
21. C. Daniel C., Pfammatter W., Kehrli P. and E. Wyss E. (2005) 'Processed kaolin as an alternative insecticide against the European pear sucker, *Cacopsylla pyri* (L.)'. *JEN* 129 (7). Research Institute of Organic Agriculture, Ackerstrasse, Postfach, Switzerland. p.363. http://citeseerx.ist.psu.edu/viewdoc/download?doi=10.1.1.565.666&rep=rep1&type=pdf
22. Nassar R.M.A, Yasser Y.M., Nassar D.M.A (2011) 'Effect of Foliar Spray with Active Yeast Extract on Morphological, Anatomical and Yield Characteristics of Kidney Bean (*Phaseolus vulgaris* L.)'. *Australian Journal of Basic and Applied Sciences* 5(5), p.1071. http://ajbasweb.com/old/ajbas/2011/1071-1079.pdf
23. Asmaa M.R., EL-Desuki M., Abdel-Mouty M.M. and Aisha H, Ali H. (2013) 'Effect of Compost Levels and Yeast Extract Application on the Pea Plant Growth, Pod Yield and Quality'. *Journal of Applied Sciences Research* 9(1): 149-155. In El Sagan M.A.M. (2015) 'Effect of Some Natural Extracts on Growth and Productivity of Cucumber under Sandy Soil Conditions'. *International Journal of Advanced Research* (2015), Volume 3, Issue 9. p.677.
24. Fawzy Z.F. (2007) 'Increasing productivity of head lettuce by foliar spraying of some bio- and organic soil conditions'. *Australian Journal of Basic and Applied Science* 2(2): 296-300. In El Sagan M.A.M. (2015) 'Effect of Some Natural Extracts on Growth and Productivity of Cucumber under Sandy Soil Conditions'. *International Journal of Advanced Research* (2015), Volume 3, Issue 9. p.677.
25. El Sagan M.A.M. (2015) 'Effect of Some Natural Extracts on Growth and Productivity of Cucumber under Sandy Soil Conditions'. *International Journal of Advanced Research* (2015), Volume 3, Issue 9. p.677.
26. El-Tohamy, W.A., H.M. El-Abagy and N.H.M. El-Greadly, (2008) 'Studies on the effect of Putrescine, Yeast'. In El Sagan M.A.M. (2015) 'Effect of Some Natural Extracts on Growth and Productivity of Cucumber under Sandy Soil Conditions'. *International Journal of Advanced Research* (2015), Volume 3, Issue 9. p.677.
27. El-Fathy S.L., Farid S. and El-Desouky S.A. (2000) 'Induce cold tolerance of out door tomatoes during early summer season by using adenosine triphosphate (ATP), yeast, other material and chemical treatments to improve their fruiting and yield'. *J. Agric. Sci., Mansoura Univ.* 25(1), 377-401. In Nassar R.M.A, Yasser Y.M., Nassar D.M.A (2011) 'Effect of Foliar Spray with Active Yeast Extract on Morphological, Anatomical and Yield Characteristics of Kidney Bean (*Phaseolus vulgaris* L.)'. *Australian Journal of Basic and Applied Sciences* 5(5), p.1072. http://ajbasweb.com/old/ajbas/2011/1071-1079.pdf

28. Volker K. (2014) 'Surround for Insect Management and Protection against Heat and Light Stress'. TKI NovaSource. p.14. www.abim.ch/fileadmin/abim/documents/presentations2014/3_Kurt_Volker_ABIM2014.pdf
29. Chambre D'agriculture (2012) 'Réussir son projet de plantation'. (Programme Eva 17) Chambre D'agricultur, 3 bd Vladimir, 17100 Saintes. France. p.18. www.charente-maritime.chambagri.fr/fileadmin/ publication/CA17/25_EVA/Documents/dossier_projet_BD.pdf
30.0. Jack A.L.H. and Nelson E.B. (2010) 'Suppression of Pythium damping off with compost and vermicompost'. Final report to the Organic Farming Research Foundation. Department of Plant Pathology and Plant-Microbe Biology. Cornell University. 334 Plant Science Building. Ithaca, NY 14853. p.5.
30.1. Jack A.L.H. (2012) 'Vermicompost Suppression of Pythium Aphanidermatum Seedling Disease: Practical Application and an Exploration of the Mechanism of Diseases Suppression'. Cornell University. p.18. www.researchgate.net/publication/255702218
31. Fukuoka M. (1978) *The One-Straw Revolution*. Rodale Press. [Re: Protection from animals, p.43. Low Maintenance, p.51. Reduced Rotting, p.43. Difficult seeds, p.67. Arthropod Behaviour, p.67.]
32. Van Veen A.J., Overbeek L.S., Van Elsas J.D.(1997) 'Fate and Activity of Microorganisms Introduced into Soil'. *Microbiol. Mol. Biol. R. June 1997.* Vol 61. No.2. p.124. www.ncbi.nlm.nih.gov/pmc/articles/PMC232604/pdf/610121.pdf
33. Chandrashekara C. and Bhatt J.C., Kumar R., Chandrashekara K.N. (2012) 'Supressive Soils in Plant Disease Management'. p.241. www.researchgate.net/publication/23295760
34. Jack A.L.H. and Nelson E.B. (2010) 'Suppression of Pythium damping off with compost and vermicompost'. Final report to the Organic Farming Research Foundation'. Department of Plant Pathology and Plant-Microbe Biology. Cornell University. 334 Plant Science Building. Ithaca, NY 14853. p.3.
35. Heil M. and Bostock R.M. (2002) 'Induced systemic resistance (ISR) against pathogens in the context of induced plant defenses'. *Ann. Bot.* 89(5), 503-12. In Diánez F., Santos M. and Tello J.C. (2005) 'Suppression of Soilborne Pathogens by Compost: Suppressive Effects of Grape Marc Compost on Phytopathogenic Oomycetes'. p.446. Acta horticulturae 697:441-460· January 2005. p.446. www.researchgate.net/publication/284354688
36. Diánez F., Santos M. and Tello J.C. (2005) 'Suppression of Soilborne Pathogens by Compost: Suppressive Effects of Grape Marc Compost on Phytopathogenic Oomycetes'. Departamento Producción Vegetal Universidad de Almería. Spain. *Acta Hort.* 697 ISHS. p.442.
37. Scheuerell S.J.(2004) 'Compost Tea Production Practices, Microbial Properties and Plant Disease Suppression'. Oregon State University. Dept. of Botany and Plant Pathology, 2082 Cordley Hall, Corvallis. p.45.
38. Coleman D.C., Reid C.P.P. and Cole C.V. (1983) 'Biological strategies of nutrient cycling in soil systems'. *Advances in Ecological Research* 13, 1-55. In Harris A. and Hill R. (2007) 'Carbon-Negative Primary Production: Role of Biocarbon and challenges for Organics in Aoteroa/New Zealand'. Tauranga, New Zealand, *Journal of Organic Systems* Vol.2, No.2. Lincoln University, New Zealand. p.3 and p.7.
39. Lind K., Lafer G. , Schloffer K., Innerhofer G., Meister H. (2003) *Organic Fruit Growing*. CABI Publishing. CAB International, Wallingford Oxon OX10 8DE, UK. p.126.
40. Van Zwieten M., Stovold G., Lukas Van Zwieten L. (2007) 'Alternatives to Copper for Disease Control in the Australian Organic Industry'. *Environmental Centre of Excellence, NSW Agriculture, Wollongbar.* pp. Forward (iii).
41. Dumestre A., Sauve S., McBride M., Baveye P., Berthelin. J. (1999) 'Copper speciation and microbial activity in long-term contaminated soils'. *Archives of Environmental Contamination & Toxicology* 36(2), 124-131. In Van Zwieten M., Stovold G., Lukas Van Zwieten L. (2007) *Alternatives to Copper for Disease Control in the Australian Organic Industry. Environmental Centre of Excellence*, NSW Agriculture, Wollongbar. p.7.
42. Van Zwieten M., Stovold G., Lukas Van Zwieten L. (2007) *Alternatives to Copper for Disease Control in the Australian Organic Industry*. Environmental Centre of Excellence, NSW Agriculture, Wollongbar. p.7.
43. Van Zwieten L., Rust J., Kingston T., Merrington G., Morris S. (2004) 'Influence of copper fungicide residues on occurrence of earthworms in avocado orchard soils'. *The Science of the Total Environment* 329, 29-41. In Bünemann E.K., Schwenke G.D., and Van Zwieten L. (2006) 'Impact of agricultural inputs on soil organisms – a review'. *Australian Journal of Soil Research* 2006, 44. pp.394.
44. Van Zwieten M., Stovold G., Lukas Van Zwieten L. (2007) *Alternatives to Copper for Disease Control in the Australian Organic Industry*. Environmental Centre of Excellence, NSW Agriculture, Wollongbar. p.47.
45. Ma W. (1984) Sublethal toxic effects of copper on growth, reproduction and litter breakdown activity in the earthworm *Lumbricus rubellus*, with observations on the influence of temperature and soil pH'. *Environmental Pollution*, A 33(3), 207-219. In Van Zwieten M., et al. (2007) *Alternatives to Copper for Disease Control in the Australian Organic Industry*. Environmental Centre of Excellence NSW Agriculture. p.7.
46. Donahue S. (2001) 'Agricultural Management Effects on Earthworm Populations'. Soil Quality Agronomy. Technical Note No.11. Natural Resources Conservation Service. Soil Quality Institute (NRCS, USDA). p.7. Soil Quality Institute. 411 Auburn, AL 36832 https://directives.sc.egov.usda.gov/OpenNonWebContent.aspx?content=18543.wba
47.0. Sinkevičienė A., Jodaugienė D., Pupalienė R., Urbonienė M. (2009) 'The influence of organic mulches on soil properties and crop yield'. *Agronomy Research* 7 (Special issue I). [Re: Crop Yield, p.485.]
47.1. Hill D.E., Hankin L. and Stephens (1982) 'Mulches: Their Effect on Fruit Set, Timing and Yields of Vegetables'. Bulletin 805. The Connecticut Agricultural Experiment Station, New Haven. [Re: Vegetable crops, p.9.] www.ct.gov/caes/lib/caes/documents/publications/bulletins/b805.pdf
47.2. Hassan S.A., Abidin R.Z., Ramlan M.F. (1995) 'Growth and Yield of Chilli (*Capsicum annuum* L.) in Response to Mulching and Potassium Fertilization'. *PertanikaJ. Trap. Agric. Sci.* 18(2). [Re: Vegetable crops, p.113.]
48. Phillips M.(2012) *The Holistic Orchard: Tree Fruits and Berries the Biological Way*. Chelsea Green Publishing. p.110.
49. Wiedemeier T.H., Lucas M.A, Haas P.E. (2000) 'Designing Monitoring Programs to Effectively Evaluate the Performance of Natural Attenuation'. Parsons Engineering Science, Inc. Pasadena, California. pp.2-12. www.environmentalrestoration.wiki/images/8/8f/AFCEE_Long_Term_Monitoring_Protocol_2000.pdf
50. USDA (2006) 'Agricultural Management Effects on Earthworm Populations'. Soil Quality Agronomy. Technical Note No.11. p.6. https://directives.sc.egov.usda.gov/OpenNonWebContent.aspx?content=18543.wba
51. USDA (2017) 'Soil Quality Test Kit Guide'. United States Department of Agriculture (USDA). Natural Resources Conservation Service. [Re: Compaction, p.9. Nutrient Availability, p.66.] www.nrcs.usda.gov/wps/portal/nrcs/detail/soils/health/assessment/
52. USDA (2017) 'Soil Texture Calculator'. United States Department of Agriculture. www.nrcs.usda.gov/wps/portal/nrcs/detail/soils/survey/?cid=nrcs142p2_054167
53. Pumphrey F.V., Klepper B.L., Rickman R.W., Hane D.C. (1980) 'Sandy Soil and Soil Compaction' (1980). Circular of Information 687. Agricultural Experiment Station. Oregon State University, Corvallis. p.4. (Note : References 54-69 inc. on p.143 – Tillage Comparison.)
54. Domínguez A., Bedano J.C., Becker A.R., Arolfo R.V. (2014) 'Organic farming fosters agroecosystem functioning in Argentinian temperate soils: Evidence from litter decomposition and soil fauna'. National Council for Scientific and Technical Research, Argentina. p.170. www.sciencedirect.com/science/article/pii/S0929139313002916
55. Kladivko E.J. (2001) 'Tillage systems and soil ecology'. *Soil and Tillage Research* Volume 61, Issue 1. Elservier. p.61. www.sciencedirect.com/science/article/pii/S0167198701001799
56. Kassam A., Friedrich T., Shaxson F., Pretty J. (2009) 'The Spread of Conservation Agriculture: Justification, Sustainability and Uptake'. *International Journal of Agricultural Sustainability* p.14 and p.316. www.researchgate.net/publication/232861957
57. Benites J., Vaneph S., Bot A. (2002) 'Planting concepts and harvesting good results'. *LEISA Magazine*, October. 18(3): 6. In Derpsch R., Theodor Friedrich T., Kassam A., Hongwen L. (2010) 'Current status of adoption of no-till farming in the world and some of its main benefits'. *Int J Agric & Biol Eng.* Vol1. No.1. p.16. www.fao.org/ag/ca/CA-publications/China_IJABE.pdf
58. Montgomery D.R. (2007) 'Soil erosion and agricultural

sustainability'. Proceedings of the National Academy of Sciences of the United States of America. Department of Earth and Space Sciences, University of Washington, Seattle, WA 91895. pp. Abstract. www.pnas.org/content/104/33/13268.long

59. Ibendahl G.(2016) 'A Profitability Comparison of No-Till and Tillage Farm'. Kansas State University Department of Agricultural Economics. [Re: Profitability, p.2. Yield, p.1.] www.agmanager.info/sites/default/files/Profitability_NoTill-Tillage.pdf
60. Tebrügge F., Böhrnsen A. (1977) 'Crop yields and economic aspects of no-tillage compared to plough tillage: Results of long-term soil tillage field experiments in Germany'. In Derpsch R., et al. (2010) 'Current status of adoption of no-till farming in the world and some of its main benefit'. *Int J Agric & Biol Eng.* Vol1. No.1. p.7. www.fao.org/ag/ca/CA-publications/China_IJABE.pdf
61. Baker C.J., Ritchie W.R., Keith E., Saxton K.E. (2010) 'The Nature of Risk in No-tillage'. In Baker C.J., Saxton K.E., Ritchie W.R., Chamen W.C.T., Reicosky D.C., Ribeiro F., Justice S.E. and P.R. Hobbs P.R. (2010) *No-tillage Seeding in Conservation Agriculture.* 2nd Ed. Food and Agriculture Organisation of the United Nations. p.40. www.fao.org/3/a-al298e.pdf
62. Soane B. D., Ball B. C., Arvidsson J., Basch G. Moreno F., Roger-Estrande J. (2012) 'No-till in northern, western and south-western Europe: A review of problems and opportunities for crop production and the environment'. *Soil Tillage and Research* 118: 66-87. In Godwin T.J. (2014) *Potential of "No-till" Systems for Arable Farming.* Soil and Water Management Centre, Harper Adams University. Newport. Shropshire. TF108NB. [Re: Best Yields for Dry, p.7. Equal, p.21.]
63. Peign J., Ball B.C., Roger Estrade J., David C. (2007) 'Is conservation tillage suitable for organic farming'. *Soil use and Management.* British Society of Soil Science. [Re: Perennial Weeds, p.9. Herbicides and Transition, p.11.] www.isara.fr/en/content/download/1075/7506/version/1/file/Article+S-CAB-4.pdf
64. Gill G. (2002) 'Ecological management of agricultural weeds'. *Agriculture, Ecosystems and Environment* 90(1), 106-107. In Kristiansen P., Taja A., Reganold J. (Eds) (2006) 'Organic Agriculture: A Global Perspective'. Csiro Publishing. PO Box 1139 (150 Oxford St) Collingwood VIC 3066. Australia. p.1.
65. Hobbs P.R. and Govaerts B. (2010) 'How Conservation Agriculture can Contribute to Buffering Climate Change'. (Chapter 10) In Reynolds M.P. (Ed)(2010) *Climate change and Crop Production.* Volume 13. CABI. p.180.
66. Jardin D., and DeWolf E. (2009) 'Disease Factors to Consider in No-till'. No Till Series. Kansas State University. p.1. www.bookstore.ksre.ksu.edu/pubs/MF2909.pdf
67. Friedrich T. and Kassam A. 'No-Till Farming and the Environment'. p.156. In (2012) August. *Outlooks on Pest Management.* www.fao.org/ag/ca/ca-publications/outlook_pestidices_no_till_and_inputs.pdf
68. Bedano J.C., and Domínguez A. (2016) 'Large-Scale Agricultural Management and Soil Meso- and Macrofauna Conservation in the Argentine Pampas'. *Sustainability* 2016, 8, 653. MDP1, Basel Switzerland. p.11. www.mdpi.com/2071-1050/8/7/653/pdf
69. Bolton S.(2014) 'Ploughing versus reduced tillage'. NFU Cymru. AHDB-Agriculture and Horticulture Development Board. p.6. https://cereals.ahdb.org.uk/media/329443/susannah-bolton.pdf
70. Kassam A., Friedrich T., Derpsch R. and Kienzle J. (2015) 'Overview of the Worldwide Spread of Conservation Agriculture'. *Field Actions Science Reports.* p.6. www.rolf-derpsch.com/fileadmin/templates/main/downloads/Overview-of-the-worldwide-spread-of_CA_2015_Kassam_et_al.pdf
71. Fageria N.K. and Moreira A. (2011) 'The Role of Mineral Nutrition on Root Growth of Crop Plants'. In Sparks D.L. (Ed) *Advances in Agronomy,* Vol. 110. Elsevier. p.276.
72. Derpsch R., Friedrich T. (2009) 'Global Overview of Conservation Agriculture Adoption'. p.2. ResearchGate. www.researchgate.net/publication/238743255
73. Kemper W.D., Trout T.J., Segeren A., Bullock M. (1987) 'Worms and Water'. *Journal of Water and Soil Conservation.* Vol 42. No.6. pp.403-404. https://eprints.nwisrl.ars.usda.gov/492/1/644.pdf
74. Atkinson J. (2000) 'Soil description and classification'. City University, London/University of the West of England. p.6. http://environment.uwe.ac.uk/geocal/SoilMech/classification/default.htm
75. Carpenedo & Mielniczuk (1990) 'J. Estado de agregação e qualidade de agregados de Latossolos Roxos, submetidos a diferentes sistemas de manejo'. *R. Bras. Ci. Solo* 14, 99-105, 1990. In Da Veiga M., Reinart D.J. and Reichert J.M. (2009) 'Aggregate Stability as Affected by Short and Long-Term Tillage Systems and Nutrient Sources of a Hapludox in Southern Brazil'. *R. Bras. Ci. Solo* 33, p.772.
76. Mitchell C.C, Pinkston C.B. and Caylor A. (2003) 'Garden Tillage research and demonstrations'. Agronomy Series. Agriculture & Natural Resources. Department of Agronomy and Soils, Auburn University. [Re: Compaction, p.2. Crop Production, p.6-7. Tillage Equipment and Compaction, p.2. Double Digging Best, p.4. Subsoiling Effective, p.5.] www.aces.edu/timelyinfo/Ag%20Soil/2003/November/s-03-03-TILLAGE.pdf
77. Murdock L.W., Call D. and James J. (2008) 'Compaction, Tillage Method, and Subsoiling Effects on Crop Production'. University of Kentucky, College of Agriculture [Re: Subsoiling Effective, p.2. Subsoiling Improved No-Till, pp.2-3.]
78. Gowariker V., Krishnamurthy V.N., Gowariker S., Dhanorkar M., and Paranjape K. (2009) *The Fertiliser Encyclopedia.* John Wiley & Sons, Inc., Hoboken, New Jersey. p.655.
79. Capowiez Y., Cadoux S, Bouchand P., Roger-Estrade J., Richard G., Boizard H. (2009) 'Experimental evidence for the role of earthworms in compacted soil regeneration based on field observations and results from a semi-field experiment'. *Soil Biology & Biochemistry* 41, p.711. www.academia.edu/23798023
80. Bertrand M., Barot S., Manuel Blouin M., Whalen J., Tatiana De Oliveira T., Roger-Estrade J. (2015) 'Earthworm services for cropping systems. A review'. *Agronomy for Sustainable Development.* Springer. 35 (2). p.560. https://link.springer.com/article/10.1007/s13593-014-0269-7
80. Wolkowski R.P. (2011) 'Impact of Tillage on Soil Properties'. Extension Soil Scientist, Department of Soil Science, University of Wisconsin-Madison. p.3. http://citeseerx.ist.psu.edu/viewdoc /download?doi=10.1.1.604.3101&rep=rep1&type=pdf
81. Pfiffner L. (2014) 'Earthworms – Architects of Fertile Soils'. Research Institute of Organic Agriculture FiBL. Ackerstrasse 113, P.O. Box 219, CH-5070 Frick, Switzerland. p.6. www.tilman-org.net
82. Roberson J.D. (1936) 'The Function of the Calciferous Glands of Earthworms'. *Journal of Experimental Biology* Vol. 13., No. 3. [Re: Buffer, p.294.] http://jeb.biologists.org/content/jexbio/13/3/279.full.pdf
83. Willie J.G.M., Vijver P. and Vijver M.G.(2015) 'Earthworms and Their Use in Eco(toxico)logical Modeling'. Laboratory for Ecological Risk Assessment, National Institute of Public Health and the Environment, P.O. Box 1, 3720 BA Bilthoven, The Netherlands. p.183. www.researchgate.net/publication/225878540
84. Briones M.J.I., Ostle N.J., Piearce T.G. (2008) 'Stable isotopes reveal that the calciferous gland of earthworms is a CO_2-fixing organ'. *Soil Biology and Biochemistry,* Volume 40, Issue 2, February. p.554. www.sciencedirect.com/science/article/pii/S0038071707003872
85. Six J., Bossuyt H., Degryze S., Denef K. (2004) 'A history of research on the link between (micro)aggregates, soil biota, and soil organic matter dynamics'. *Soil & Tillage Research* 79, p.17. 'www.sciencedirect.com/science/article/pii/S0167198704000881
86. Pfiffner L. (2014) *Earthworms – Architects of Fertile Soils.* Research Institute of Organic Agriculture FiBL. Ackerstrasse 113, P.O. Box 219, CH-5070 Frick, Switzerland. p.4. https://shop.fibl.org/CHen/mwdownloads/download/link/id/624/?ref=1
87. Schoenholzer F. (2000) *Automated image analysis as a tool to study abundance of fungi and bacteria during leaf litter decomposition.* Swiss Federal Institute of Technology. [Re: Consume Decomposed Leaves, p.6. Preference/Preparation, pp.8-9 and pp.93-94.] http://e-collection.library.ethz.ch/eserv/eth:24091/eth-24091-02.pdf
88. RHS (2017) 'Clay Soils'. The Royal Horticultural Society. www.rhs.org.uk/advice/profile?PID=620
89. VON (2005) 'Growing on Clay Soils'. VON, 58 High Lane, Chorlton, Manchester M21 9DZ. http://veganorganic.net/growing-on-clay-soils/
90. Cleveland T.E. (2011) U*tilization of waste products from sustainable forestry in the Pinelands National Reserve.* Urbana, Illinois. [Re: Hardwood Bark, p.3. Softwood Bark, p.4.] http://citeseerx.ist.psu.edu/viewdoc/download?-

doi=10.1.1.227.7872&rep=rep1&type=pdf
91. USDA (2013) 'Mulching. Natural Resources Conservation Practice Standard'. Code 484. pp.484-CPD-2. www.nrcs.usda.gov/Internet/FSE_DOCUMENTS/stelprdb1249892.pdf
92. Holmes J. and Kassel P. (2006) 'Can ground eggshell be used as a liming source'. Iowa State University. [Re : Effective, p.238. Comparison, pp.236-237.] www.agronext.iastate.edu/soilfertility/info/eggshell-lime.pdf
93. Funderburg E. (2017) 'Personal Correspondence'. The Samuel Roberts Noble Foundation. www.noble.org/staff/ag/eddie-funderburg/publications
94. Ball J. (1999) 'Understanding and Correcting Soil Acidity'. The Sam Roberts Noble Foundation. 2510 Sam Noble Parkway, Ardmore, Ok 7340. www.noble.org/news/publications/ag-news-and-views/1999/january/understanding-and-correcting-soil-acidity/
95. USDA (2010) 'Soil Glue'. United States Department of Agriculture. www.nrcs.usda.gov/Internet/FSE_DOCUMENTS/nrcs142p2_051280.pdf
96. Darwish L. (2013) *Earth Repair : a grassroots guide to healing toxic and damaged landscapes*. New Society Publishers P.O. Box 189, Gabriola Island, BC V0R 1X0, Canada. www.newsociety.com
97. Perry M. (2007) 'Gentilly residents guide to do it yourself soil clean up using natural processes'. Common Ground. [Re: Heavy Metal and Petrochemicals, p.7. Petrochemicals, p.5.] https://organizingforpower.files.wordpress.com/2009/06/bioremediation-handbook.pdf
98. Sparks D.L. and Ginder-Vogel M. (2012) 'The Role of Synchrotron Radiation in Elucidating the Biogeochemistry of Metal(loids) and Nutrients at Critical Zone Interfaces'. In Huang P.M, Li T., Sumner M.E.(Eds) 2nd Ed. *Handbook of Soil Sciences: Resource Management and Environmental Impacts*. CRC Press, Taylor & Francis Group, 6000 Broken Sound Parkway NW, Suite 300, Boca Raton, FL 33487-2742. p.43 (1-15).
99. Kodjo-Wayo L. (2006) *Biodegradation and Phytoremediation of Polycyclic Aromatic Hydrocarbons Using Mushroom Compost*. The University of Georgia. https://getd.libs.uga.edu/pdfs/kodjo-wayo_lina_k_200605_phd.pdf
100. Eggen T., Sasek V. (2002) 'Use of Edible and Medicinal Oyster Mushroom [*Pleurotus ostreatus* (Jacq.:) Kumm.] Spent Compost in Remediation of Chemically Polluted Soils'. *International Journal of Medicinal Mushrooms* Vol 4, p.255. www.researchgate.net/publication/231180793
101. Hodder M. (2008) University of Reading. In Mitchell E. (2008) 'Earthworms to aid soil clean-up'. BBC News. http://news.bbc.co.uk/2/hi/science/nature/7611522.stm
102. Karen R., Taber H.G, Taber M. (2003) 'Control of the Foliar Disease, *Septoria lycopersici*, in Organic Tomato Production'. Department of Horticulture, Iowa State University, Ames, IA 50011. www.public.iastate.edu/~taber/Extension/Progress%20Rpt%2003/OrgTom.pdf
103. McGrath M.T. (2005) 'Evaluation of Compost Tea for Managing Foliar Diseases in Tomato'. Dept of Plant Pathology, Cornell University, LIHREC, 3059 Sound Avenue, Riverhead, NY 11901.
104. Weltzien, H.C. and Ketterer, N. (1986) 'Control of Downy Mildew, *Plasmopara viticola* (de Bary) Berlese et de Toni, on grapevine leaves through water extracts from composted organic wastes'. *Phytopath* 76,1104. In Diánez F., Santos M., Tello J.C. (2006) 'Suppression of Soilborne Pathogens by Compost: Suppressive Effects of Grape Marc Compost on Phytopathogenic Oomycetes'. *Acta horticulturae* 697:441-460· January 2005. p.447. ResearchGate.www.researchgate.net/publication/284354688
105. Carlson B. (2016) 'Using compost tea in organic farming'. Seeds Farm 201 Lincoln St. Northfield. MN 55057. [Re : Poor Results, p.14. Poorly Ventilated Compost, p.17.] Example of Poor Ventilation: www.cropservicesintl.com/shop/minerals-microbes/instant-compost-tea

Chapter Ten

1.0. Radin D., Michel L., Johnston J., Delorme A. (2012) 'Psychophysical interactions with a double-slit interference pattern'. *Physics Essays* 26, 4, p.157. www.deanradin.com/evidence/RadinPhysicsEssays2013.pdf
1.1. PEAR (2010) 'Princeton Engineering Anomalies Research. Scientific Study of Consciousness-Related Physical Phenomena: Implications and Applications'. http://pearlab.icrl.org/implications.html
2.0. Radin D. , Michel L., Johnston J., Delorme A. (2012) 'Psychophysical interactions with a double-slit interference pattern'. *Physics Essays* 26, 4, p.157. www.deanradin.com/evidence/RadinPhysicsEssays2013.pdf
2.1. PEAR (2010) 'Princeton Engineering Anomalies Research. Scientific Study of Consciousness-Related Physical Phenomena: Implications and Applications'. http://pearlab.icrl.org/implications.html
3.0. Schlosshauer M., Kofler J.,and Zeilinger A.(2013) 'A Snapshot of Foundational Attitudes Toward Quantum Mechanics'. Department of Physics, University of Portland, 5000 North Willamette Boulevard, Portland, Oregon 97203, USA. https://arxiv.org/pdf/1301.1069.pdf
3.1. Ball P. (2013) 'Experts still split about what quantum means'. *Nature*, January 11th. www.nature.com/news/experts-still-split-about-what-quantum-theory-means-1.12198
4.0. CERN (2017) 'Dark Matter'. [Re: Indirect evidence only, p.1.] Conseil Européen pour la Recherche Nucléaire,.CH-1211, Geneva 23, Switzerland. https://home.cern/about/physics/dark-matter
4.1. Lewin S. (2016) 'Dark Matter Stays Dark'. *Scientific Amercian*. [Re: 80% of the Universe and Indirect evidence only, p.1.] www.scientificamerican.com/article/dark-matter-stays-dark
5. Heisenberg W. (1990) *Across The Frontiers*. Publishers Ox Bow Pr. www.amazon.co.uk/Across-Frontiers-Werner-Heisenberg/dp/0918024803/ref=la_B001IOBIQCBIQC
6. De Foe A. (2015) *Hearts in Transcendance*. Author House, 1663 Liberty Drive, Bloomington, IN 47403. p.291. www.AuthorHouse.com
7. PEAR (2010) 'Princeton Engineering Anomalies Research. Scientific Study of Consciousness-Related Physical Phenomena: Implications and Applications'. http://pearlab.icrl.org/experiments.htm
8. Boeing G. (2016) 'Visual Analysis of Nonlinear Dynamical Systems: Chaos, Fractals, Self-Similarity and the Limits of Prediction'. *Systems* 4 (4), p.37. www.mdpi.com/2079-8954/4/4/37/htm
9. Lee J. and Khitrin A.K. (2004) 'Quantum amplifier: Measurement with entangled spins'. *Journal of Chemical Physics*. 121 (9), p.3949. http://aip.scitation.org/doi/pdf/10.1063/1.1788661
10. Jenson H., Guilaran L., Jaranilla R. and Garingalao G. (2006) *Natural Farming Manual*. National Initiative on Seed and Sustainable Agriculture in the Philippines Batong Malake, Los Baños Laguna, The Philippines (PABINHI-Pilipinas) and Resource Efficient Agricultural Production-Canada Ste-Anne de Bellevue, Québec, Canada. (REAP-Canada). pp.1,2,5 and 7.
11. Thun M. (1964). In Kollerstrom N. (2012) *Gardening and Planting by the Moon 2013*. W. Foulsham & Co. p.29.
12. Van Steensel F., Nicholas P., Bauer-Eden H., Kenny G., Campbell H., Ritchie M., Macgregor A.N., Koppenol M., Blake G. and Bacchus P. (2002) *A Review of New Zealand and International Organic Land Management Research*. The Research and Development Group of the Bio Dynamic Farming and Gardening Association in New Zealand: PO Box 39045, Wellington. p.133. www.biodynamics.com/research/review-new-zealand-and-international-organic-land-management-research
13. Tipa G. (2001) 'The Maori Perspective: Customary and traditional freshwater and riparian values'. In MacGibbon R. (2001) 'Managing Waterways on Farms. A guide to sustainable water and riparian management in rural New Zealand'. Ministry for the Environment. Wellington, New Zealand. p.8. https://mfe.govt.nz/sites/default/files/managing-waterways-jul01.pdf
14. Marsden Rev. Maori (2003) In Hook G.R. and Raumati L.P. (2011) 'Cultural Perspectives of Fresh Water'. *MAI Review* 2, p.8. www.review.mai.ac.nz/MR/article/download/442/442-3300-1-PB.pdf
15. Institut für Strömungswissenschaften Herrischried (2013) [Photo Credit]. Stutzhofweg 11, 79737 Herrischried DE. sekretariat@stroemungsinstitut.de http://stroemungsinstitut.de/english-publications
16. Silverstone M. (2011) *Blinded by Science*. Lloyd's World Publishing. p.77.
17.0. Ikramul H., Hamad A., Javed I. and Qadeer M.A. (2003) 'Production of alpha amylase by *Bacillus licheniformis* using as economical medium'. *Bioresource Technology* 83, 57-61.

17.1. Yang S.S. (1996) 'Antibiotics production of cellulosic waste with solid state fermentation by Streptomyces'. *Renewable Energy* 9, 976-979.

18. Zadrazil F. (1978) 'Cultivation of Pleurotus'. In *The Biology and Cultivation of Edible Mushroom* (Eds.). Chang S.T and Hayes W.A. Academic Press, New York. pp.521-554.

19. Steiner R. (1924) 'Agriculture Lectures'. In Kollerstrom N. (2012) *Gardening and Planting by the Moon 2013*. W. Foulsham & Co. Ltd. p.36.

20. Willis G. (1995) *Lost Secrets of the Garden*. Peasant Wisdom Publishing. GW Agriculture, POB 3362, McKinney, TX 75070. USA. p.18.

21. Zürcher E. (1998) 'Chronobiology of trees: Synthesis of traditional phytopractices and scientific research as tools of future forestry'. p.257. www.researchgate.net/publication/237773086

22. Zürcher E. and Rogenmoser C. (2010) 'Considering reversible Variations in Wood Properties: possible Applications in the Choice of the Tree-Felling Date?' Proceedings of the International Convention of Society of Wood Science and Technology and United Nations Economic Commission for Europe - Timber Committee. October 11-14, 2010, Geneva, Switzerland. Bern University of Applied Sciences BFH / Architecture. Wood and Civil Engineering. Biel, Switzerland. [Re: Synodic duration, p.2. Rotting, p.2.]

23. Oregon Biodynamics Group (2005) 'Lunar Growth Cycle'. In 'Ecosystem Management'. Oregon Biodynamics Group, c/o Carol Ach, 44382 McKenzie Hwy, Leaburg, OR 97489. http://oregonbd.org/Class/Mod5.htm

24. Brown F.A. (1958) 'Monthly Cycles in an Organism in Constant Conditions During 1956 and 1957'. *Proc Natl Acad Sci* USA. 1958 Apr; 44(4): 290–296. www.ncbi.nlm.nih.gov/pmc/articles/PMC335411

25. Lai T.M. (1976) 'Phosphorous and Potassium Uptake by Plants Relating to Moon Phases'. *Biodynamics*. Summer 1976. pp.1-15. In Kollerstrom (1980) 'Plant Response to the Synodic Lunar Cycle'. *Cycles*. Foundation for the Study of Cycles. South Highland Avenue, Pittsburgh, Pennsylvania 15206. Vol 31. No3. p.63.

26. Brown F.A. and Chow C.S. (1973) 'Lunar Correlated Variations in Water Uptake by Bean Seeds'. *Biol. Bull.* 145. October. p.277.

27. Naylor E. (2010) 'Chronobiology of Marine Organism'. Cambridge University Press. The Edinburgh Building, Cambridge CB2 8RU, UK. [Re: Highest Tides, p.11. Lunar Day, p.5.]

28.0. Skov M.W., Hartnoll R.G., R.K. Ruwa, Shunula J.P., Vannini M. and Cannicci S. (2005) 'Marching to a different drummer: Crabs synchrononize reproduction to a 14-Month Lunar-Tidal Cycle'. *Ecology* 86(5). Ecological Society of America. [Re: Crabs, p.1164.]

28.1. Engelmann W. (2007) 'Rhythms of Life. An introduction using selected topics and examples'. March edition. Institut für Botanik. Physiologische Ökologie der Pflanzen Universität Tübingen. Auf der Morgenstelle 1. D72076 Tübingen (Germany). [Re: Palolo worm, p.205.]

29.0. Vignoli L. and Luiselli L. (2013) 'Better in the dark: two Mediterranean amphibians synchronize reproduction with moonlit nights'. *Web Ecol.*, 13, 1-11, 2013. Copernicus Publications/EEF. [Re: Mammals, Birds and Amphibians, p.1.]

29.1. Grant R., Halliday T. and Chadwick E. (2012) 'Amphibians' response to the lunar synodic cycle – a review of current knowledge, recommendations, and implications for conservation'. *Behavioral Ecology*. 7th November. Oxford University Press. p.53.

30. Kollerstrom N. and Staudenmaier G. (2001) 'Evidence for Lunar-Sidereal Rhythms in Crop Yield: A Review'. *Biological Agriculture and Horticulture* Vol. 19. A B Academic Publishers. Printed in Great Britain. [Re: Example of Equivocal Results, p.247. Greater Yields, p.249.]

31.0. Bockemühl J. (1985) 'Elements and Ethers: Modes of Observing the World'. In Bockemühl J. (Ed)(1985) *Toward a Phenomenology of the Etheric World*. Anthroposophical Press Inc. [Re: Four Elements, p.5.]

31.1. Bockemühl J. (1985) 'The Formative Movements of Plants'. In Bockemühl J. (Ed) (1985) *Toward a Phenomenology of the Etheric World*. Anthroposophical Press Inc. [Re: Formative Forces, p.137.]

32. Kollerstrom N. (2012) *Gardening and Planting by the Moon 2013*. W. Foulsham & Co. Ltd. [Re: Varying Divisions, p.22. Ascending and Descending Moon, p.17. Apogee-Perigee Cycle, p.15. Sidereal Cycle, p.15. Nodal Points, p.32. Plutarch, p.39.]

33.0. Kollerstrom N. 'The Lunar Calendar' (2012) In *Gardening and Planting by the Moon 2013*. W. Foulsham & Co. Ltd. pp.64-138.

33.1. Crawford E.A. (2000) *The Lunar Garden: Planting by the Moon Phases*. 2nd Ed. Cynthia Parzych Publishing Inc. New York. pp.55-127.

34.0. Thun M. (2001) 'Working with the Stars, a Biodynamic Sowing and Planting Calendar'. 26th Ed. Lanthorne Press. Launceston, UK.

34.1. Wildfeuer S. (2012) *Stella Natura 2013 Biodynamic Planting Calendar*. Stella Natura. PO Box 783. Kimberton PA 19442. Email: info@stellanatura.com

35. Shaw C. (2009) 'Under the Moon'. In *Grow it*. December 2009. Findhorn Press Ltd. p.78.

36. Gros M. (2013) *In Tune With the Moon 2013*. Findhorn Press. [Re: Watering, p.18. Hemispheres, p.4. Rotting/Harvesting, p.27. Ascending moon, p.5.] www.findhornpress.com

37. Kollerstrom N. (2005) 'The Bishop Radish 1978 experiment'. 9 Primrose Gardens, London. NW3 4UJ. p.4.

38. Daly M., Behrends P.R., Wilson M.I. and Jacobs L.F. (1992) 'Behavioural modification of predation risk: moonlight avoidance and crepuscular compensation in nocturnal desert rodent, *Dipodomys merriami*'. *Anim. Behav* 44, 1-9. In Griffin P.C., Suzanne C., Griffin S.C., Waroquiers C. and Scott-Mills L. (2005) 'Mortality by moonlight: predation risk and the snowshoe hare'. *Behavioral Ecology*. 16(5). June 2005. p.939. DOI:10.1093/beheco/ari074

39. Speiss H. (1999) 'Lunar Rhythms and Plants'. In *Biodynamics* May/June 2000. [Re: Perigee/All Plants, p.22. Potatoes, p.21. Beans, p.21 and p.22. Radish, pp.21-22. Bark Beetle, p.19.]

40. Podolinsky A. (1990) 'Bio-Dynamic Agriculture Introductory Lectures'. Vol.1.Australia. In Kollerstrom N. (2012) *Gardening and Planting by the Moon 2013*. W. Foulsham & Co. Ltd. p.39.

41. Vogt K.A., Beard K.H, Hammann S., Palmiotto J.O., Vogt D.J., Scatena F.N. and Hecht B.P (2002) 'Indigenous Knowledge Informing Management of Tropical Forests: The Link between Rhythms in Plant Secondary Chemistry and Lunar Cycles'. *Ambio* Vol. 31 No. 6, Sept. 2002. p.489. DOI: 10.1579/0044-7447-31.6.485

42.0. Neiskins A. (2013) 'Full Worm Moon' – March. www.youtube.com/watch?v=tT_aE6PlGuQAlmanac.com

42.1. Thomas R.B. (1792) *The Old Farmer's Almanac*. In Neiskens A. (2011) Full Moon: March Full Worm Moon. www.youtube.com/watch?v=tT_aE6PlGuQ&list=UU6coL3mCKGgdGTj780i7Q0g&index=9

Taken by the Author: Selected photos on p.30: Annual Sow Thistle, Black Bindweed, Chickweed, Coltsfoot, Lamb's Quarters, Hedge Mustard. pp.45,61,73,75,77,78,79,90,91,95, 99, 113,116,118,123,124,127,137,157,162,164,165. Chapter and pre-chapter pages for Chapters Six, Eight, Nine and Ten.

Appendix Two

1. Steiner R. (1924) 'Agriculture Course'. The Rudolf Steiner Archive. http://wn.rsarchive.org/Lectures/Agri1958/Ag1958_index.html

2. International Federation of Organic Agriculture Movements.IFOAM Head Office : Charles-de-Gaulle-Str. 53113 Bonn, Germany.

3. The Rudolf Steiner Archive. www.rsarchive.org

4. 4. Grijalava B. (2013) 'Horns in Spring'. www.facebook.com/newurbanfarmers/photos

5. Steiner R. (1924) 'Agriculture Course. Lecture 4'. The Rudolf Steiner Archive. http://wn.rsarchive.org/Lectures/Agri1958/19240612p01.html

6. Steiner R. (1924) 'Agriculture Course. Lecture 6'. The Rudolf Steiner Archive. http://wn.rsarchive.org/Lectures/Agri1958/19240614p01.html

7. Steiner R. (1924) 'Agriculture Course. Lecture 2'. The Rudolf Steiner Archive. http://wn.rsarchive.org/Lectures/Agri1958/19240610p01.html

8. Steiner R. (1924) 'Agriculture Course. Lecture 3'. The Rudolf Steiner Archive http://wn.rsarchive.org/Lectures/Agri1958/19240611p01.html

9. Decoction. From the Latin decoquere to boil down. Mash up, boil approximately eight minutes in this context, and strain.

10. Steiner R. (1924) 'Agriculture Course. Lecture 5'. The Rudolf Steiner Archive. http://wn.rsarchive.org/Lectures/Agri1958/19240613p01.html

11. Mesentery. By the original meaning, which would be the way Steiner used it: The flexible sack that connect the parts of the small intestine.

Index